Groundwater for Sustainable Livelihoods and Equitable Growth

International Contributions to Hydrogeology
Series Editor: Dr. Nick S. Robins, Editor-in-Chief IAH Book Series, British Geological Survey, Wallingford, UK

International Association of Hydrogeologists

22 **Managing Water Well Deterioration**
   Robert G. McLaughlan

23 **Understanding Water in a Dry Environment**
   Edited by Ian Simmers

24 **Urban Groundwater Pollution**
   Edited by David Lerner

25 **Introduction to Isotope Hydrology**
   Willem G. Mook

26 **Methods in Karst Hydrogeology**
   Edited by Nico Goldscheider and David Drew

27 **Climate Change Effects on Groundwater Resources**
   A Global Synthesis of Findings and Recommendations
   Edited by Holger Treidel, Jose Luis Martin-Bordes and Jason J. Gurdak

28 **History of Hydrogeology**
   Edited by Nicholas Howden and John Mather

29 **Investigating Groundwater**
   Ian Acworth

30 **Groundwater for Sustainable Livelihoods and Equitable Growth**
   Edited by V. Re, R.L. Manzione, T.A. Abiye, A. Mukherji, and A. MacDonald

For more information about this series, please visit: www.routledge.com/IAH—International-Contributions-to-Hydrogeology/book-series/TFIAHICH

Groundwater for Sustainable Livelihoods and Equitable Growth

Edited by

V. Re, R.L. Manzione, T.A. Abiye,
A. Mukherji, and A. MacDonald

CRC Press
Taylor & Francis Group
Boca Raton London New York

CRC Press is an imprint of the
Taylor & Francis Group, an **informa** business

Cover image: Photo taken by Viviana Re, 2006.

First published 2022
by CRC Press/Balkema
Schipholweg 107C, 2316 XC Leiden, The Netherlands
e-mail: enquiries@taylorandfrancis.com
www.routledge.com – www.taylorandfrancis.com

CRC Press/Balkema is an imprint of the Taylor & Francis Group, an informa business

© 2022 Taylor & Francis Group, LLC

All rights reserved. No part of this book may be reprinted or
reproduced or utilised in any form or by any electronic, mechanical,
or other means, now known or hereafter invented, including
photocopying and recording, or in any information storage or retrieval
system, without permission in writing from the publishers.

Although all care is taken to ensure integrity and the quality of this
publication and the information herein, no responsibility is assumed
by the publishers nor the author for any damage to the property or
persons as a result of operation or use of this publication and/ or the
information contained herein.

Library of Congress Cataloging-in-Publication Data
Names: Re, Viviana, editor.
Title: Groundwater for sustainable livelihoods and equitable growth /
edited by Viviana Re, Rodrigo Lilla Manzione, Tamiru A. Abiye,
Aditi Mukherji and Alan MacDonald.
Description: First edition. | Boca Raton, FL : CRC Press, 2022. |
Includes bibliographical references and index.
Identifiers: LCCN 2021047650 (print) | LCCN 2021047651 (ebook) |
ISBN 9780367903862 (hardback) | ISBN 9781032199511 (paperback) |
ISBN 9781003024101 (ebook)
Subjects: LCSH: Groundwater. | Water resources development.
Classification: LCC TD403 .G1325 2022 (print) | LCC TD403 (ebook) |
DDC 628.1/14—dc23/eng/20211116
LC record available at https://lccn.loc.gov/2021047650
LC ebook record available at https://lccn.loc.gov/2021047651

ISBN: 978-0-367-90386-2 (hbk)
ISBN: 978-1-032-19951-1 (pbk)
ISBN: 978-1-003-02410-1 (ebk)

DOI: 10.1201/9781003024101

Typeset in Times New Roman
by codeMantra

Contents

Preface	*ix*
Acknowledgments	*xi*
Contributors	*xiii*

Introduction: Groundwater, sustainable livelihoods and equitable growth *xvii*
V. Re, R.L. Manzione, T.A. Abiye, A. Mukherji, and A. MacDonald

1 Groundwater and livelihood in Gunungsewu karst area, Indonesia 1
E. Haryono, T.N. Adji, A. Cahyadi, M. Widyastuti, U. Listyaningsih, and E. Sulistyowati

2 Groundwater resources development for livelihoods enhancement in the Sahel Region: a case study of Niger 25
A.E. Cheo, B. Ibrahim, and E.G. Tambo

3 Groundwater, informal abstraction, and peri-urban dwellers in the Techiman Municipality of Ghana 63
L. Kwoyiga

4 Urban development and intensive groundwater use in African coastal areas: the case of Lomé urban area in Togo 77
R. Barry, F. Barbecot, M. Rodriguez, A. Djongon, and W. Akakpo

5 Contribution of groundwater towards urban household water security 95
N. Mujere

vi Contents

6 Sustainable and resilient exploitation of small alluvial aquifers in the Brazilian semi-arid region: the experience of Sumé 101
J.C. Rêgo, J.P. Albuquerque, J.D. Pontes Filho, B.B. Tsuyuguchi, T.J. Souza, and C.O. Galvao

7 Stubble burning in northwestern India: is it related to groundwater overexploitation? 123
D. Saha, M. Chakraborty, and A. Chowdhury

8 Groundwater recharge through landscape restoration and surface water harvesting for climate resilience: the case of upper Tekeze river basin, Northern Ethiopia 135
K. Woldearegay, L. Tamene, F. van Steenbergen, and K. Mekonnen

9 The Quaternary aquifer: an affordable resource to address water scarcity in the northern part of the Lake Chad basin 159
B. Collignon, C. Estienne, C. Masse, and I.A. Nassour

10 An overview of Karst groundwater springs in Al Jabal Al Akhdar region (North East Libya) 179
S.M. Hamad and A. El Hasia

11 The governance and water security of groundwater obtained from private domestic wells in periurban areas in Brazil: a case study on the Guandu river basin in the metropolitan region of Rio de Janeiro, Brazil 195
D. Tubbs Filho, A.S. Schueler, and S.Y. Pereira

12 Groundwater policy, legal and institutional framework situation analysis: gaps and action plan: the case of Malawi 213
J. Sauramba, T. Mkandawire, B. Munyai, and M. Majiwa

13 Groundwater: a juggernaut of socio-economic development and stability in the arid region of Kachchh 231
P.M. Patel and D. Saha

14 The role of groundwater in economic and social development of Mato Grosso do Sul State, Midwest of Brazil 253
S.G. Gabas, G.F. Dourado, D.A. Uechi, G.H. Cavazzana, and G. Lastoria

Contents vii

15 Valuing groundwater use: resolving the potential of groundwater in the Upper Great Ruaha River Catchment of Tanzania 275
D.B. Mosha, J.L. Gudaga, D. Gama, and J.J. Kashaigili

16 Conjunctive use of surface and groundwater: operational and water management strategies to build resilience, water security, and adaptation 295
G.F. Marques, C.D.P. Mattiuzi, S.D. Cota, and M. Pulido-Velazquez

17 The role of groundwater in rural water supply: the case of six villages of Taunggyi District, Southern Shan State, Myanmar 315
S.Y. May, K.K. Khaing, and J.S.T. Ward

18 Groundwater-driven paddy farming in West Bengal: how a smallholder-unfriendly farm power policy affects livelihoods of farmers 337
M. Shah, T. Shah, and S. Daschowdhury

19 Assessment of options for small-scale groundwater irrigation in Lao PDR 347
P. Pavelic, D. Suhardiman, O. Keovilignavong, C. Clément, J. Vinckevleugel, S.M. Bohsung, K. Xiong, L. Valee, M. Viossanges, S. Douangsavanh, T. Sotoukee, K.G. Villholth, B.R. Shivakoti, and K. Vongsathiane

Index 365

Preface

For many people, groundwater is fundamental to everyday life. This hidden component of the water cycle provides one-third of the world's water supply and plays a pivotal role in alleviating poverty and sustaining and growing livelihoods. However, case studies examining these complex relationships are almost entirely absent from the academic literature, particularly those that demonstrate the impact that groundwater development has on households, communities, and regions. This missing voice has hampered the ability to demonstrate the positive role that groundwater development can have and therefore the importance of developing groundwater sustainably so that future generations can also benefit.

Groundwater for Sustainable Livelihoods and Equitable Growth helps to bridge this knowledge gap, providing a significant body of evidence for how groundwater has contributed to reducing poverty, increasing resilience to climate and environmental change, and helping to develop equitable growth. The nineteen contributions presented in this book are from sixteen different countries in Africa, Asia, and South America, and despite the local peculiarities, they are all united by the central role that groundwater plays in supporting environmental and human development. The idea for the book arose from the Burdon Groundwater Network for International Development (BGID), which is an official network of the International Association of Hydrogeologists (IAH) created to support the sustainable development and management of groundwater for reducing poverty. Discussions at the BGID sessions at the IAH Congresses in Dubrovnik (2017) and Malaga (2019) were instrumental in getting the book off the ground. The book also benefited from the support of many international researchers who gave their time in reviewing the chapters – a reminder of the dedication of many in the international hydrogeological community to support each other and the study and protection of global groundwater resources.

We hope that this book will shed more light on the fundamental, but often hidden, role that groundwater plays in sustaining livelihoods, particularly for the least well off, and to encourage more interdisciplinary research in this area. With greater understanding of these complex relationships, a commitment to sharing knowledge, and increased investment in groundwater characterization and governance, groundwater will continue to improve people's lives for centuries to come.

August 2021
V. Re, R.L. Manzione, T.A. Abiye,
A. Mukherji, and A. MacDonald

Acknowledgments

The Burdon Groundwater Network for International Development (BGID) is an official network of the International Association of Hydrogeologists (IAH) created to support the sustainable development and management of groundwater for reducing poverty.

The Network is named in memory of the internationally renowned Irish hydrogeologist David Burdon, who had a long-time interest and work in global groundwater, and IAH operates a fund in his name to support these activities. This book is part of the IAH-BGID mission of creating an international groundwater knowledge base contributing to sustainable groundwater management and development in low-income countries.

The editors wish to thank all the chapter authors who contributed to the realization of the book. The COVID-19 pandemic outbreak in 2020 has challenged all of us in completing the book in a timely manner, and we are grateful to each of them for their dedication to sharing their knowledge and their patience with us during this challenging time. We are also thankful to all the external reviewers who gave their time and expertise to improve the book chapters: Alvar Closas, Anderson Luiz Ribeiro de Paiva, Antonio Meira Alves Neto, Archisman Mitra, Bruno Pirilo Conicelli, Carlos Maldaner, Daniel Nkhuwa, Davi de Carvalho Diniz Melo, Dolores Fidelibus, Donald John MacAllister, Fabio Fussi, Jade Ward, James Sorensen, Marie Charlotte Buisson, Matthys Dippenaar, Molla Demlie, Richard Taylor, Rim Trabelsi, Roberto Eduardo Kirchener, Robyn Johnston, Tushaar Shah, and Thokozani Kanyerere.

A special thanks to Nick Robins for his precious support for the realization of the book and to Janjaap Blom at Taylor and Francis for his patience.

Contributors

T.A. Abiye University of the Witwatersrand, Johannesburg, South Africa

T.N. Adji Department of Environmental Geography, Universitas Gadjah Mada, Yogyakarta, Indonesia

W. Akakpo Direction des Ressources en Eau (DRE), Ministère de l'Eau et de l'Hydraulique villageoise, Lomé, Togo

J.P. Albuquerque Federal University of Campina Grande, Campina Grande, Brazil

F. Barbecot GEOTOP, Département des sciences de la Terre et de l'atmosphère, Université du Québec A Montréal (UQAM), Montréal, Canada

R. Barry GEOTOP, Département des sciences de la Terre et de l'atmosphère, Université du Québec A Montréal (UQAM), Montréal, Canada

S.M. Bohsung International Water Management Institute, Vientiane, Lao PDR

A. Cahyadi Karst Research Group, Faculty of Geography, Universitas Gadjah Mada, Yogyakarta, Indonesia, Department of Environmental Geography, Universitas Gadjah Mada, Yogyakarta, Indonesia

G.H. Cavazzana Dom Bosco Catholic University, Campo Grande, Brazil

M. Chakraborty Partners in Prosperity, New Delhi, India

A.E. Cheo Institute for Environment and Human Security, United Nations University, Bonn, Germany

A. Chowdhury Partners in Prosperity, New Delhi, India

C. Clément International Water Management Institute, Vientiane, Lao PDR

B. Collignon Urbaconsulting, Paris, France

S.D. Cota Nuclear Technology Development Center (CDTN/CNEN), Belo Horizonte, Brazil

S. Daschowdhury Consultant, International Water Management Institute, Anand, India

xiv Contributors

A. Djongon GEOTOP, Département des sciences de la Terre et de l'atmosphère, Université du Québec A Montréal (UQAM), Montréal, Canada, Département des sciences de la terre, Université Paris Saclay, laboratoire Géosciences Paris sud, Orsay, France

S. Douangsavanh International Water Management Institute, Vientiane, Lao PDR

G.F. Dourado School of Engineering, University of California, Merced, California

C. Estienne Hydroconseil, Avignon, France

D. Tubbs Filho Departamento de Geociências, Instituto de Agronomia, Universidade Federal Rural do Rio de Janeiro – UFRRJ, Seropédica, Brasil

J.D. Pontes Filho Federal University of Campina Grande, Campina Grande, Brazil, Federal University of Ceará, Fortaleza-CE, Brazil

S.G. Gabas School of Engineering, Architecture and Urbanism and Geography, Federal University of Mato Grosso do Sul, Campo Grande, Brazil

C.O. Galvao Federal University of Campina Grande, Campina Grande, Brazil

D. Gama Department of Forest and Environmental Economics, Sokoine University of Agriculture, Morogoro, Tanzania

J.L. Gudaga Department of Community Development, Amani College of Management and Technology, Njombe, Tanzania

S.M. Hamad Faculty of Natural Resources & Environmental Sciences, University of Omar Al-Mukhtar, Al Baydah, Libya

E. Haryono Karst Research Group, Faculty of Geography, Department of Environmental Geography, Universitas Gadjah Mada, Yogyakarta, Indonesia

A. El Hasia Faculty of Engineering, University of Omar Al-Mukhtar, Baydah, Libya

B. Ibrahim Department of Geology, University of Abdou Moumouni, Niamey, Niger

J.J. Kashaigili Department of Forest Resource Assessment and Management, Sokoine University of Agriculture, Morogoro, Tanzania

O. Keovilignavong International Water Management Institute, Vientiane, South Africa

K.K. Khaing Pathein University, Pathein, Myanmar

L. Kwoyiga Department of Environment, Water and Waste Engineering, University for Development Studies, Tamale, Ghana

G. Lastoria School of Engineering, Architecture and Urbanism and Geography, Federal University of Mato Grosso do Sul, Campo Grande, Brazil

U. Listyaningsih Department of Environmental Geography, Universitas Gadjah Mada, Yogyakarta, Indonesia, Centre for Population Studies and Policy, Universitas Gadjah Mada, Yogyakarta, Indonesia

A. MacDonald British Geological Survey, United Kingdom

M. Majiwa Southern African Development Community-Groundwater Management Institute (SADC-GMI), University of the Free State, Institute for Groundwater Studies, Bloemfontein, South Africa

R.L. Manzione Biosystens Engineering Department (DEB), School of Sciences and Engineering (FCE), São Paulo State University (UNESP), Tupã, São Paulo, Brazil

G.F. Marques Institute of Hydraulic Research, Federal University of Rio Grande do Sul (IPH/UFRGS), Porto Alegre, Brazil

C. Masse Urbaconsulting, Paris, France

C.D.P. Mattiuzi Brazilian Geological Survey (CPRM/SGB), Porto Alegre, Brazil

S.Y. May Department of Water and Environmental Studies, University of Yangon, Yangon, Myanmar

K. Mekonnen International Livestock Research Institute (ILRI), Addis Ababa, Ethiopia.

T. Mkandawire Department Civil Engineering, Faculty of Engineering, Malawi University of Business and Applied Sciences, Chichiri, Malawi

D.B. Mosha Institute of Continuing Education, Sokoine University of Agriculture, Morogoro, Tanzania

N. Mujere Department of Geography, Geospatial Science and Earth Observation, University of Zimbabwe, Harare, Zimbabwe

A. Mukherji International Water Management Institute, New Delhi

B. Munyai Southern African Development Community-Groundwater Management Institute (SADC-GMI), University of the Free State, Institute for Groundwater Studies, Bloemfontein, South Africa

I.A. Nassour Hydroconseil, Avignon, France

P.M. Patel International Water Management Institute, Anand, India

P. Pavelic International Water Management Institute, Vientiane, Lao PDR

S.Y. Pereira Instituto de Geociências, Universidade Estadual de Campinas-UNICAMP, Campinas, Brasil

M. Pulido-Velazquez Institute of Water and Environmental Engineering, IIAMA, Universitat Politècnica de València, Valencia, Spain

V. Re Earth Sciences Department, University of Pisa, Italy

J.C. Rêgo Federal University of Campina Grande, Campina Grande, Brazil

M. Rodriguez École supérieure d'aménagement du territoire et de développement régional (ESAD), Université Laval, Québec, Canada

D. Saha, Central Ground Water Board, Ministry of Jal Shakti, Government of India, New Delhi, India

J. Sauramba Southern African Development Community-Groundwater Management Institute (SADC-GMI), University of the Free State, Institute for Groundwater Studies, Bloemfontein, South Africa

A.S. Schueler Departamento de Arquitetura, Instituto de Tecnologia, Programa de Pós Graduação em Desenvolvimento Territorial e Políticas Públicas, Universidade Federal Rural do Rio de Janeiro – UFRRJ, Seropédica, Brasil

M. Shah International Water Management Institute, Anand, India

T. Shah International Water Management Institute, Anand, India

B.R. Shivakoti Institute for Global Environmental Strategies, Hayama, Japan

T. Sotoukee International Water Management Institute, Vientiane, Lao PDR

T.J. Souza Federal University of Campina Grande, Campina Grande, Brazil

D. Suhardiman International Water Management Institute, Vientiane, Lao PDR

E. Sulistyowati Graduate School on Environmental Science, Universitas Gadjah Mada, Faculty of Science and Technology, UIN Sunan Kalijaga, Yogyakarta, Indonesia

E.G. Tambo Institute for Environment and Human Security, United Nations University, Bonn, Germany

L. Tamene International Centre for Tropical Agriculture (CIAT), Addis Ababa, Ethiopia

B.B. Tsuyuguchi Federal University of Campina Grande, Campina Grande, Brazil

D.A. Uechi School of Engineering, Architecture and Urbanism and Geography, Federal University of Mato Grosso do Sul, Campo Grande, Brazil

L. Valee International Water Management Institute, Vientiane, Lao PDR

F. van Steenbergen MetaMeta Research, The Netherlands

K.G. Villholth International Water Management Institute, Pretoria, South Africa

J. Vinckevleugel International Water Management Institute, Vientiane, Lao PDR

M. Viossanges International Water Management Institute, Vientiane, Lao PDR

K. Vongsathiane Department of Irrigation, Vientiane, Lao PDR

J.S.T. Ward British Geological Survey, London, United Kingdom

M. Widyastuti Karst Research Group, Faculty of Geography, Department of Environmental Geography, Universitas Gadjah Mada, Yogyakarta, Indonesia

K. Woldearegay School of Earth Science, Mekelle University, Mekelle, Ethiopia

K. Xiong International Water Management Institute, Vientiane, Lao PDR

Introduction

Groundwater, sustainable livelihoods and equitable growth

V. Re
University of Pisa

R.L. Manzione
São Paulo State University (UNESP)

T.A. Abiye
University of the Witwatersrand

A. Mukherji
International Water Management Institute

A. MacDonald
British Geological Survey

I INTRODUCTION

From the earliest times, the use of groundwater has been critical for human life and sustaining and growing livelihoods. Perennial springs in east Africa are thought to have helped early humans survive extended drought (Cuthbert et al., 2017). Access to groundwater through springs and shallow wells enabled early civilizations to develop settlements away from rivers and extend the growing season (Angelakis et al., 2020; Tzanakakis et al., 2020). The development of groundwater facilitated growing urbanization in the eastern Mediterranean during Greek and Roman times (Angelakis et al., 2016; Crouch, 1993); however, it is in the last two centuries, and particularly since 1970, that the world has come to rely more on groundwater to underpin growth. For example, India's Green Revolution, which transformed the country from a food deficit to a food surplus, was only made possible due to intensive groundwater irrigation (Mukherji, 2020; Pingali, 2012). Deep boreholes supplied many cities with clean drinking water and provided a reliable water supply for rapid industrial expansion throughout the 19th and 20th centuries, and for the green revolutions in agriculture in the latter half of the 20th century (Siebert et al., 2010). It is clear that groundwater continues to solve many water security issues today, but equally, the invisibility of groundwater means that its critical role is often overlooked. This book aims to provide an initial evidence base for how groundwater is contributing to reducing poverty, increasing resilience to climate and environmental change, and helping to develop equitable growth.

The benefits of groundwater are well understood by users: storage is several orders of magnitude greater than any other freshwater resource (Gleeson et al., 2016).

Groundwater also buffers water supplies against droughts (MacDonald et al., 2019) and has been instrumental in agricultural growth and poverty alleviation in regions that make intensive use of groundwater for irrigation (Salem et al., 2018; Sekhri, 2014). Natural quality is generally good (Lapworth et al., 2020) (with some notable exceptions in some areas for geogenic contamination such as arsenic and fluoride; Selinus et al., 2013), and groundwater can be found in many different environments, often close to the point of need. These benefits can deliver water, food, and livelihood security, and help in reducing current poverty. However, although the benefits are enjoyed, groundwater itself is poorly understood, and continued unregulated development can lead to overexploitation (Gleeson et al., 2020) and degradation in certain hotspots (Elshall et al., 2020). Degradation reduces the benefits, for example, in the form of reduced access to reliable irrigation (Asoka & Mishra, 2020; Jain et al., 2021); reduced adaptive capacity of rural populations (Blakeslee et al., 2020; Fishman, 2018); quality deterioration (Mas-Pla & Menció, 2019); higher costs for pumping and treatment (McDonough et al., 2020; Turner et al., 2019); land subsidence (Shirzaei & Bürgmann, 2018); and poor ecosystem services (WLE, 2015). However, in many parts of the world, and particularly in Africa, groundwater is still under-used (Cobbing & Hiller, 2019), and there is still considerable potential for new groundwater development to contribute to sustainable livelihoods. What is clear is that groundwater resources and the sustainability of water supplies need to be better understood and managed for the benefits of groundwater to continue to accrue to society.

Recent studies examining how to ensure long-term and equitable benefits from groundwater development suggest the importance of strengthening the interactions between the science–policy–practice interface in groundwater systems (Milman & MacDonald, 2020), advancing groundwater governance (Villholth et al., 2019) and promoting transdisciplinary approaches, integrating the social dimension into hydrogeological assessments (Hynds et al., 2018; Re, 2015). Regardless of the angle of observation, all the recent debates on groundwater sustainability have two elements in common: fostering the connection and engagement with water users and managers; and promoting information and data sharing. Groundwater information (also critical for the achievement of SDG6) is only of use if shared in a timely manner; otherwise, possible remedial actions may already have become obsolete (Re et al., 2018).

Key to success is sharing lessons and experiences of groundwater development. For this reason, this book aims to provide a significant body of case studies in which groundwater has contributed to reducing poverty, increasing resilience to climate and environmental change, and helping to develop equitable growth. Such case studies and examples are lacking in the academic literature, particularly those that demonstrate the impact that groundwater development has had on households, communities, and regions. This missing voice has hampered the ability to demonstrate the positive role that groundwater development can have, and therefore, the importance of developing groundwater sustainably so that future generations can also benefit.

2 CONTRIBUTIONS TO THE BOOK

In this book, there is an attempt to cover different views and perspectives of groundwater development and management focusing on supporting livelihoods and reducing poverty worldwide. There are nineteen contributions from four different continents,

Africa, Asia, South America, and Oceania, and sixteen different countries: Brazil, Cameroon, Chad, Ethiopia, Ghana, India, Indonesia, Laos, Libya, Myanmar, Niger, Pakistan, South Africa, Tanzania, Togo, and Zimbabwe.

From South America, the studies encompass the four corners of Brazil. Marques et al. (2022) established operational and water management strategies to build resilience, water security, and adaptation for the conjunctive use of surface and groundwater in the south. From the mid-western region, Gabas et al. (2022) presented the role of groundwater in economic and social development, focusing on agricultural activities and minorities such as indigenous and quilombolas (former slaves) communities. Rêgo et al. (2022) highlighted the importance of alluvial aquifers in the Brazilian semiarid region. Tubbs et al. (2022) reflect upon an important water source of Rio de Janeiro city, the Guandu River basin, the management of private domestic wells, and its implications on water governance for coastal areas.

In Asia, Haryono et al. (2022) showed how community-based karst tourism recovered the Gunungsewu region in Indonesia, after the creation of a geopark. Pavelic et al. (2022) featured small-scale groundwater irrigation solutions in Laos, and how access to groundwater irrigation provided farmers with options to irrigate additional crops and improve livelihoods in the non-rainy season. From Myamar, May et al. (2022) presented the case of the Six Villages of Taunggyi, Shan State, where water is taken directly from springs without any treatment. Several contributions came from India: Shah et al. (2021) discussed the impact of energy tariff changes from a field pilot in Birbhum district (West Bengal state), showing the close linkages between electricity pricing and tariffs and groundwater use incentives; Patel and Saha (2022) explained the importance of groundwater and its rational use in the post-2001 earthquake in the Kutch district (Gujarat state); and Saha et al. (2022) examined the links between an Act that promulgates postponement of the date of paddy sowing to reduce groundwater extraction and increase in stubble burning later in the season, and found a complex relationship with other factors such as access to technology.

In the African continent, Cheo et al. (2022) compared two case studies from Cameroon and Niger, drawing synergies between the two countries for groundwater interventions in specific livelihood and agro-ecological zones. Woldearegay et al. (2022) focused on the Tigray region in northern Ethiopia and the continuous improvements in land and water management methods over the years to ensure food and water security. Hammad and Hasia (2022) showed the long-term importance of the Al Jabal Al Akhdar area karst springs, a source of water for local communities since ancient Greek and Roman civilizations. Collignon et al. (2022) created an atlas based on an extensive groundwater inventory, aiming to optimize investments to ensure water security in the northern part of Lake Chad. From Ghana, Kwoyiga (2022) explored how informal groundwater development has filled the gap between demand and municipal supply and enabled peri-urban growth in the Techiman municipality. Mosha et al. (2022) showed how groundwater development improves agricultural livelihoods by surveying three villages in Usangu Plains, Tanzania by undertaking an economic analysis to demonstrate the returns to groundwater investment. In Togo, Barry et al. (2022) presented how groundwater is essential to the population of Lomé, projecting its use in urban and coastal areas and the vulnerability to climate change. In Zimbabwe, Mujere (2022) studied the groundwater available in a suburb of Harare and the value of groundwater quality to prevent outbreaks of water-borne and water-related diseases.

3 CHALLENGES AND FUTURE PERSPECTIVES

Finally, in Malawi, Sauramba et al. (2022) presented the legal framework established between 2017 and 2019 in order to identify gaps and challenges to develop an action plan for groundwater.

3 CHALLENGES AND FUTURE PERSPECTIVES

The papers in this collection cover many different perspectives using a diverse set of methodologies and discuss successes, challenges, and opportunities common to many groundwater-dependent communities worldwide. Common to many were the tradeoffs required for society to both benefit from groundwater and protect the quality and quantity of the resource to sustain future use. This requires a good understanding of aquifers, the dynamics of the groundwater system, and the current abstraction rates and contaminants while placing groundwater in specific legal, economic, and social contexts. Many of the studies reflected on the paucity of sound data on groundwater and aquifer features at local, regional, and transboundary levels, despite the high dependency of local communities to this resource, which limited the ability to reliably forecast future opportunities and threats.

Data scarcity, or limited access to existing data, is often the result of limited long-term investments and a lack of meaningful interactions at the science– policy – practice interface for groundwater (Milman & MacDonald, 2020). Indeed, it is clear that improved monitoring is key for the successful achievement of SDG6, as evidenced by the recently launched UN-Water SDG 6 Data Portal (UN Water, 2021), although even here, groundwater is under-represented given the scale of groundwater use. New advances in data and information sharing offer opportunities to coordinate and plan effective groundwater monitoring and aquifer characterization, and make this information widely available and accessible. The rapid advance of open-access publications is providing opportunities to level up access to information, and facilitate lesson learning and knowledge transfer.

In some areas, the lack of qualified groundwater professionals was also highlighted as a threat to the continued use of groundwater for growth and livelihoods. Groundwater is often informally developed and poorly regulated with the result that expertise in inexpensive drilling and pump installation is valued over expertise in understanding the resource or developing it sustainably (Healy et al., 2020). Increasing the knowledge and capacity of users to understand more how groundwater behaves and how its sustainable use is linked to ongoing livelihood security may help develop demand for increased groundwater professionalization and regulation. Looking to the future, the implementation of transdisciplinary approaches may help to foster the dialogue among scientists, policymakers, and users, thus resulting in improved groundwater governance and sustainable benefits for all.

There is compelling evidence that the development of groundwater has profoundly improved many people's lives and continues to lift people out of poverty today. The examples in this book provide a wide variety of case studies from Asia, Africa, and South America that show how groundwater, often invisibly, improves people's lives and livelihoods and promotes equitable growth. However, the studies also demonstrate how groundwater can be over-used and contaminated, and how ignorance of the nature of groundwater is one of the greatest threats to its sustainable use. It is, therefore, of paramount importance to help an interdisciplinary community grow to facilitate sharing knowledge and to increase investment in characterizing, monitoring, and governing groundwater, so that groundwater will continue to improve people's lives for centuries to come.

REFERENCES

Angelakis, A. N., Voudouris, K. S. & Mariolakos, I. (2016) Groundwater utilization through the centuries focusing on the Hellenic civilizations. *Hydrogeol. J.* **24**(5), 1311–1324. Springer Verlag. doi:10.1007/s10040-016-1392-0.

Angelakis, A. N., Zaccaria, D., Krasilnikoff, J., Salgot, M., Bazza, M., Roccaro, P., Jimenez, B., et al. (2020, May 1) Irrigation of world agricultural lands: Evolution through the Millennia. *Water (Switzerland)*. MDPI AG. doi:10.3390/W12051285.

Asoka, A. & Mishra, V. (2020) A strong linkage between seasonal crop growth and groundwater storage variability in India. *J. Hydrometeorol.* **22**(1), 125–138. American Meteorological Society. doi:10.1175/JHM-D-20–0085.1.

Blakeslee, D., Fishman, R. & Srinivasan, V. (2020) Way down in the hole: Adaptation to long-term water loss in rural India. *Am. Econ. Rev.* **110**(1), 200–224. American Economic Association. doi:10.1257/aer.20180976.

Cobbing, J. & Hiller, B. (2019) Waking a sleeping giant: Realizing the potential of groundwater in Sub-Saharan Africa. *World Dev.* **122**, 597–613. Elsevier Ltd. doi:10.1016/j.worlddev.2019.06.024.

Crouch, D. P. (1993) *Water Management in Ancient Greek Cities. Water Manag. Anc. Greek Cities*. Oxford University Press. doi:10.1093/oso/9780195072808.001.0001.

Cuthbert, M. O., Gleeson, T., Reynolds, S. C., Bennett, M. R., Newton, A. C., McCormack, C. J. & Ashley, G. M. (2017) Modelling the role of groundwater hydro-refugia in East African hominin evolution and dispersal. *Nat. Commun.* **8**. Nature Publishing Group. doi:10.1038/ncomms15696.

Elshall, A. S., Arik, A. D., El-Kadi, A. I., Pierce, S., Ye, M., Burnett, K. M., Wada, C. A., et al. (2020, September 1) Groundwater sustainability: A review of the interactions between science and policy. *Environ. Res. Lett.* IOP Publishing Ltd. doi:10.1088/1748–9326/ab8e8c.

Fishman, R. (2018) Groundwater depletion limits the scope for adaptation to increased rainfall variability in India. *Clim. Change* **147**(1–2), 195–209. Springer Netherlands. doi:10.1007/s10584-018-2146-x.

Gleeson, T., Befus, K. M., Jasechko, S., Luijendijk, E. & Cardenas, M. B. (2016, February 1) The global volume and distribution of modern groundwater. *Nat. Geosci.* Nature Publishing Group. doi:10.1038/ngeo2590.

Gleeson, T., Cuthbert, M., Ferguson, G. & Perrone, D. (2020) Global groundwater sustainability, resources, and systems in the anthropocene. *Annu. Rev. Earth Planet. Sci.* **48**(1), 431–463. Annual Reviews Inc. doi:10.1146/annurev-earth-071719-055251.

Healy, A., Upton, K., Capstick, S., Bristow, G., Tijani, M., Macdonald, A., Goni, I., et al. (2020) Domestic groundwater abstraction in Lagos, Nigeria: A disjuncture in the science-policy-practice interface? *Environ. Res. Lett.* **15**(4), 045006. Institute of Physics Publishing. doi:10.1088/1748-9326/ab7463.

Hynds, P., Regan, S., Andrade, L., Mooney, S., O'Malley, K., DiPelino, S. & O'Dwyer, J. (2018) Muddy waters: Refining the way forward for the "sustainability science" of socio-hydrogeology. *Water* **10**(9), 1111. MDPI AG. doi:10.3390/w10091111.

Jain, M., Fishman, R., Mondal, P., Galford, G. L., Bhattarai, N., Naeem, S., Lall, U., et al. (2021) Groundwater depletion will reduce cropping intensity in India. *Sci. Adv.* **7**(9), eabd2849. American Association for the Advancement of Science. doi:10.1126/sciadv.abd2849.

Lapworth, D. J., MacDonald, A. M., Kebede, S., Owor, M., Chavula, G., Fallas, H., Wilson, P., et al. (2020) Drinking water quality from rural handpump-boreholes in Africa. *Environ. Res. Lett.* **15**(6), 064020. Institute of Physics Publishing. doi:10.1088/1748–9326/ab8031.

MacDonald, A. M., Bell, R. A., Kebede, S., Azagegn, T., Yehualaeshet, T., Pichon, F., Young, M., et al. (2019) Groundwater and resilience to drought in the Ethiopian highlands. *Environ. Res. Lett.* **14**(9), 095003. Institute of Physics Publishing. doi:10.1088/1748–9326/ab282f.

Mas-Pla, J. & Menció, A. (2019) Groundwater nitrate pollution and climate change: learnings from a water balance-based analysis of several aquifers in a western Mediterranean region (Catalonia). *Environ. Sci. Pollut. Res.* **26**(3), 2184–2202. Springer Verlag. doi:10.1007/s11356-018-1859-8.

McDonough, L. K., Santos, I. R., Andersen, M. S., O'Carroll, D. M., Rutlidge, H., Meredith, K., Oudone, P., et al. (2020) Changes in global groundwater organic carbon driven by climate change and urbanization. *Nat. Commun.* **11**(1), 1–10. Nature Research. doi:10.1038/s41467-020-14946-1.

Milman, A. & MacDonald, A. (2020, September 1) Focus on interactions between science-policy in groundwater systems. *Environ. Res. Lett.* Institute of Physics Publishing. doi:10.1088/1748-9326/aba100.

Mukherji, A. (2020) Sustainable groundwater management in India needs a water-energy-food nexus approach. *Appl. Econ. Perspect. Policy.* John Wiley & Sons, Ltd. doi:10.1002/AEPP.13123.

Pingali, P. L. (2012) Green revolution: Impacts, limits, and the path ahead. *Proc. Natl. Acad. Sci.* **109**(31), 12302–12308. National Academy of Sciences. doi:10.1073/PNAS.0912953109.

Re, V. (2015) Incorporating the social dimension into hydrogeochemical investigations for rural development: The Bir Al-Nas approach for socio-hydrogeology. *Inc. la Dimens. Soc. en las Investig. hidrogeológicas para el Desarro. Rural el enfoque Bir Al-Nas para la socio-hidrogeología.*

Re, V., Thin, M. M., Setti, M., Comizzoli, S. & Sacchi, E. (2018) Present status and future criticalities evidenced by an integrated assessment of water resources quality at catchment scale: The case of Inle Lake (Southern Shan state, Myanmar). *Appl. Geochem.* **92**(March), 82–93. Elsevier. doi:10.1016/j.apgeochem.2018.03.005.

Salem, G. S. A., Kazama, S., Shahid, S. & Dey, N. C. (2018) Groundwater-dependent irrigation costs and benefits for adaptation to global change. *Mitig. Adapt. Strateg. Glob. Chang.* **23**(6), 953–979. Springer Netherlands. doi:10.1007/S11027-017-9767-7.

Sekhri, S. (2014) Wells, Water, and welfare: The impact of access to groundwater on rural poverty and conflict. *Am. Econ. J. Appl. Econ.* **6**(3), 76–102. American Economic Association. doi:10.1257/APP.6.3.76.

Selinus, O., Alloway, B., Centeno, J. A., Finkelman, R. B., Fuge, R., Lindh, U. & Smedley, P. (2013) *Essentials of medical geology: Revised edition. Essentials Med. Geol. Revis. Ed.* Springer Netherlands. doi:10.1007/978-94-007-4375-5.

Shirzaei, M. & Bürgmann, R. (2018) Global climate change and local land subsidence exacerbate inundation risk to the San Francisco Bay Area. *Sci. Adv.* **4**(3). American Association for the Advancement of Science. doi:10.1126/sciadv.aap9234.

Siebert, S., Burke, J., Faures, J. M., Frenken, K., Hoogeveen, J., Döll, P. & Portmann, F. T. (2010) Groundwater use for irrigation - A global inventory. *Hydrol. Earth Syst. Sci.* **14**(10), 1863–1880. doi:10.5194/hess-14-1863-2010.

Turner, S. W. D., Hejazi, M., Yonkofski, C., Kim, S. H. & Kyle, P. (2019) Influence of groundwater extraction costs and resource depletion limits on simulated global nonrenewable water withdrawals over the twenty-first century. *Earth's Futur.* **7**(2), 123–135. John Wiley and Sons Inc. doi:10.1029/2018EF001105.

Tzanakakis, V. A., Paranychianakis, N. V. & Angelakis, A. N. (2020, August 21) Water supply and water scarcity. *Water (Switzerland).* MDPI AG. doi:10.3390/w12092347.

UN Water. (2021) *SDG6 Data*. Retrieved February 4, 2021. Retrieved from https://www.sdg6data.org/.

Villholth, K.G., Lopez-Gunn, E., Conti, K., Garrido, A. & Gun, J. Van Der (Eds.). (2019) *Advances in Groundwater Governance - 1st Edition*. CRC Press. Retrieved from https://www.routledge.com/Advances-in-Groundwater-Governance/Villholth-Lopez-Gunn-Conti-Garrido-Gun/p/book/9780367890100.

WLE. (2015) *Groundwater and ecosystem services: A framework for managing smallholder groundwater-dependent agrarian socio-ecologies - applying an ecosystem services and resilience approach. Groundw. Ecosyst. Serv. a Framew. Manag. Smallhold. groundwater-dependent Agrar. socio-ecologies - Appl. an Ecosyst. Serv. Resil. approach*. International Water Management Institute (IWMI). CGIAR Research Program on Water, Land and Ecosystems (WLE). doi:10.5337/2015.208.

FULL DETAILS OF THE BOOK CHAPTERS (CITED IN PARAGRAPH 2)

Barry, R., Barbecot, F., Rodriguez, M., Djongon, A. & Akakpo, W. (2022) Urban development and intensive groundwater use in African coastal areas: The case of Lomé urban area in Togo.

Cheo, A. E., Ibrahim, B. & Tambo, E.G. (2022) Groundwater resources development for livelihoods enhancement in the Sahel Region - Case Study of Niger.

Collignon, B., Estienne, C., Masse, C. & Nassour, I.A. (2022) The Quaternary aquifer: an affordable resource to address water scarcity in the northern part of the Lake Chad basin.

Gabas, S. G. Dourado, G. F., Uechi, D. A, Cavazzana, G. H. & Lastoria, G. (2022) The role of groundwater in economic and social development of Mato Grosso do Sul State, Midwest of Brazil.

Hamad, S. M., & El Hasia, A. (2022) An overview of Karst groundwater springs in Al Jabal Al Khader region (North East Libya).

Haryono, E., Adji, T. N., Cahyadi, A., Widyastuti, M., Listyaningsih, U. & Sulistyowati, E. (2022) Groundwater and livelihood in Gunungsewu Karst Area, Indonesia.

Kwoyiga, L. (2022) Groundwater, informal abstraction and peri-urban dwellers in the Techiman Municipality of Ghana.

Marques, G. F., Mattiuzi, C. D. P., Cota, S. D. S & Pulido-Velazquez, M. (2022) Conjunctive use of surface and groundwater: Operational and water management strategies to build resilience, water security and adaptation.

May, S Y, Khaing K. K. & Ward J. S. T. (2022) The role of groundwater in rural water supply: The case of six villages of Taunggyi District, Southern Shan State, Myanmar.

Mosha, D. B., Gudaga, J. L., Gama, D. & Kashaigili J. J. (2022) Valuing groundwater use: Resolving the potential of groundwater in the Upper Great Ruaha River Catchment of Tanzania.

Mujere N (2022) Contribution of groundwater towards urban household water security.

Patel, P. M. & Saha D. (2022) Groundwater: A juggernaut of socio-economic development and stability in the Arid Region of Kachchh.

Pavelicl, P., Suhardiman, D., Keovilignavong, O., Clément, C., Vinckevleugel, J., Bohsung, S. M., Xiong, K., Valee, L., Viossanges, M., Douangsavanh, S., Sotoukee, T., Villholth, K.G., Shivakoti, B. R. & Vongsathiane, K. (2022) Assessment of options for small-scale groundwater irrigation in Lao PDR.

Rêgo, J. C., Albuquerque, J. P., Pontes Filho, J. D., Tsuyuguchi, B. B., Souza, T. J. & Galvao, C. O. (2022). Sustainable and resilient exploitation of small alluvial aquifers in the Brazilian semi-arid region: The experience of Sumé.

Saha, D., Chakraborty, M., & Chowdhury, A. (2022) Stubble burning in northwestern India- is it related to groundwater over-exploitation?

Sauramba, J., Mkandawire, T., Munyai, B. & Majiwa, M. (2022) Groundwater policy, legal, and institutional framework situation analysis: Gaps and action plan: the case of Malawi.

Shah, M., Shah, T. & Daschowdhury, S. (2022) Groundwater-driven paddy farming in West Bengal. How a smallholder-unfriendly farm power policy affects livelihoods of farmers.

Tubbs D. F., Soares de Schueler, A. & Pereira, S. Y. (2022) The Governance and Water Security of Groundwater Obtained from Private Domestic Wells in Periurban Areas in Brazil: The Case Study of the Guandu River Basin in the Metropolitan Region of Rio de Janeiro, Brazil.

Woldearegay, K., Tamene, L., van Steenbergen, F., & Mekonnen, K. (2022) Groundwater Recharge through Landscape Restoration and Surface Water Harvesting for Climate Resilience: The case of Upper Tekeze River Basin, Northern Ethiopia.

Chapter 1

Groundwater and livelihood in Gunungsewu karst area, Indonesia

E. Haryono, T.N. Adji, A. Cahyadi, M. Widyastuti, and U. Listyaningsih
Universitas Gadjah Mada

E. Sulistyowati
UIN Sunan Kalijaga

CONTENTS

1.1 Introduction...1
1.2 Hydrogeology and water resource situation 4
1.3 Groundwater extraction and water supply development 9
1.4 Livelihood and water scarcity adaptation strategy............................10
    1.4.1 Living condition and adaptation strategy 11
        1.4.1.1 Seasonal crop farming 11
        1.4.1.2 Agroforestry..13
        1.4.1.3 Livestock raising...14
        1.4.1.4 Migration 15
        1.4.1.5 Groundwater-related tourism 16
    1.4.2 Water consumption pattern...17
1.5 Impact of water supply development 18
1.6 Summary 19
Acknowledgments...20
References..20

1.1 INTRODUCTION

Gunungsewu karst covers $1,300\,km^2$, extending $84\,km$ from Parangtritis Beach, Bantul Regency-Special Province of Yogyakarta to Telengria Beach, Pacitan Regency – East Java Province (Figure 1.1). This area comprises the Wonosari-Punung Formation, a Neogen Limestone composed of massive coral limestone and bedded chalky limestone. The basement of the limestone formation is mostly a Miocene volcanic clastic sedimentary rocks of Wuni Formation, Sambipitu Formation, Semilir Formation, Nglanggran Formation, and Nampol Formation (Table 1.1).

The karst of the Gunungsewu karst area has conical/Kegel karst morphology with cockpit-type closed depression (Haryono and Day, 2004; Tjia, 2013). The cockpit is the area where people are inhabiting and growing crops for staple food (Haryono et al., 2016). The prevailing climate of the area is monsoon, with annual precipitation from 1,600 to $2,000\,mm$. Unfortunately, the people of these areas are suffering from water shortages due to difficult access to surface and groundwater resources. Most of the

DOI: 10.1201/9781003024101-1

2 Groundwater for Sustainable Livelihoods

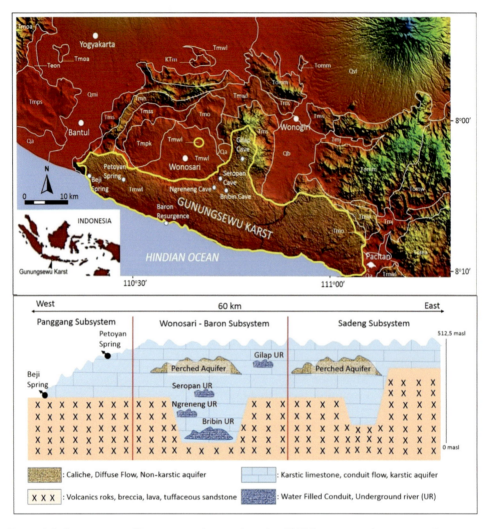

Figure 1.1 Gunungsewu Karst area, depicted in the SRTM image with stratigraphic units. (Crossed section (no scale) is modified from Kusumayudha (2015), showing the basement rock structure governing the main underground river system of the area. Stratigraphic units are extracted from the Geological map of Yogyakarta and Surakarta Quadrangle (Surono et al., 1992; Rahardjo et al., 1995). Springs and underground river extraction sites are shown in the figure in a small solid white circle.)

rainwater sinks to the underground through ponors and sinking streams. The underground rivers are at a depth of 60–100 m or more and are found only in a few places. Therefore, local people neither have access to the groundwater nor are they able to make their own water supply system. Even, the local government and water supply companies facing difficulties in providing water supply. The government is still exploring the possible new caves with underground rivers in search of water supply.

Table 1.1 Geological formation in Gunungsewu karst area and its surrounding

Geological formation	Description
Wonosari-Punung formation (Tmwl)	Bedded limestone, massive limestone, and reefs
Quartenary alluvium (Qa)	Alluvium
Baturetno formation (Qb)	Quaternary deposits, consisting of sand, gravel, clay (dominated by black clay)
Lawu volcanic rock (Qvl)	Tuff and volcanic breccia, andesitic lava
Merapi volcanic rock (Qmi)	Tuff and volcanic breccia, lava, Lahar deposits
Nglanggran formation (Tmn)	Volcanic breccias, agglomerates and several basaltic andesite intrusions, pillow lavas
Semilir formation (Tms)	Andesitic tuffs, sandstones, agglomerates, claystone, siltstone, shale, and andesitic to basaltic breccias
Bemmelen formation (Tmoa)	Andesite
Jaten formation (Tmj)	Quarzt sandstone, tuffaceous sandstone, shale
Watu Patok formation (Tomw)	Lava, intercalation of sandstone, claystone, chert
Oyo formation (Tmo)	Mixed marine-volcanic facies of very well bedded tuffaceous marl, clays, andesitic tuffs, conglomeratic limestones
Metamorphic rock (Ktm)	Schist, marble, meta volcanic rock, metasediment, slate
Kepek formation (Tmpk)	Claystone, sandy marl, calcarenite, bedded limestone
Sentolo formation (Tmps)	Agglomerates and marl, interbedded limestone
Arjosari formation (Toma)	Agglomerate, sandstone, siltstone, limestone, claystone, volcanic breccia, lava, and tuff
Sambipitu formation (TmSS)	Mudstone, interbedded calcareous sandstones, marl, tuffaceous sandstones
Nanggulan formation (Teon)	Sandstone, quartz sandstone, calcareous sandstone, claystone, fossiliferous claystone, calcareous claystone, siltstone, and coal seam intercalations
Mandalika fromation (Tomm)	Volcanic breccia, tuff, lava, sandstone, siltstone, claystone
Wuni formation (Tmw)	Volcanic breccia, tuff, tuffaceous sandstone, lithic sandstone, mudstone, lignite, tuff, carbonaceous shale

Source: Compiled from Surono et al. (1992) and Rahardjo et al. (1995).

Surface water is available as doline ponds (*telaga*) and springs. The ponds occur when the ponors are clogged by clayed sediment in the base of cockpits. However, most of the doline ponds are drying up due to deforestation, sedimentation, and leakages (Haryono et al., 2009; Widyastuti and Haryono, 2016). But karst springs emerge only in some localities and fulfill only the water supply for the surrounding villages. Therefore, the primary water source is groundwater as an underground river. Four caves with underground rivers are already in use for water supply, i.e., Seropan, Bribin, Ngobaran, and Baron. However, the water supply is not able to meet water demand for the whole karst area.

Villages are distributed with a population density of 496 people/km^2. The people make a living from agriculture, tourism, agroforestry, and mining sectors. The agricultural carrying capacity of the karst area is 3.14 (Haryono, 2011), indicating that the population number is already 3.14 times higher than food-related agricultural production. People grow rainfed paddy, maize, groundnut, and cassava for daily needs. Besides, they raise cattle for medium-term savings and grow teak for long-term savings. Not to mention, agriculture productivity is low compared to the surrounding areas. Rainfed agriculture makes people more vulnerable to climate- and water-related

4 Groundwater for Sustainable Livelihoods

issues. The rice productivity is only 4 quintals/ha, much lower than the average rice production of the surrounding regency, which is up to 62 quintals/ha.

Low agricultural productivity and difficult water access push the people to migrate to other cities, islands, or even other countries (Purnomo, 2004, 2009). Migration is one of the strategies for livelihood in the Gunungsewu karst. The productive people, usually after completing secondary school, migrate circularly to the other cities. During the last 20 years, farmers extract limestone for livelihood. The other recent livelihood diversification is tourism by utilizing caves and underground rivers. They use underground rivers for cave-tubing activities, show caves, and water-related activities. This chapter herein summarizes the livelihoods of the Gunungsewu karst concerning drought due to difficult access to groundwater. Most of the data are from Gunungkidul Regency (GK) covering 60% of the Gunungsewu karst area.

1.2 HYDROGEOLOGY AND WATER RESOURCE SITUATION

The hydrological conditions in the Gunungsewu karst area are not much different from other karst areas. The main water flow system is through an underground river network that is recharged by diffuse infiltration, allogenic rivers, and autogenic percolation. Karstic aquifers in Gunungsewu are characterized by secondary porosity governed by the enlargement of voids resulting from the dissolution process. However, local variations can still be distinguished based on geological structures and lithological variations. Haryono (2000) suggested that the porosity of Gunungsewu karst in epikarst zones varies in several regions that have different surface morphology due to differences in the level of karst development. Chalky limestone and mudstone have lower secondary porosity compared to hard reef limestone. The distribution of porosity in the epikarst layer in the Gunungsewu karst region is shown in Table 1.2.

Table 1.2 Distribution of porosity in the epikarst zone in the three karst areas of Gunungsewu, Gunungkidul Regency

Sample area	Characteristics	Porosity (%)		
		Rocks	Solutionally cavities	Infilled material
Polygonal karst of Panggang	Coral limestone, hard and shallow, severe karren and secondary porosity, springs abundant.	1.1–14.0	22.52	40.0–58.9
Labyrinth karst of Paliyan and Tanjungsari	Deep and hard coralline limestone, well-developed karren and secondary porosity, intensive surface dry valley, few springs are found.	13.0–16.6	22.0–52.0	36.6–40.2
Residual cone of Bedoyo	Bedded limestone, soft and deep, karren does not develop well, planation is extensive, and springs are rare.	23.1–48.2	<10	20.6–31.9

Source: From Haryono (2000).

Allogenic recharge from surface rivers and autogenic recharge through a sinkhole/ponor are also widespread in the Gunungsewu karst region. The recharging point mostly comes from the sinkhole that drains water from closed depression during the rainy season. Recharge from the ponor brings suspended load to the springs and underground rivers after rain events. Interviews with residents indicate that, in general, water will return to clear in at least 2 days after the rain stop. Small springs with a discharge of fewer than 5 L/s are usually less affected by the sinkhole recharge to have clear water throughout the year. Adji (2012) explains that autogenic recharge through the ponor at Bribin underground river has time to peak (Tp) about 3 hours in the upper stream (Gilap cave) and 5.5 hours downstream (Bribin cave). The fact also shows that autogenic recharge only contributes a small percentage of the total discharge in the Bribin underground river, which is around 18%–22% (Adji, 2011), while what contributes significantly is percolation water from the epikarst zone in the form of diffuse and fissure flow.

The hydrogeological system in the Gunungsewu karst is controlled by the major underlying geological structures that make up this area. The interpretation of 1:50,000 scale aerial photographs and 1:1,000,000 scale European Remote-Sensing Satelite (ERS) imagery by Kusumayudha (2005) shows that the crack structure consisting of faults and fractures has a general direction of northwest–southeast and northeast–southwest. These structures then divide this area into blocks bounded by faults that also regulate the hydrogeological system due to the creation of low and high configurations (Figure 1.1).

The Panggang Hydrogeology System lies in the Polygonal karst (the western part of the Gunungkidul karst area, GK). The limestone bed in the system is 15–30 m thick, so that karst springs with small discharge and doline pond become the main features of this region. The Wonosari-Baron Hydrogeological System is located in the middle part of the Gunungkidul karst, which is classified as Labyrinth-cone karst. The limestone bed is around 600 m in depth forming vertical caves, allogenic rivers, and underground rivers. Bribin-Baron underground river is an essential system in this area. Tracer test studies by McDonald and Partners (1984) showed that the Bribin underground river (1,736 L/s) was recharged by the Pentung allogenic river (23 L/s), Jombangan cave (80 L/s), Jomblangbanyu cave, and Gilap cave (40.1 L/s). This flow has leaked into Ngreneng underground river (311 L/s), which according to Sidauruk et al. (2018), was also recharged from Seropan underground river (875.9 L/s). This system then empties into the Indian Ocean (Baron spring – 8,000 L/s) (Figure 1.2).

Furthermore, the Ponjong Hydrogeological Subsystem (the upstream part of the Bribin-Baron system) is characterized by the appearance of springs along the fault that separates the karst area from the Wonosari basin (Haryono, 2000). This system is also affected by allogenic rivers originating from the old volcanic catchment, so that has several springs with large discharge, such as Beton (1,192 L/s) and Gremmeng (1,870 L/s). Finally, Agniy et al. (2017) explains the Pindul Hydrogeological System's existence in the physiography of the Wonosari basin. Tracer tests at this location indicate that the system was recharged by an underground flow network originating from some caves (Asri, Greng, Emas, Candi, Suruh, and Sioyot caves), which are interconnected from East to West, and are also being recharged by the allogenic river of Kedungbuntung (Figure 1.2).

Recent research tries to examine more closely the hydrological characteristics in Gunungsewu, which include flow and hydrogeochemical characteristics, to detail the

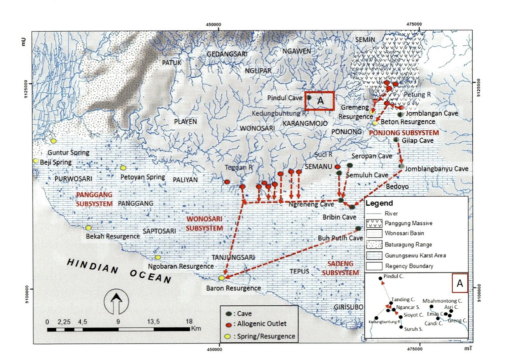

Figure 1.2 Ponjong-Bribin-Baron system. (Modified from MacDonalds and Partners, 1984.)

relationship with the level of karst development in Gunungsewu, Gunungkidul Regency. Adji et al. (2017a) explain that the Bribin and Ngreneng underground rivers and the Beton spring (located in the eastern part of Gunungsewu) have the most developed karst hydrological system. This fact is evidenced by the Master Recession Curve (MRC) analysis, which shows that the aquifer storage releases two types of conduit flow (turbulent). Meanwhile, the release of water storage in the Seropan and Gilap underground rivers shows that there is one type of conduit flow reflecting that the karst in both locations has developed sufficiently. In contrast, the Petoyan spring (located in the western part of Gunungsewu) is one of the springs whose recharge predominantly originates from the epikarst zone in the Polygonal karst (Adji and Bahtiar, 2016). This spring only has a diffuse (laminar) flow type in the release of its water storage, so it can be concluded that the catchment area has a low karst hydrological system development.

Besides, the hydrogeochemical analysis conducted by Adji et al. (2017b) reinforces these findings. Petoyan springs, which are dominated by diffuse flow, have a strong relationship between fluid conductivity and calcium (Ca^{2+}) and bicarbonate (HCO_3^-) throughout the year. In contrast, Gilap underground river and Jomblangan cave (eastern part of Gunungsewu) have a low relationship between hydrochemical parameters. This is caused by the input from autogenic, allogenic, and rainwater entering through the conduit void that has already developed (point-recharged type). Meanwhile, the analysis of the flow duration curve (FDC) conducted by Nurkholis et al. (2019) states that developed karst in Gunungsewu is still capable of having ample groundwater storage, as found in the Pindul Hydrogeological System. The same condition is also found in the Bribin underground river, which has a large and stable average discharge, even though conduit-type voids have developed (Adji, 2012). In conclusion, the karst hydrological system in Gunungsewu, Gunungkidul Regency in the eastern and central parts is more complex and more developed when compared to that found in the western part of Gunungsewu (Figure 1.2).

Underground rivers in the Gunungsewu karst area are concentrated in the Wonosari subsystem. This is due to the shallow underlying rock of the Wonosari subsystem, as shown in Figure 1.1. Besides, the Wonosari subsystem has the most allogeneic recharge, mainly originating from the Suci River and Sumurup River. Both allogenic rivers are perennial rivers. This allogenic recharge enters the underground river system in the Wonosari subsystem and emerges as a resurgence in the Baron spring (Figures 1.2 and 1.3).

People in Gunungsewu karst use four different water sources, rainwater, doline pond, karst spring, and underground river (Figure 1.4). Rainwater is collected from the roof and is diverted to a water tank. People use rainwater for about 4 months during the rainy season. The volume of the water tank is $3 m^3$. The rainwater is run out 3 weeks after the rain stopped. Those who do not have access to piped water, buy water from the nearest karst spring and store it in the rainwater tank. The average daily water usage per capita in the area is 50 L/s. Therefore, a family with five members must buy water every 10–15 days during the dry season. During the dry season, the average price of water at water vendors is about Rp. 150,000 (US$ 11.047) per $8 m^3$. This amount is higher than the water tariff from the Regional Water Company (PDAM) Handayani, which costs Rp. 45,000 (US$ 4.31) for the monthly connection fee and Rp. 3,000 (US$ 0.224) per $5 m^3$ of water, and thus customers only need to pay around Rp. 48,000 (US$ 3.53) to Rp. 50,000 (US$ 3.68) per month, depending on how much water they use. Karst spring is the only water resource most people of the areas rely on. However, karst spring only emerges in the foot slope of the Gunungsewu karst in

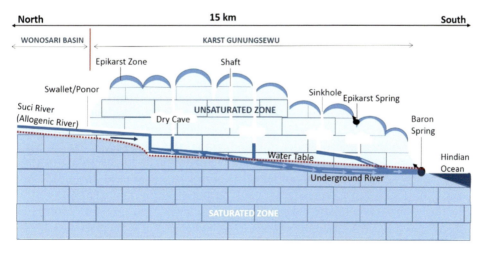

Figure 1.3 North-south crossed section showing the hydrogeological situation of the Gunungsewu karst.

8 Groundwater for Sustainable Livelihoods

Figure 1.4 Water resources in Gunungsewu karst area. (a) Rainwater tank, to which people divert rainwater from the roof. (Photo: Haryono, E.) (b) Doline pond used for drinking water for people who do not access to pipeline water. (Photo: Yuwono, JSE.) (c) Karst spring, used for water tank, distributed to the local people by trucks. (Photo: Haryono, E.) (d) Underground river dam, built to lift underground river and distributed to the surrounding communities. (Batan/IWG-KIT.)

the west, north, and east slope. Some springs/resurgences in the southern shoreline of the area are in the base of a coastal cliff. The center of the Gunungsewu karst, in the plateau part, does not have springs. The water resource in this part relies on an underground river that is difficult to locate and access. Doline ponds are the only water resource in the plateau part during the dry season. However, after the government provides pipeline water, the people do not use doline ponds for drinking water because pipe water is considered to be much cheaper and cleaner. Hence, people use doline water just for washing and bathing cattle. A government-owned company, PDAM Handayani, operates the water provider in Gunungkidul Regency. People mostly rely on this connection, although the company can only provide water for a few days in a week during the dry season. In some households, rainwater is also used during the rainy season. People mainly use rainwater for cooking and drinking, as arguably they prefer the taste of rainwater compared to piped water.

Karst water in the Gunungsewu has good quality for water supplies (MacDonald and Partner, 1984; Haryono et al., 2009; Sudarmadji et al., 2009). The electric conductivity of most water samples shows good conditions (197–559 µS/cm) and is still below the quality standard allowed for drinking water based on WHO requirements.

Suspension sediment measurements show large ranges, depending on the season and the type of aquifer recharge from the underground river spring catchment area. The highest suspended sediment content is found in the rainy season, especially in springs with developed conduit systems such as the Baron spring, Bribin, and Seropan underground river springs. Due to these springs, the high suspended sediment has a recharge area of the allogenic systems such as Kali Tegoan, Kali Suci, and Kali Petung. In contrast, Haryono (2000) found that most karst springs that are always clear water throughout the year do not have a catchment area in the form of allogenic rivers and sinkholes in large numbers (the majority located in the western part of the Gunungsewu karst).

Organic matter content in springs and underground rivers, as shown by organic fluorescence in the karst area of Gunungkidul Regency, also suggests good water quality. Measurements made by MacDonald and Partner (1984) show that there are slight differences between water samples, which generally have low organic content. Also, only 2 of the 29 samples were contaminated by coliform bacteria. More than half of these samples are of good quality and have a total coliform number below WHO's quality standards. Analysis of the quality of other springs in the Ponjong karst area was also carried out by Haryono et al. (2009), where coliform bacteria were also found in only two of the eleven analyzed samples, with the quality of the water still considered safe for drinking after boiling. Comparison between the analysis carried out in 1984 and the most recent analysis showed an increase in the concentration of coliform bacteria. Sudarmadji et al. (2009) revealed that coliform bacteria in the groundwater depends on the season, which shows an increase during the rainy season with concentrations reaching >2,400 MPN/100 mL. Nitrate in the Seropan underground river extraction site is considered high, reaching 323 mg/L (Eiche et al., 2016).

1.3 GROUNDWATER EXTRACTION AND WATER SUPPLY DEVELOPMENT

The first groundwater exploration was conducted in 1982 and 1983 by public work in cooperation with MacDonald and Partners under the Greater Yogyakarta Groundwater Development Project (P2AT) scheme. The UK government funded the project through the Overseas Development Administration (ODA). P2AT made some 112 exploration boreholes, of which 31 have been used for irrigating 985 ha agricultural land (McDonald and Partners, 1984). The project involved many partners, including the National Planning Board (BAPPENAS) and local government planning board (BAPPEDA). These planning boards coordinated other related water supply institutions, i.e., Water Supply Project (PAB), PAB-IKK, Drinking Water Supply Institution (BPAM), Perusahaan Daerah Air Minum Tirtamarta Kotamadya Yogyakarta (PDAM), P2AT, Healthcare Office, Irrigation Affair Office, PERUMNAS, Dian Desa, and other donor agencies (McDonald and Partners, 1984).

Further water supply projects in the Gunungkidul area were started following the development planning of the Sub-District Capital City (IKK) project in 1982. The project was funded by the central government, which included IKK Rongkop, IKK Paliyan, IKK Playen, and IKK Semanu. The management was under the Drinking Water Supply Bureau of Gunungkidul Regency (BPAM). BPAM was then transformed into the Drinking Water Supply Company (PDAM) in 1989, and in 2009 was named PDAM

10 Groundwater for Sustainable Livelihoods

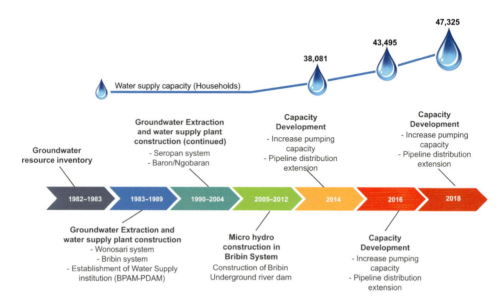

Figure 1.5 History of groundwater extraction and water supply development.

Tirta Handayani Gunungkidul. Further developments of the groundwater-based water supply have occurred during the 2009–2014 and 2014–2019 periods (personal communication with Engineering Director of PDAM Tirta Handayani). Currently, PDAM Tirta Handayani is supplying drinking water to eighteen sub-districts, 144 villages, 1,416 neighborhoods, or 50.32% of the total Gunungkidul population. Figure 1.5 shows the history of groundwater extraction and water supply development in the karst area of Gunungkidul Regency. The Indonesian government has already established five groundwater extraction plants and drinking water distribution (Table 1.3). Cave surveys and geophysical surveys are still undergoing in search of a new underground river to fulfill the need for water demand.

1.4 LIVELIHOOD AND WATER SCARCITY ADAPTATION STRATEGY

Gunungkidul Regency, of which most of the area is karst land, has the lowest Human Development Index (HDI) among other regencies in the Special Province of Yogyakarta. The HDI is a macro indicator to show how groundwater (water scarcity in general) influences people living in the area. The HDI of Gunungkidul Regency is 68.73, which is lower than the other four regencies which are situated in a volcanic area and alluvial plain with ample groundwater or surface water resources. The HDI average of the province is 78.9 (BPS, 2019). The other two regencies situated in Gunungsewu karst, i.e., Wonogiri and Pacitan, are also the poorest regions in Central Java and East Java. Groundwater access also governs the inhabitant density. Within the Gunungsewu karst, karst areas have a lower population density (<400 people/km^2) compared to non-karst areas, which are higher than 500 people/km^2 (Figure 1.6). Girisubo is the least populated sub-district (255 people/km^2). Girisubo lies in the center part of the

Table 1.3 Water supply system in the Gunungkidul Regency, Gunungsewu karst area

No	Water supply system	Description	Services
1.	Wonosari system	Borewells, water depth, is less than 5 m. The water is pumped using a diesel generator. The wells are in the Wonasari basin. The aquifer is non-karstified limestone. Capacity 426 L/s. Funded and operated by PDAM Gunungkidul	Eleven sub-districts, 42 villages, 270 neighborhoods, covering 14,493 households (31%)
2.	Seropan system	An underground river, the cave depth is 60 m from the land surface. The water is pumped up by the generator set and electricity power. The capacity is 205 L/s. Funded and operated by PDAM Gunungkidul.	Seven sub-districts, 26 villages, and 282 neighborhoods. Total services are 14,580 households (22%)
3.	Bribin system	An underground river, 100 m depth. Pumped up using electric power and micro-hydro. The water is pumped up to the highest point on top of conical karst hills and distributed gravitationally to the pipeline system without treatment. Capacity 126, 60 L/s by an electricity-powered pump and 66 L/s by micro-hydro pump. Funded by the central government, UNICEF, the German government (Karlsruhe Univ.) and Gunungkidul government. Operated by PDAM Gunungkidul	Nine sub-districts, 26 villages, and 325 neighborhoods. 8,598 household connections (30%) as of total household
4.	Ngobaran/ Baron system	Resurgences of an underground river, pumped up using electric power. Capacity 120 L/s. Funded by JICA-Japan. Operated by PDAM Gunungkidul	Seven sub-district, 27 villages, 187 neighborhoods, and 9,654 households
5.	Springwater system	Pumped from the karst spring by local people or managed by village authority. Capacity varies depending on the spring discharge. Funded by various parties. Mostly less than 10 L/s	

Note: Summarized from PDAM Tirta Handayani reports.

karst plateau with deep groundwater and limited surface water. Figure 1.6 is a population density comparison between the karst area with difficult groundwater access and the non-karst area/partial karst area with better access to groundwater.

1.4.1 Living condition and adaptation strategy

1.4.1.1 Seasonal crop farming

The living condition of the inhabitant of Gunungsewu karst is affected by groundwater scarcity and limited land resources. Dry field cultivation is the main livelihood and adaptation strategy in the karst area of Gunungsewu (Suryanti et al., 2010; Lukas and Teinhilper, 2005). People grow seasonal crops at the bottom of the close depression

12 Groundwater for Sustainable Livelihoods

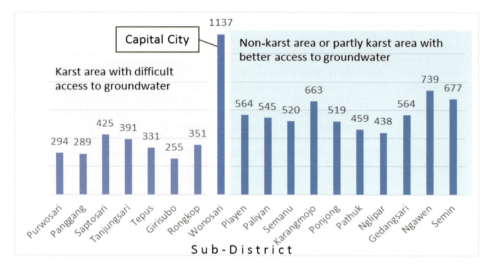

Figure 1.6 Population density in Gunungkidul Regency (person/km²), analyzed from the Statistic Bureau (BPS) data. Karst area with limited or difficult access to groundwater access has a lower population density.

Figure 1.7 Drone and ground views of agricultural land in the Gunungsew karst area. The figures show how local people utilize limited land to grow crops. In the right picture, karst slope are used for teak plantation, and the bottom of the depression is used for seasonal crops. (Drone photo by Fathoni, ground photo by Haryono.)

and the slope of karst hills. They grow multiple crops for subsistence living, i.e., rainfed rice, cassava, groundnuts, maize, soybeans, and vegetables. Rice and vegetables are generally used for self-consumption, while cassava, maize, soybeans, and groundnuts are sold for additional rice, water, and other primary foodstuffs or for buying cattle feed. To optimize the limited land resource, farmers make terraces on the slope of karst hills by pilling limestone boulder (Figure 1.7).

People grow rainfed paddy in the rainy season mixed with maize or cassava. To keep up with rain, farmer sow paddy seeds before the rain come, the so-called *awur-awur*. They hope that they do not lose time to optimize water availability. However,

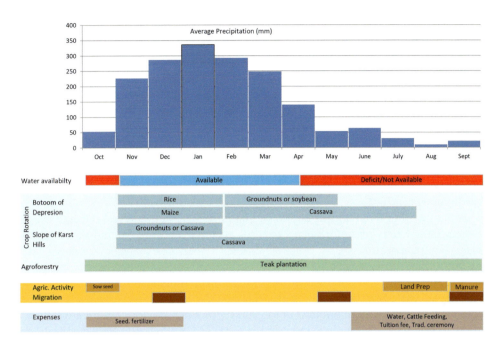

Figure 1.8 Crops rotation and agricultural activities in Gunungsewu karst as an adaptation strategy to water resource availability. Rainfall data is from the Paliyan rain gauge station 1981–2008 periods. (Crop rotation, agricultural activities, migration pattern, expenses pattern are summarized from Lukas and Teinhilper (2005); Suryanti (2010), and field survey 2019.)

climate variability has become a severe problem for farmers in applying such practices due to climate change. Sometimes, after the paddy seed grows at the beginning of the rainy season, the rain stops for about 2 weeks or more, causing paddy to dry up. Farmers then have to spend more expenses on seed replacement. Groundnuts or soybeans are grown afterward after rice harvesting in the spacing between cassava crops. Cassava is the crop that people harvest at the end of the cropping season. Cassava was dried up on the farmland at the end of the dry season while farmers prepare the land for the next growing season. Cassava is mostly used for making tapioca or cattle feeding ingredients (Figure 1.8).

1.4.1.2 Agroforestry

In addition to seasonal crops, local people also employ agroforestry practices. Agroforestry, mainly teak plantation, is the second essential livelihood adaptation strategy in the Gunungsewu karst area (Fujiwara et al., 2018; Roshetko et al., 2013; Nibbering, 1999). Teak is often mixed with maize, especially in the early teak growing period, to increase land productivity (Khasanah et al., 2015). Besides, people also plant exotic species such as acacia and mahogany. Acacia and mahogany are planted for supplying firewood and building materials, as well as occasional animal fodder. However, teak

14 Groundwater for Sustainable Livelihoods

receives its popularity due to its superior quality and exotic wood pattern (Sunkar, 2008). The earliest effort for reforestation in Gunungkidul Regency was the establishment of the Wanagama forest in 1964. As this program became successful, in the 1990s, the Regent of Gunungkidul encouraged teak planting and rehabilitation program in karst area of Gunungkidul Regency [Whitten et al. (1996) in Wardhana et al. (2012)]. Since then, teak planting became more popular among villagers. Farmers plant woody species either in the slope of karst hill, in the home garden, or along the border of crops, while flat areas are used for cash crops such as maize, cassava, soybean, and groundnuts. Although most households have woody species plantations, they are not commonly called forests because they cover only a small amount of land, quite often in small patches. The Statistic Bureau (BPS) does not attribute these small patches as forest, although these lands can also be considered to be a type of agroforestry (Wardhana et al., 2012).

Teak and other tree planting are for saving purposes. They harvest the teak timber for secondary needs or unforeseen expenses such as wedding ceremonies, building houses, buying a motorcycle, or other expenses. Farmers cop the teak tree down when they do not have anything to sell (such as cattle or jewelry) for unforeseen expenses, the so-called *tebang butuh*. *Tebang butuh* literally means cutting tree planting whenever farmers need cash. Smallholder teak agroforestry in Gunungsewu karst was started in the 1960s and getting more popular in the late 1990s when the Indonesian government promoted the reforestation movement in the area. In Gunungkidul Regency itself, the annual timber production in 2017 was 187,557 m^3, and the area of timber is 35,704 ha, approximately 24% of the area (Maryudi et al., 2017).

Nowadays, *sengon laut* (*Albizia falcataria*) is also popular because of its fast growth and the structure of the canopy that enables understory cash crops to flourish, such as groundnuts, maize, and soybeans. Although the price of *sengon laut* is lower than teak, it only takes 4 years to harvest, compared to teak that needs more than 10 years to be attributed to a significant price. Sellers usually sell timbers as trees in their field; hence, buyers still need to allocate a budget for cutting the trees and transporting them to the nearest pool. As a result, the price of wood varies depending on the size of the tree and the logging location.

1.4.1.3 Livestock raising

Raising livestock is the third side business of the people in the karst region (Triyanto et al., 2018). Cattle are usually for medium-term (1 or 2 years) savings. People sell their cattle for predictable and regular expenses (such as school tuition fees and health insurance/BPJS) or other secondary needs such as making gifts during relatives' essential events such as weddings or birth. Expressing money for a gift during those events is considered to be a social duty that must be fulfilled. Hence, people often sell their goods or find a loan in the neighborhood community loan (kelompok arisan) or local bank to meet this need. Livestock is considered to be a valuable asset. Hence, villagers mostly have more than one animal. Each family either has up to five cows or up to seven goats. In the market, the price of cows and goats may depend on the season. During the dry season, when plant fodder is rare, the price of livestock falls considerably. The price of Livestock in Gunungsewu Local Market is shown in Table 1.4.

Table 1.4 Prices of livestock in local markets of Semanu District, Gunungkidul Regency in the rainy season

Livestock	Average ownership (in a household)	Average price per individual (IDR/USD)
Cow	2	Rp. 18,000,000/$1,325.62
Male (adult)	2	Rp. 17,000,000/$1,251.98
Female (adult)	1	Rp. 8,000,000/$589.17
Offspring		
Goat	2	Rp. 2,000,000/$147.29
Male (adult)	3	Rp. 1,500,000/$110.47
Female (adult)	2	Rp. 700,000/$51.55
Offspring		
Chicken	3	Rp. 100,000/$7.36
Rooster	4	Rp. 75,000/$5.52
Hen		

Source: Field survey 2020.
Note: The local currency is Indonesian Rupiah (IDR/Rp), USD 1 ($) equals to IDR 13,578 (Rp). Currency rate is based on Bank Indonesia's published rate. This conversion will be used throughout the text.

Cows are usually brought by farmers to the field in the morning while working in the farmland, brought back to their home in the afternoon, or just kept in the farmland. The farmland is generally far (up to 4 km or more); therefore, farmers spend their whole day in the field and sheltering in a hut during midday. There is also a growing trend that a group of farmers build communal animal rearing. The facility is usually located in the hamlet's perimeter (*dusun*), and each hamlet can have more than one farmer group. In the rainy season, farmers feed livestock from hays or any fodder on the farmland. In addition, to supply plant fodders for livestock, people often grow *kolonjono* (*Braciaria mutica*), *kleresede* (*Gliricidia sepium*), and *odot* (*Pennisetum purpureum*). These vegetations are planted along the borders of crops or in the villagers' yards. In the prolonged dry season, when fodder on the farmland runs out, the farmer has to spend more additional money to buy fodder from other areas. Sometimes, they have to sell their belongings (jewelry, etc.), or often the livestock itself, to buy cattle fodders.

Livestock is not only an important asset for saving, but villagers also take advantage of the manure. This manure is often piled on the ground with other fodder residues such as hays and is transformed into compost. The compost is used to fertilize crops and home gardens and hence contributes to organic farming. The use of organic fertilizer can reduce the dependency on chemical fertilizer which price is relatively high. For example, based on field data in 2020, the price for nitrogen fertilizer is Rp. 120,000 ($8.84)/bag (bag size: 50 kg).

1.4.1.4 Migration

Migration is the other adaptation strategy for Gunungsewu karst inhabitants to cope with difficult access to water resources and limited land resources. People move to other cities temporarily or permanently. The temporary migration usually happens after

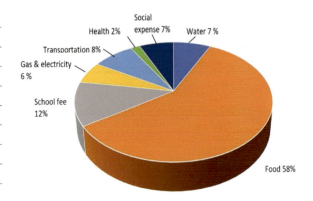

Expenditure	Value (in IDR/USD)
Water	Rp.70,581/ $5.20
Food	Rp.566,774/ $41.00
School fee	Rp.115,000/ $8.47
Gas and electricity	Rp.57,016/ $4.20
Transportation	Rp.77,097/ $5.68
Health	Rp.17,935/ $1.32
Social expense	Rp.65,161/ $4.80

Figure 1.9 Farmers' expenditure in Gunugsewu karst. (Data was from an interview in Candirejo Village and Tileng Village, Gunungkidul Regency.)

planting season (in the rainy season), after harvesting season, and after land preparation in the dry season (Figure 1.9). The migration culture in the area was already started in the colonialization era when the area was severely degraded due to forest disruptions and uncontrolled logging. As a result, the economic situation of Gunungsewu occupants became complicated that led to outmigration. The earliest record of outmigration was in 1810 (Nibbering, 1991). As the exploitation of the teak forest continued to increase during the Republic of Indonesia's early independence, so was the migration to the urban area became a common culture for seeking better livelihood. The latter migration is more permanent and farther cities or other islands. Relatively old literature suggested that economic factors became a major driving force for the outmigration of villagers, with adult males were most likely to seek job opportunities outside the village whereas women often stayed at home tendering the lands and cattle (Mantra, 1988).

1.4.1.5 Groundwater-related tourism

Recently, the new living in some parts of the Gunungsewu karst is in the tourism sector. Among the tourist destination are underground rivers, namely Kalisuci and Pindul caves. Local communities surrounding the caves make use of the underground river for cave tubing activities. The cave tubing at Pindul cave was first introduced in 2010 by the students of Universitas Gadjah Mada who were doing community services in the area. Kalisuci cave, on the other hand, was used for cave tubing earlier, introduced by the Federation of Indonesian Speleology president. The two tourist destinations are getting more popular since the Gunungsewu karst was designated as geopark area in 2012 and then accepted as a UNESCO Global Geopark in 2015 (Table 1.5).

Cave tubing and other related activities considerably increase the income and livelihood condition of the surrounding two caves, especially young people. They work as cave tubing guides, operators and selling souvenirs, and other local products for tourists. The income of households working in the tourism sector is much better than the surrounding households involved in the tourism business. All the cave tourism operator's household income is higher than the local minimum wage ruled by the local

Table 1.5 Income and assets of the villager in Gunungsewu (based on field data in Candirejo and Tileng Villages)

Income and assets	Average value per household (in IDR/USD)	Notes
Income		
Agriculture (total harvest in a year)	Rp. 5,528,871/$407.18	Mostly from rice, maize, cassava, and groundnuts
Labor	Rp. 500,806/$36.88	Informal works that are available in the village
Remittance	Rp. 320,968/$23.64	Temporary migrant in nearby cities
Trading	Rp. 80,645/ $5.94	Informal shop or food stall
Total income	Rp. 1,334,126/$98.25	
Assets		
Cattle	Rp. 4,998,387/$368.11	One cow and three goats (average per household)
Timber	Rp. 4,819,355/$354.92	Mostly teak, mahogany, acacia, and sengon
Jewelry		Not significant

government (Indriyaningsih et al., 2016; Aji and Makfatih, 2014). In contrast, 4,473 of the households who do not work in the tourism sector are under minimum wage (Aji and Makfatih, 2014).

1.4.2 Water consumption pattern

Gunungsewu inhabitants use different water sources conjunctively, depending upon the water resource availability in the surrounding areas. It is either rainwater, tanker water, *telaga* water (doline pond), or piped water. Local people use rainwater during the rainy season. Some families prefer using rainwater, although piped water is already available. They argue that rainwater is tastier than piped water. This situation makes the water supply company suffer losses, results from an operational cost deficit. Some people said that the quality of piped water is not satisfying, referring to the murky appearance of water, especially during the rainy season. To overcome this situation, people in Candirejo and Tileng Village in Gunungkidul Regency mix rainwater and piped water in their water tanks. Households insisting on using rainwater have a separate tank for rainwater and use the water only for cooking and drinking, especially during the rainy season. At the time of water scarcity, people also use reclaimed water. After washing rice and kitchen utensils, excessive water can be used to water home gardens or provide drinking water for cattle (Table 1.6).

In general, water consumption in Gunungsewu follows a relatively regular pattern according to the season. However, piped water can be a reliable water source sometimes can only be drawn 3–4 times a week due to regular maintenance by the water company. In the rainy season (November to February), it is also reported that the water is not clean with suspended mud and limestone. From May to October, some people are observed to buy water from water vendors. The price of water varies depending on the village location, between Rp. 100,000 and Rp. 150,000 (±$7.36 and $11.05) per $5\,m^3$.

Table 1.6 Average water used in a month in Candirejo and Tileng Villages

Source of water	Average volume (per month) (m^3)	Average expense (IDR/USD)
Piped Water	8.5	Rp. 70,501/$5.19
Tank water (bought from water vendors)	5	Rp. 110,000/$8.10
Rainwater (mostly in the rainy season)	1.5	-

Note: Based on the field survey in 2020.

Table 1.7 Water consumption pattern in Gunungsewu karst

	Jan	Feb	Mar	Apr	May	Jun	Jul	Aug	Sep	Oct	Nov	Dec
Rainwater	■										■	
Tanker water (from spring)				■	■	■	■	■	■	■	■	
Doline pond water (*telaga*)			■									
Piped water (from underg. River or spring)	■	■	■	■	■	■	■	■	■	■	■	■

However, in some cases, water sharing between neighbors is normal practice because not all households have a private water connection. In this case, a water connection (water meter) is shared between 3 and 4 households. Social agreement plays a vital role in this regard, including an agreement on cost-sharing, water volume, and time for fulfilling each household's water tank. Although it remains slow, the PDAM continues to add more water connections across the karst area. Thankfully, although the water from PDAM is often not available during a certain time, people can still buy tank water very easily. The improved infrastructure, such as roads, enables easy transportation and water access (Retnowati, 2014) (Table 1.7).

In general, villagers' water consumption ranges between 3 and $12\,m^3$ per month, with an average value of $8.5\,m^3$ per month and an average expense of Rp. 70,501 ($5.19). This expense contributes 7% of the total expenditure of villagers, with the highest expense is for food (buying vegetables and dishes), at about 58% of total expenditure (Figure 1.9).

1.5 IMPACT OF WATER SUPPLY DEVELOPMENT

Although it is difficult to separate the impact of water supply development and other infrastructure development, water supply development likely has a positive impact on the Gunungsewu karst area's welfare. The development of access to piped water in Gunungsewu has considerably increased during the past 10 years. If it is to compare poverty data and water availability, the number of poor people has steadily decreased since 2011–2018. This decrease is comparable with the increase in water supply, as is shown in Figure 1.10.

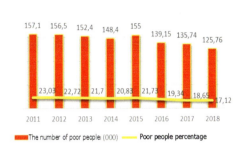 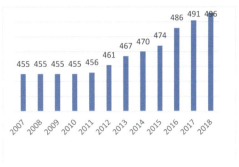

Figure 1.10 The left figure is the number of poor people, and right is the population density since 2012.

The development of groundwater supply increases population density. Since the area's birth rate is constant during the last 10 years, the increase of the population density in the Gunungsewu karst area (inferred from Gunungkidul Regency) results from in-migration. We are not sure yet whether the in-migration is from local inhabitants who return to their home town or newcomers from other areas. It is most probable that the in-migration is from the first case because as Gunungsewu karst is more promising and has better livelihood opportunities, people who formerly migrated to other cities return.

A qualitative survey regarding the perception of welfare among villagers proved that people view their economic situation as much better due to water supply availability. People indicated that, before the availability of piped water, they had to buy tank water that may cost between Rp. 100,000 and Rp. 150,000 (±$7.36–$11.05), with a volume of 5,000 L and could only last for 3 weeks (based on a household member of four). Using piped water, they only spent Rp. 25,000 ($1.58) for the same quantity of water. However, further research is needed to figure out the contribution of groundwater and water supply development in Gunungsewu karst compared to other infrastructure development.

1.6 SUMMARY

Gunungsewu karst extends from Yogyakarta Province to East Java Province and is known as Indonesia's most populated karst area. Water resources in Gunungsewu karst are in the form of an underground river, doline pond (*telaga*), and spring. Gunungkidul Regency, as the largest administrative area in Gunungsewu karst, hosts underground rivers that provide water for the local water company (PDAM). Four caves with underground rivers become the main sources of water supplies, i.e., Seropan, Bribin, Ngobaran, and Baron. However, it is sometimes not sufficient due to a temporary shutdown during the water shortage in the dry season and dissatisfaction with quality (high suspended load). In the rainy season, people rely on rainwater for drinking and cooking, whereas piped water is widely used for other purposes such as washing clothes, watering home gardens, and cleaning utensils and equipment. The spending on the water made up 7% of their monthly income. In some parts, the expenditure to buy water from a water vendor exceeds school fees and health/medicine. People adopt several strategies in using water, such as collecting rainwater, taking animals to drink in *telaga* (doline pond), connecting to piped water, accessing public water storage, and buying water from water vendors.

Dry field cultivation is the main livelihood and adaptation strategy in the karst area of Gunungsewu. To secure necessities, people make a living by cultivating cash crops on hills, slopes, and karst valleys/karst depression, migrating to nearby cities during the dry season and working on informal jobs nearby whenever available. They grow multiple crops for subsistence living, i.e., rainfed rice, cassava, groundnuts, maize, soybeans, and vegetables. In addition to seasonal crops, local people also practice agroforestry. Farmers mainly have teak plantations. Besides, acacia and mahogany are planted for supplying firewood and building materials, and occasional animal fodder. Raising livestock is the third livelihood of the people in the karst region. Cattle are usually raised for medium-term (1 or 2 years) saving. People sell their cattle for predictable and regular expenses or other secondary needs such as making gifts during relatives' wedding parties or giving birth. Migration is the other adaptation strategy for Gunungsewu karst inhabitants to cope with difficult access to water resources and limited land resources. People move to other cities temporarily or permanently. The migration pattern is usually seasonal; when agricultural work is idle in the dry season, people move to nearby urban areas. Recently, the new living in some parts of the Gunungsewu karst is in the tourism sector. Among the tourist destination are underground river caves. Local communities make use of the underground river for cave-tubing activities.

ACKNOWLEDGMENTS

Thanks to Pratomo Hadi ST, the Engineering Director of Drinking Water Company of Gunungkidul Regency (PDAM Tirta Handayani), for providing data on the water supply development; the people of Candirejo Village, Semanu District, and the people of Tileng Village, Girisubo District for their participation during Focus Group Discussion.

REFERENCES

Adji, T.N. (2011) Upper catchment of Bribin underground river hydrogeochemistry (Gunung Sewu Karst, Java, Indonesia). *Proceedings of Asian Trans-Disciplinary Karst Conference 2011*, Yogyakarta, 7–10 January 2011, pp. 117–133.

Adji, T.N. (2012) Wet season hydrochemistry of Bribin River in Gunung Sewu Karst, Indonesia. *Environ Earth Sciences*, 76, 1563–1572. Available from https://link.springer.com/article/10.1007%2Fs12665-012-1599-x [Accessed 6th June 2020].

Adji, T.N., Bahtiar, I.Y. (2016) Rainfall–discharge relationship and karst flow components analysis for karst aquifer characterization in Petoyan Spring, Java, Indonesia. *Environmental Earth Sciences*, 75 (735). Available from https://link.springer.com/article/10.1007/s12665-016-5553-1 [Accessed 27th January 2020].

Adji, T.N., Haryono, E., Fatchurohman, H., Oktama, R. (2017b) Spatial and temporal hydrochemistry variations of karst water in Gunung Sewu, Java Indonesia. *Environmental Earth Sciences*, 76, 709. Available from https://link.springer.com/article/10.1007/s12665-017-7057-z [Accessed 9th June 2020].

Adji, T.N., Haryono, E., Mujib A, Fatchurohman, H., Bahtiar, I. (2017a) Assessment of aquifer karstification degree in some karst sites on Java Island, Indonesia. *Carbonates Evaporites*, 34, 53–66. Available from https://link.springer.com/article/10.1007%2Fs13146-017-0403-0 [Accessed 2nd December 2019].

Agniy, R.F, Cahyadi, A., Nurkholis, A. (2017) Analisis karakteristik akuifer karst dengan uji perunutan dan pemetaan gua. *Proceeding of 2nd Annual Conference of Groundwater Expert Association (PIT-PAAI)*, 13–15 September 2017, Yogyakarta, Indonesia. Vol. 1. Available from https://doi.org/10.31219/osf.io/dfxjh [Accessed 5th January 2020].

Aji, D.N., Makfatih, A. (2014) Analisis dampak obyek wisata Gua Pindul terhadap peningkatan pendapatan masyarakat Desa Bejiharjo. [*Thesis*]. [Online] Universitas Gadjah Mada. Available from http://etd.repository.ugm.ac.id/home/detail_pencarian/71813 [Accessed 20th December 2019].

BPS. (2019) *Data Statistik Makro Ekonomi Kabupaten Gunungkidul*. Wonosari, BPS.

Eiche, E., Hochschild, M., Haryono, E., Neumann, T. (2016) Characterization of recharge and flow behaviour of different water sources in Gunung Kidul and its impact on water quality based on hydrochemical and physico-chemical monitoring. *Applied Water Science.* doi:10.1007/s13201-016-0426-z.

Fujiwara, T., Awang, S.A., Widayanti, W.T., Septiana, R.M., Hyakumura, K., Sato, N. (2018) Socioeconomic conditions affecting smallholder timber management in Gunungkidul District, Yogyakarta Special Region, Indonesia. *Small-Scale Forestry*, 17 (1), 41–46. Available from https://doi.org/10.1007/s11842-017-9374-1 [Accessed 5th January 2020].

Haryono, E. (2000) Some properties of epikarst drainage system in Gunung Kidul Regency, Yogyakarta, Indonesia. *The Indonesian Journal of Geography*, 32, 10–15.

Haryono, E. (2008) Model perkembangan karst berdasarkan morfometri jaringan lembah di Karangbolong, Gunungsewu, Blambangan, dan Rengel. [*Dissertation*] Fakultas Geografi, Universitas Gadjah Mada.

Haryono, E. (2011) Introduction to Gunungsewu karst, Java-Indonesia: Field guide Asian Trans Disciplinary Karst Conference 2011. Yogyakarta. [Online] Available from https://osf. io/preprints/inarxiv/7w2sh/ [Accessed 13th August 2018].

Haryono, E., Adji, T.N., Widyastuti, M. (2009) Environmental problems of telaga (doline ponds) in Gunungsewu Karst, Java Indonesia. In: White, W.B. (ed) *Proceeding 15th International Congress of Speleology 2009*, 19–26 July 2009 Kerrville, Texas. Vol. II, pp. 1112–1116.

Haryono, E., Danardono, M.S., Putro, S.T, Adji, T.N. (2016) The nature of carbon flux in Gunungsewu karst, Java Indonesia. *Acta Carsologica*, 45 (1), 173–185.

Haryono, E., Day, M.J. (2004) Landform differentiation within the Gunung Kidul karst, Java-Indonesia. *Journal of Cave and Karst Studies*, 66 (2), 62–68.

Indriyaningsih, D., Armaidy, A., Muhammad. (2016) Peran pemuda dalam pengembangan ekowisata kawasan Gua Pindul dan implimentasinya terhadap ketahanan ekonomi masyarakat. *Jurnal Ketahanan Nasional*, 22 (2). Available from https://jurnal.ugm.ac.id/jkn/article/ view/11986 [Accessed 9th June 2020].

Khasanah, N, Perdana, A., Rahmanullah, A., Manurung G., Roshetko, J.M., van Noordwijk, M. (2015) Intercropping teak (*Tectona grandis*) and maize (*Zea mays*): Bioeconomic trade-off analysis of agroforestry management practices in Gunungkidul, Java. *Agroforestry Systems*, 89 (6), 1019–1033. Available from https://doi.org/10.1007/s10457-015-9832-8 [Accessed 7th January 2020].

Kusumayudha, S.B. (2005) *Hidrogeologi karst dan geometri fractal di daerah Gunungsewu*. Yogyakarta, Adicita Karya Nusa.

Lukas, M., Teinhilper, D. (2005) *Living Condition in the Gunungsewu karst Region*. Verlag Johannes Herrmann. http://www.johannes-herrmann-verlag.de/files/9783937983035.html

MacDonald and Partners. (1984) *Greater Yogyakarta Groundwater Study. Volume 3C Cave Survey*. Directorate General of Water Resources Development of the Republic of Indonesia. London, Overseas Development Administration.

Mantra, I.B. (1988) Population mobility and the links between migrants and the family back home in Ngawis Village, Gunung Kidul Regency, Yogyakarta Special Region. *The Indonesian Journal Geography*, 18 (55). Available from https://jurnal.ugm.ac.id/ijg/article/view/2180 [Accessed 9th June 2020].

Maryudi, A., Nawir, A.A, Sekartaji, D.A., Sumardamto, P., Purwanto, R.H., Sadono, R., Suryanto, P., Soraya, E., Soeprijadi, D., Affianto, A., Rohman, R., Riyanto, S. (2017) Smallholder farmers' knowledge of regulations governing the sale of timber and supply chains in Gunungkidul District, Indonesia. *Small-Scale Forestry*, 16 (1), 119–131. Available from https://doi.org/10.1007/s11842-016-9346-x [Accessed 6th January 2020].

Nibbering, J.W. (1991) Hoeing in the hills: Stress and resilience in an upland farming system in Java. *[Unpublished thesis]* Australia, Australian National University.

Nibbering, J.W. (1999) Tree planting on deforested farmlands, Sewu Hills, Java, Indonesia: Impact of economic and institutional changes. *Agroforestry Systems*, 46 (1), 65–82. Available from https://doi.org/10.1023/A:1006202911928 [Accessed 3rd January 2020].

Nurkholis, A., Adji, T.N., Haryono, E., Cahyadi, A., Suprayogi, S. (2019). Time series analysis application for karst aquifer characterisation in Pindul Cave karst system, Indonesia. *Acta Carsologica*, 48, 69–84. Available from https://doi.org/10.3986/ac.v48i1.6745 [Accessed 2nd June 2020].

Purnomo, D. (2004). Studi tentang niatan menetap migran sikuler (Kasus migran sirkuler asal Wonogiri ke Jakarta). *Jurnal Ekonomi Pembangunan*, 5 (2), 135–46.

Purnomo, D. (2009) Fenomena migrasi tenaga kerja dan perannya bagi pembangunan daerah asal: studi empiris di Kabupaten Wonogiri. *Jurnal Ekonomi Pembangunan*, 10 (1), 84. Available from https://doi.org/10.23917/jep.v10i1.810 [Accessed 3rd January 2020].

Rahardjo, W., Rumidi, S., Rosidi, H.M.D. (1995) *Geological map of Yogyakarta Sheet, Jawa*, 2nd edition, 1:100,000. Bandung, Pusat Penelitian dan Pengembangan Geologi.

Retnowati, A. (2014) Culture and risk based water and land management in karst areas: An understanding of local knowledge in Gunungkidul, Java, Indonesia. *[Thesis]*. Germany, Justus Liebig University Giessen.

Roshetko, J.M., Rohadi, D., Perdana, A., Sabastian, G., Nuryartono, N., Pramono. A.A., Widyani, N., Manalu, P., Fauzi, M.A., Sumadamto, P., Kusumowardani, N. (2013) Teak agroforestry systems for livelihood enhancement, industrial timber production, and environmental rehabilitation. *Forests Trees and Livelihoods*, 22 (4). Available from https://www.tandfonline.com/doi/abs/10.1080/14728028.2013.855150 [Accessed 8th June 2020].

Sidauruk, P., Prasetio, R., Satrio. (2018) Hydraulic interconnections study of Seropan-Ngreneng-Bribin underground rivers in Gunung Kidul karst area using tracer technique. *International Journal Water*, 12 (1), 39–53.

Sudarmadji, W.M., Adji, T.N., Haryono, E. (2009) Pengembangan metode konservasi air bawahtanah di kawasan karst sistem bribin-baron, Kabupaten Gunungkidul. *Research Report of World Class Research University Program*. Yogyakarta, Universitas Gadjah Mada.

Sudarmadji, W.M., Haryono, E. (2005) Pengembangan metode konservasi air bawah tanah di kawasan karst sistem Bribin-Baron, Kabupaten Gunungkidul. *Reserach Report*. Yogyakarta, Universitas Gadjah Mada.

Sunkar, A. (2008). Sustainability in karst resources management: The case of the Gunungsewu in Java. *[Thesis]*. New Zealand, The University of Auckland.

Surono, T.B., Sudarno, J. (1992) *Geological map of Surakarta-Giritontro Sheet, Jawa*. 1:100,000. Bandung, Pusat Penelitian dan Pengembangan Geologi.

Suryanti, E.D. (2010) Strategi adaptasi ekologis masyarakat di kawasan karst Gunungsewu dalam mengatasi bencana kekeringan. *[Thesis]*. Yogyakarta, Universitas Gadjah Mada. Available from https://doi.org/10.1017/CBO9781107415324.004 [Acessed 18th December 2020].

Suryanti, E.D., Sudibyakto, Baiquni, M. (2010) Strategi adaptasi ekologis masyarakat di kawasan karst Gunungsewu dalam mengatasi bencana kekeringan (Studi kasus Kecamatan Tepus, Kabupaten Gunungkidul). *Jurnal Kebencanaan Indonesia*, 2 (3), 658–674.

Tjia, H.D. (2013). Morphostructural development of Gunungsewu karst, Jawa Island. *Indonesian Journal of Geology*, 8 (2), 75–88.

Triyanto, E.S., Rahayu, Purnomo, S.H. (2018) Strategy of beef cattle livestock development in Gunungkidul District, Indonesia. *Russian Journal of Agricultural and Socio-Economic Sciences*. [Online] 82 (10), 209–218. Available from https://doi.org/10.18551/rjoas.2018-10.23 [Accessed 3rd December 2019].

Wardhana, W., Sartohadi, J., Rahayu, L., Kurniawan, A. (2012) Analisis transisi lahan di Kabupaten Gunungkidul dengan citra penginderaan jauh multi temporal. *Jurnal Ilmu Kehutanan*, 2 (14), 89–102.

Widyastuti, M., Haryono, E. (2016) Water quality characteristics of jonge telaga (doline pond) as water resources for the people of Semanu District Gunungkidul Regency. *Indonesian Journal of Geography*, 48 (2), 157–167.

Chapter 2

Groundwater resources development for livelihoods enhancement in the Sahel Region

A case study of Niger

A.E. Cheo
Institute for Environment and Human Security, United Nations University

B. Ibrahim
University of Abdou Moumouni

E.G. Tambo
Institute for Environment and Human Security, United Nations University

CONTENTS

2.1 Introduction 25
2.2 Materials and methods 28
2.3 Rural livelihood and groundwater supply 29
2.4 Concept and theories for analyzing livelihood 31
    2.4.1 Sustainable livelihoods framework 31
    2.4.2 Livelihood zoning 35
2.5 Case study – Niger 36
    2.5.1 Institutional arrangement and water-related policy in Niger 39
    2.5.2 Groundwater supply and rural livelihood in Niger 41
        2.5.2.1 Groundwater supply for irrigation 42
        2.5.2.2 Groundwater supply for other uses 43
    2.5.3 Livelihood zoning in Niger 43
    2.5.4 Rural livelihoods: groundwater interventions and agriculture 46
2.6 Discussions 51
2.7 Conclusions 56
References 57

2.1 INTRODUCTION

The Sahel region in Africa marks the continental physical and cultural transition between the more fertile tropical regions in the south and the desert in the north. The region is home to about 135 million people and considered as one of the world's climate change hotspots. Commonly, the Sahel stretches from Senegal on the Atlantic coast, through parts of Mauritania, Mali, Burkina Faso, Niger, Nigeria, Chad, and Sudan to Eritrea on the Red Sea coast (Figure 2.1). The geographic definition of the Sahel region

DOI: 10.1201/9781003024101-2

26 Groundwater for Sustainable Livelihoods

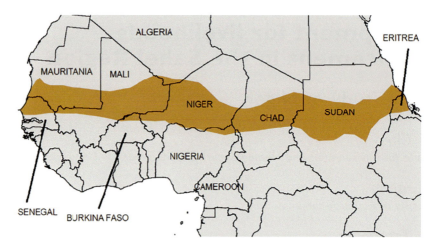

Figure 2.1 The Sahel region of Africa is highlighted in dark orange color (Stephen, 2014).

varies. In agro-ecological terms, the Sahel constitutes areas that receive between 200 and 800mm of mean annual rainfall (UN Office for the Coordination of Humanitarian Affairs, 2011). In another definition, the region is a semi-arid grassland and shrubland transition zone stretching across the African continent between the Sahara Desert to the north and the tropical savannas to the south (Herrmann et al., 2005).

> The Sahel region has always been highly integrated into the wider West and North Africa region, with long-established patterns of migration south to better watered coastal countries, and northwards to the oases of the Sahara and onto the Maghreb coast. (UNEP, 2011)

In recent decades, the region has suffered from extreme drought, affecting agriculture and causing widespread hunger. In addition, the region has been affected by wars (till date the situation is still unstable), increasing population pressures, and pervasive poverty and aid dependency.

> The Sahel region is better understood as disequilibrium environments, where climatic variability – characterized by rainfall that is highly scattered and unpredictable over time and space, and droughts that are unpredictable but periodic and expected – are the norm rather than the exception. (Hesse et al., 2013)

With a larger part of the population depending on subsistence agriculture, the region comprises the most vulnerable production environments and populations in the world. In addition, the risks posed by the impacts of climate change threaten crop production and yield, and lead to a decrease in food availability. Any increase in temperature and/or modification in rainfall quantities and distribution will significantly affect the natural resource on which agriculture depends, thus affecting local livelihoods. Figure 2.2 shows the distribution of the different livelihoods and production systems in the Sahel region. Agriculture is the main source of livelihood for the majority of people living in the Sahel region.

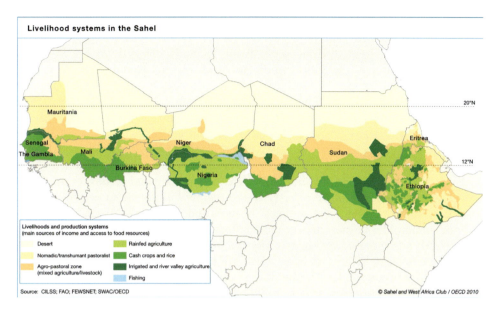

Figure 2.2 Livelihood systems in the Sahel. (Sahel and West Africa Club (SWAC)/OECD, 2010.)

Farming, herding, and fishing remain the dominant livelihoods in the region, despite the high seasonal variation in rainfall, erratic crop yields, poor soil, and depleted fish stocks (UNEP, 2011). According to World Bank (2009) statistics in 2008, 46% of the land area of the nine countries in the region was categorized as agricultural land as defined by the Food and Agriculture Organization (FAO). Farming accounts for more than 50% of the Gross Domestic Product (GDP) derived from agricultural practices (ECOWAS-SWAC/OECD, 2007) and pastoralism accounts for approximately 40% share of the GDP derived from agriculture, making it the second largest contributing source of the GDP from agriculture (SWAC-OECD/ECOWAS, 2008). There is therefore no doubt that climate change will most likely exacerbate and accelerate the current vulnerability of the livelihoods based on agriculture (UNEP, 2011).

In the Sahelian region, farm households have developed a range of strategies to cope with the difficult climatic situation. Some of the coping strategies include selling animals, on-farm diversification, and farm specialization. According to Sissoko et al. (2011), early warning systems including an operational agro-meteorological information system to provide crucial information to framers in the western Sahel have already been installed at the regional level. Further coping strategies were the recognition of the fact that the efforts from political, institutional, and financial bodies at the national and international levels are indispensable for the sustenance of millions of lives in the region (Sissoko et al., 2011). "In terms of development, priority needs to be given to adaptation and implementation of comprehensive programs on water management and irrigation, desertification control, development of alternative sources of energy and the promotion of sustainable agricultural practices by farmers" (Sissoko et al., 2011). Water is the main limitation to improved agricultural production in the region.

In the broader context of economic transformation and environmental challenges, the need for affordable, safe, and reliable supplies of water is paramount and important in reducing rural poverty.

Without many rivers to draw water from, groundwater systems are the main source of fresh water in the Sahel region. In rural Africa, groundwater has proven to be cheaper to develop, offers reliable supply, provides a buffer against drought and the aquifer provide natural protection from contamination (even though natural occurring fluoride [F] and arsenic [As] are problematic in some areas as well as the growing concern of anthropogenic pollution). In addition, yields from irrigated farms by groundwater have proven to be much higher than surface water schemes (Burke et al., 1999). This is mainly because groundwater supply can be easily adjusted and this allows for efficiency and flexible in-field application on demand (Calow et al., 2010). "Efforts to reduce or eradicate poverty in the region will not be successful without substantial gains in agricultural incomes" (Faurès and Santini, 2008).

Poor smallholder farmers are the most vulnerable to water scarcity. Groundwater is less directly and more slowly impacted by climate change when compared with other water sources such as rivers and lakes. Only after prolonged droughts because of climate change, groundwater levels show a significant declining trend. In this context, future intervention using groundwater should be able to respond to challenges of climate change so as to protect and enhance rural livelihoods. No existing datasets and reports can show the exact status and trend of groundwater access in the Sahel region. It would be therefore too simplistic to assume that simply increasing access to well-managed and reliable groundwater services will have a direct benefit to the poor (UPGro, 2017) and improve rural livelihood without risks or unintended impacts. Therefore, the aim of this chapter was to enhance our understanding on how the benefits of groundwater intervention could be analyzed to provide policy guidance and investments in innovative measures that could lead to the improvement of rural livelihoods. Furthermore, the case study being Niger will be used as evidence-based demonstration for groundwater contribution in enhancing rural livelihoods.

2.2 MATERIALS AND METHODS

The main data and information sources for this study were collected from literature works (e.g. published, peer and non-peer reviewed, and unpublished sources). The list of reviewed literature is not exhaustive, although an effort has been made to conduct compressive coverage. The collection process has been in combination with the analyses of secondary data obtained from FAO databases, especially AQUSTAT (FAO, 2020a), FAOSTAT (FAO, 2020b), and World Bank databank (World Bank, 2020) on poverty, agricultural water resources, food production, and economic indicators. In addition, local expert consultation and involvement were valuable in strengthening the analysis.

The study framework largely builds on FAO's previous work on water and the rural poor in the Asia Pacific (Khanal et al., 2014); on water and poverty in Sub-Saharan Africa (Faurès and Santini, 2008); on the United Nations Environment Programme (2011) work on climate change, conflict and migration in the Sahel; and on FEWS NET (2011) work on livelihoods zoning "PLUS" activity in Niger, as well as on other water and livelihood initiatives. Most of the major findings of the study, especially the analysis of poverty, livelihood improvement, and water constraints in Section 5.3, have been derived using an approach developed by the previous FAO work. The challenging task for this approach was the lack of data in most of the key sectors that were analyzed in those previous works. There is therefore the need for more data in the future. However, the approach set the baseline for future research work on livelihood in Niger.

2.3 RURAL LIVELIHOOD AND GROUNDWATER SUPPLY

> An estimated 75 percent of the world's poorest people – 880 million women, children and men – live in rural areas, and the majority of them depend on agriculture and related activities for their livelihoods. (World Bank, 2007)

The rural population in the Sahel region rely mainly on crop–livestock activities and natural resources for their livelihood and food security, and to provide food for urban populations (Ickowicz et al., 2012). There is little or no diversification to other revenues sources especially for those in the agricultural sector, which makes them very vulnerable to extreme events such as droughts and floods enhanced by climate change. In cases where there is some form of diversification, it is still within the agricultural sector, which is simply the involvement in different agricultural activities such as animal rearing, crop cultivation, and provision of paid farm labor.

The role of agriculture in the reduction of rural poverty and sustaining rural livelihoods have been well recognized in recent years. It remains the primary engine for rural growth and poverty reduction (IFAD, 2010). According to World Bank (2008), the growth of agriculture in reducing rural poverty is two times more effective than other sectors. Small-scale and rainfed agriculture with little mechanical and monetary inputs are common practices in most of the rural areas in the Sahel region. However, rainfed agriculture can only be practiced in the region with a minimum rainfall of about 350 mm (UNEP, 2011), which is common in the South. There is a transition into pastoral farming went moving toward the North of the Sahel.

In the Sahel region, rural farmers typically live in permanent settlements, growing millet, maize, rice, sorghum, and raising domestic animals to provide supplementary income. The main cash crops in the region are groundnuts and cotton. Herders, conversely, raise livestock and cultivate crops along various seasonal nomadic routes, generally moving from northern to southern pastoral areas during the dry season (October–June), and back north during the wet season. Livestock production is predominant, relying on the extensive movement of herds over long distances. However, developmental investments in the agricultural sector have consistently been low and it is often accompanied by national policies that focus mostly on cash crops for export (cotton) and large-scale mechanized production (sugarcane, rice in the Niger and Senegal river valleys). Nonetheless, farming activities are vulnerable to both climate and non-climate factors.

Most smallholder farmers live in areas with poor natural resource conditions. The availability of water has been the most critical stressed natural resource and this has a negative impact on livelihoods and the level of poverty in the region. "Water-related poverty occurs when people are either denied dependable access to water, or they lack

30 Groundwater for Sustainable Livelihoods

the capacity to use it because they have insufficient or degraded land, or have poor access to markets, capital and other production factors" (Cook and Gichuki, 2006). In addition, the rising population in the region will put more pressure on water resources since there will be a need to increase agriculture productivity through more irrigated farms to meet food demands. This also offers an opportunity for reducing rural poverty through well-targeted water interventions.

The lack of control and effective management of water in the Sahel region has been one of the most critical constraints especially for rural communities located far away from surface water bodies such as the Niger River. Water plays a key role in agricultural production and contributes to livelihoods in many different ways. For example, "reliable irrigation enables farmers to adopt new technologies and intensify cultivation, leading to increased productivity, higher production and greater returns from farming" (Hussain and Hanjra, 2004). People in water-scarce areas such as the Sahel region are increasingly depending on groundwater because of its buffer capacity. In addition, groundwater is the only practical means of meeting rural community needs in the arid and semi-arid regions (Robins et al., 2006). Table 2.1 summarizes the advantages and limitations of groundwater to be suitable for meeting the water need of rural communities in the Sahel region.

"As the largest and most widely distributed store of freshwater in Africa, groundwater provides an important buffer to climate variability and change" (MacDonald et al., 2012). The Sahel sits atop some of the largest aquifers on the continent (World Bank, 2014). The long-term replenishment of groundwater resources is controlled by the long-term climate conditions. At the same time, indirect climate change impacts such as the intensification of human activities and land-use changes increase the demand for groundwater. "Groundwater in the Sahel region is under threat from a variety of sources including urban expansion and rapid development of agriculture" (World Bank, 2014) and this will have a direct or an indirect consequence to the livelihood of local communities in the region. There is therefore the need for the sustainable exploitation of groundwater, in order to increase the climate resilience of communities, which contributes to improving the livelihoods and the fight against poverty among dryland communities (World Bank, 2014).

The pathways out of poverty lie in the capabilities of the people to exploit opportunities using their own assets. Despite the significance and high potential of groundwater supply in rural communities, there is still little understanding on the concrete role of groundwater in enhancing rural economy and livelihoods. This is mainly because no systematic study has been conducted. Until now, groundwater plays an important role in domestic use, livestock sector, and agricultural production and is hence associated not only with community development, but also with poverty alleviation (Braune et al., 2008) and rural development. "The more recent evolution of livelihood framework for poverty reduction has transformed the ways in which rural poverty is perceived and addressed, especially recognizing how access to, and control over, assets impact peoples' livelihoods" (Khanal et al., 2014).

Understanding the conceptual ideas behind groundwater and poverty linkages through a livelihood lens helps design water interventions that are more specific and targeted. So far, studies (e.g. Khanal et al., 2014) have analyzed the linkages between water and poverty using the sustainable livelihoods framework and the livelihood zoning (e.g. FEWS NET, 2011) to provide solution to real-world problems such as

Groundwater and livelihoods in Niger   31

identifying the hotspots of water constraints underlying poverty (Khanal et al., 2014). It is important to note that using these frameworks to understand the complex issues surrounding poverty and the use of groundwater to enhance a sustainable rural livelihood would be challenging because of no sufficient data.

2.4 CONCEPT AND THEORIES FOR ANALYZING LIVELIHOOD

Livelihood can be best defined as the methods and means of making a living in the world (Islam and Ryan, 2015). This encompasses people's capabilities, assets, income, and activities required to secure the necessities of life (International Federation of Red Cross, 2020). These necessities include food, water, shelter, clothing, and healthcare. It is important to note that livelihoods vary within countries, from rural to urban areas, and across countries. According to FAO (2006), the household represents the unit of reference because it is by far the most important institution through which populations anywhere organize production, sharing income, and consumption. There is a growing interest in using livelihoods analysis as the basis for addressing problems ranging from emergency response to disaster mitigation and to long-term development. According to FEG Consulting and Save the Children (2008), two reasons justify this interest, which includes the fact, that (a) "information about a given area or community can only be properly interpreted if it is put into context with how people live" and (b) "interventions can only be designed and managed in ways appropriate to local circumstances if the planner knows about local livelihoods and whether or not a proposed intervention will build upon or undermine existing strategies." It is for this justification that this study is adopting the "sustainable livelihoods framework" and the "livelihood zoning" as tools for analyzing groundwater interactions that contribute to a better livelihood in the rural context, with agriculture as the main activity for subsistence.

2.4.1 Sustainable livelihoods framework

There are two approaches to secure the household necessities of life. These include (a) sustainable livelihoods framework (also known as livelihoods approach) and (b) production-based approach (also known as farming systems analysis approach). According to Khanal et al. (2014), "the adoption of a sustainable livelihoods approach (moving away from a top-down engineering-focused approach towards a more holistic, household centered one) is now widely seen as critical in ensuring success in any future water sector interventions in agricultural development". The sustainable livelihoods framework makes the household, the center of the analysis and integrates all views related to their assets or forms of capital (physical, financial, human, natural, and social). Furthermore, the framework enables us to understand poverty in the context of lack of opportunities in economic, political, and social life. Whereas, production-based approach (also known as farming systems analysis) focuses on production. Table 2.2 distinguishes between sustainable livelihoods framework and production-based approach by applying the different forms of capital, which are usually at the disposal of household.

"Livelihood strategies and outcomes at the household level depend to a large degree on the amounts and qualities of these assets owned or controlled by the household"

32 Groundwater for Sustainable Livelihoods

Table 2.1 Advantages and limitations of groundwater (MacDonald et al., 2005)

Advantage of groundwater	*Qualifying limitations*
Often available close to where it is required	Considerable effort may be needed in some situations to locate suitable sites
Can be developed relatively cheaply and progressively to meet demand with lower capital investment than many surface water schemes	As overall coverage increases, the more difficult areas which are left can become more costly to supply
Generally has excellent natural quality, and is usually adequate for potable supply with little or no treatment	Constraints on naturally occurring quality are becoming more widely observed
Generally has a protective cover provided by the soil and unsaturated zone	As development increases more rapidly, the threat of pollution from human activities needs to be assessed in relation to the nature of the protective cover

Table 2.2 Difference between two approaches in rural water development (WWAP, 2006)

Capital	*Issue*	*Production-based approach*	*Sustainable livelihoods framework*
Physical	Infrastructure for rainfed and irrigation systems	Rainfed and irrigation livelihood zones improved to increase agricultural production	Improves decision-making ability through better rainfed and irrigation livelihood zones. Removes risk and uncertainty including maintenance and management of natural capital stocks
Social	Community approach needed to raise or manage other forms of capital of crucial importance in irrigation management	Communities mobilized to establish (WUAs) to improve agricultural water management.	Identifies the poorest households and strengthens participation in, and influence on, community management systems. Creates safety nets within communities to ensure the poor have access to water. Improves rights to land and water and establishes right to access by poor households within communities.
Natural	Land and water availability	Develop new, and enhance existing water resources using physical and social assets	Enhanced through training in catchment protection and maintaining natural environment
Financial	Cash, credit, savings, animals	Develops individual or community-based tariffs and charges mechanisms for water uses	Secure through access to small-scale credit
Human	Labor, knowledge (through education, experience)	Trains people in agricultural water management and promotes gender equity	Knowledge of demand, responsive approaches, community self-assessment of needs, participatory monitoring, and gender mainstreaming

(Khanal et al., 2014). In the context of the sustainable livelihoods framework, achieving sustainable rural development will require the optimal combination of investments in the five forms of capital (Pender et al., 2004). The approach recognizes that farm household livelihoods are often diverse and using diverse assets leads to multiple priorities and strategies, and therefore multiple outcomes (Khanal et al., 2014). Furthermore, the approach analyzes problems in a holistic manner and at the same time, identifies specific interventions in response to the specific problems. Figure 2.3 summarizes the conceptual formulation of the sustainable livelihoods framework.

In Figure 2.3, the capitals also known as livelihood assets play an important role in determining the status or livelihood outcome of poor people's lives. Depending on particular context (policy, institutions), combining different capitals can enable a household to follow a variety of different livelihood strategies, which could lead to outcomes such as more income and improved food security. In some instances, the outcome can be re-used in the overall strategy either as an additional (increasing the capital stock) or as a new asset, increasing the option for the household (see Figure 2.3). There are also chances that the wrong combination of assets could lead to bad livelihood strategies that can render the household more vulnerable. This should prompt the reassessment of the household capitals and redesign a new livelihood strategy that could lead to a positive outcome. Livelihoods are sustainable only when they can recover or cope with stress and shock without undermining the existing natural resources.

According to Khanal et al. (2014), the sustainable livelihoods approach has evolved over time and there is an extensive body of literature to explain the approach and demonstrate the applicability. Other organizations (e.g. Oxfam, CARE, UNDP, and FAO) have adopted or modified the framework based on their areas of engagement. However, all approaches have shared the same basic principles and elements, which also includes the concept of "five capitals" [i.e. natural (e.g. groundwater, land, and forest), physical (e.g. infrastructure and irrigation equipment), financial (e.g. credit, saving, and incomes), social (e.g. social networks and institutions), and human (e.g. skills and knowledge)].

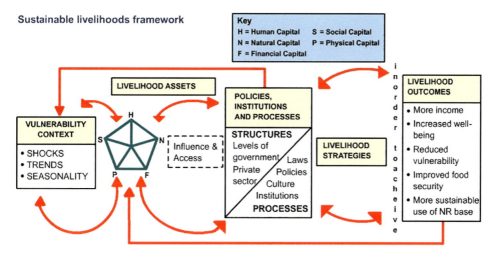

Figure 2.3 Sustainable livelihood framework (Paterson et al., 2013).

This study focuses on the potential of groundwater interventions for agricultural development to address rural poverty in the Sahel region. Usually, the sustainable livelihood framework requires a holistic approach that captures the many complexities (both constraints and opportunities) experience within livelihoods. However, this study focuses on a sectorial analysis that investigates the intervention of groundwater in the agricultural sectors. This goes contrary to the framework because the focus is only on groundwater excluding other interventions that might be contributing to local livelihood. According to Khanal et al. (2014), such study does not limit the application of the framework; rather it helps to narrow the focus, and also helps to identify options beyond water and agriculture. Besides that, rural poverty and agriculture have a wider connection to the dynamics of the economy and the environment.

"Water interventions are essential, either to create new capitals or to improve the existing capital base, but communities often lack the necessary financial capital to create or maintain the other forms of capital" (Khanal et al., 2014). In most poor communities, water projects have been realized through the injection of external financial capital to create or improve physical assets. One of the most challenging tasks in using the livelihood framework in the context of groundwater intervention is to situate how the intervention leads to an improvement in livelihood. "This requires understanding of the assets and their linkages with water and how they contribute to crop-supported livelihood outcomes" (Khanal et al., 2014). In Figure 2.4, Khanal et al. (2014) have illustrated the possible linkages between water and livelihood outcomes, which represent perfectly the same situation when using groundwater.

According to Khanal et al. (2014), there are two conditions that are essential for water interventions. These include (a) "the potential to develop water-based assets or to improve the existing asset base" and (b) "water is in reality the key limiting factor or 'binding constraint' for people's livelihoods, and water interventions can help build the needed assets for people to initiate and sustain improved livelihoods." They went further to say, the first condition is necessary but not sufficient for water interventions, as it only reveals that there is potential for water resources development. Basically, in the first condition, the scope for possible inventions is identified, whereas, in the second condition, the needs and the opportunities for intervention are

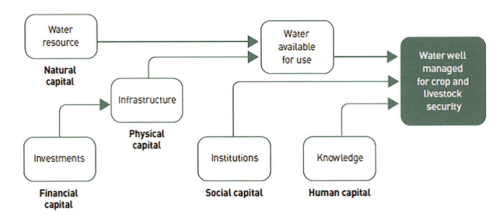

Figure 2.4 Water and the livelihood links (Khanal et al., 2014).

identified. However, both conditions are essential for water interventions to be appropriate (Khanal et al., 2014).

The outcomes of these interventions are often measured in terms of water availability and productivity, which is very essential for agriculture in the Sahel region because of the environmental conditions. For example, if groundwater is considered as the natural capital as shown in Figure 2.4, then it will need to interact with several capitals (e.g. social, human, financial, and physical capitals) to lead to a desired and sustainable outcome. The combination of these capitals helps farmers to develop specific agriculture-based livelihood strategies, which can be either mechanization, diversification, or intensification. The expectation of these livelihood strategies most often is to improve their financial capital to sustain their livelihood and can as well be used to increase their livelihood outcomes. Financial capital is very important because it can be used to create other forms of capital.

2.4.2 Livelihood zoning

"Livelihood zoning consists of identifying areas with homogeneous livelihood conditions, which are formed by considering both biophysical and socio-economic determinants" (Khanal et al., 2014). The patterns of livelihood vary from one area to another. Thus, requiring different criteria for the mapping of livelihood zones. One of the first basic steps for conducting the analysis is to map out geographical areas within which people share the same patterns of access to food and have the same access to markets. "For agriculture-based livelihood systems, zoning helps to classify areas of similar soil, productivity, climate, water resources, and land forms, enabling an assessment of the agricultural potential and constraints" (Khanal et al., 2014). Local factors such as climate, soil, and access to markets can be used to characterize of livelihood patterns. It can as well be characterized at different scales, which could be regional, country, or local levels. Table 2.3 shows the different parameters at the different scales that

Table 2.3 Main factors determining livelihood zones at different scales (Faurès and Santini, 2008)

Parameters	Regional	Country	Local (district, community, village)
Climate	high	low	n.a.
Agro-ecology	high	low	n.a.
Natural resources base	moderate/high	moderate/high	n.a
Soils	low/moderate	moderate/high	moderate
Topography	low	moderate/high	high
Cropping systems	moderate	high	moderate
Livelihood patterns	low	high	high
Population	low	high	low/moderate
Institutions	n.a.	high	moderate/high
Policies	n.a.	high	moderate/high
Infrastructures	low	moderate	high
Access to markets	n.a.	moderate	high
Access to resources	n.a.	moderate	high
Farm size	low	moderate	high
Power structure	n.a.	low	high

36 Groundwater for Sustainable Livelihoods

could be used in identifying, mapping out, and characterizing homogeneous liveli-hood zones.

Livelihood zoning often accounts for geography, production systems, and other forms of economic activities. According to Khanal et al. (2014), regional-level analyses are focused largely on production systems, country-level analyses are focused on both the geography and production systems, whereas local-level analyses focused on combining geography, production systems, and other economic activities. National and regional mapping can help in providing information about the availability of water especially during drought time. The limitation of these higher scales of analysis is that it does not contain information about the relationship between availability, access, and use. This, therefore, makes the local level very important because it helps to identify the most vulnerable areas and communities with water deficiency or food insecurity. Calow et al. (2010) contend that, such systems exist in many drought-prone countries, at least for monitoring purposes. In these countries, agriculture production is highly dependent on water and there has been increasing risk because of water availability. Groundwater has been playing an important role as a source of water, especially for the agricultural sector and for communities far away from main rivers (e.g. River Niger) in the region and these are mostly rural communities. The case study of Niger illustrates the use of groundwater in improving the lives of communities further away from River Niger.

2.5 CASE STUDY – NIGER

Niger is a landlocked Sahelian country, situated in West Africa, with an estimated population of about 19.8 million in 2016. It has a surface area of about 1,267,000 km². More than 70% of the population is mostly rural with an estimated 14 million people living in rural areas by 2011. This constitutes 15% of the country's land mass, primarily along the western and southern agropastoral zones. The capital city of the country is Niamey and it is located in the southwest corner. Niger is divided into seven regions and one capital district (Niamey) (see Figure 2.5). "With a poverty rate of 48.9% and income per head of US$420, Niger is one of the world's poorest nations" (International Fund for Agricultural Development (IFAD), 2018). Uranium and oil production are cornerstones of the economy, but agriculture is an important source of export earnings. The country is prone to political instability, long-term food insecurity (see Figure 2.5), and droughts, floods, and locust infestations. There is a high variation in the mean annual precipitation ranging from 50 mm or less in the northern desert to about 800 mm in the south of the country. Groundwater plays an important role in the water supply in Niger (Boubakar, 2010). The country, however, has immense and largely untapped fossil aquifer supplies.

Niger presents a high potential in groundwater resources distributed in geological formations (see Figure 2.6) ranging in age from Precambrian to Quaternary with some of the most productive aquifers. Figure 2.7 shows sedimentary formations as the most productive aquifers and occupies largely the western and eastern parts of the country. The excessive depths of the very productive aquifers constitute a major handicap for the exploitation of groundwater (Kimba, 2009). Groundwater resources are also found in ancient sedimentary water basins, which are non-renewable and difficult to access, reaching depths of up to 2,000 m (World Bank, 2009). "Shallower aquifers, particularly alluvial aquifers in valleys and local weathered (regolith) aquifers in basement,

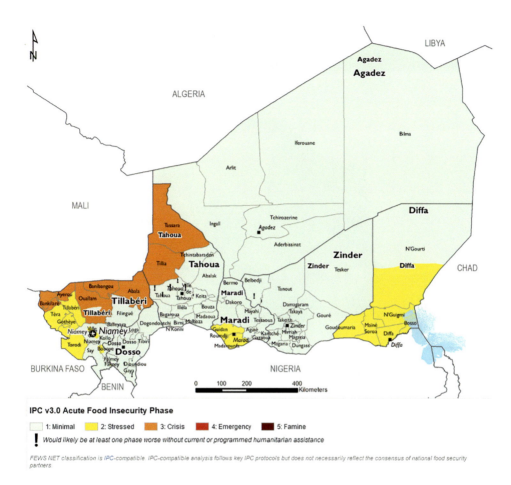

Figure 2.5 Acute food insecurity phase map of Niger (September 2020) (FEWS NET, 2020).

store much smaller amounts of groundwater, but are recharged annually by seasonal rainfall, on which they rely very heavily" (BGS Earthwise, 2020). The changes in land cover and surface water drainage patterns have led to an increase in groundwater recharge rates, and a corresponding rise in groundwater levels (Favreau et al., 2009). Groundwater levels rose by an average of ~4 m/year between 1963 and 2005. This increase in groundwater recharge was attributed to an increase in infiltration due to the increase in vegetation cover.

According to the National Environmental Council for Sustainable Development, by 2003, Niger used only about 20% of its groundwater resources. Groundwater in Niger is generally of good quality and can be used both for agriculture and for human consumption. However, in some areas, the groundwater contains chemicals (e.g. nitrates and nitrites, fluorides, and iron) that are problematic to human consumption. Furthermore, the high evaporation rate contributes to the increase in the mineralization of some Niger's groundwater especially in areas with shallow aquifers; however, this is not recognized as a widespread problem (UNICEF, 2010).

38 Groundwater for Sustainable Livelihoods

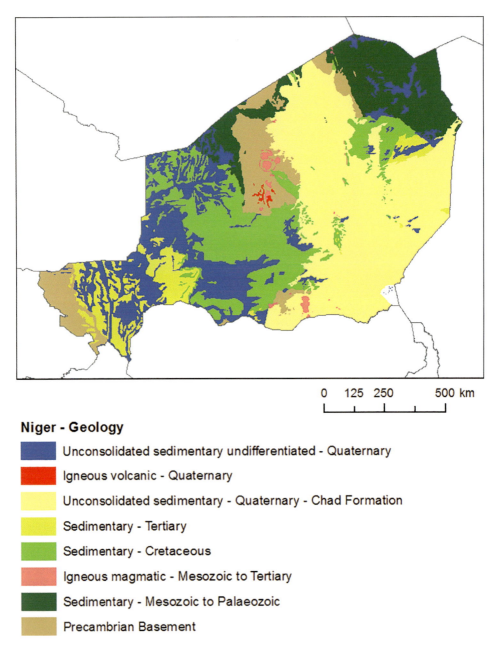

Figure 2.6 Geology of Niger at 1:5 million scale. (Based on map described by Persits et al., 2002/Furon and Lombard 1964 (BGS Earthwise, 2020).)

In addition, very little engagement has been done in terms of transboundary groundwater management. The Iullemeden Aquifer System (IAS), shared by Mali, Niger, and Nigeria is one of the most important transboundary groundwater aquifers. It consists of two major sedimentary aquifers that include the Continental Intercalaire

Groundwater and livelihoods in Niger 39

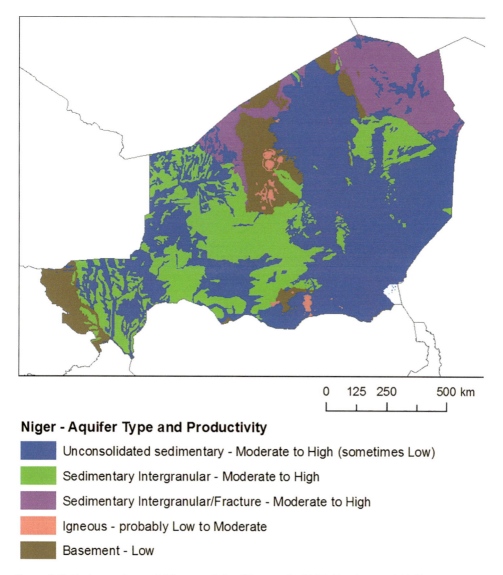

Figure 2.7 Hydrogeology of Niger at 1:5 million scale (BGS Earthwise, 2020).

and the Continental Terminal, and covers an area of approximately 500,000 km². The IAS is the main source of sustainable water for the vast majority of the populations of this region.

2.5.1 Institutional arrangement and water-related policy in Niger

In Niger, three ministries share most of the activities related to planning, mobilizing, developing, and managing water resources. These ministries are: Ministry of Water and Sanitation which has as one of its responsibilities to collect and analyze information related to surface water and groundwater resources, Ministry of Agriculture – has as one of its responsibilities to develop, implement, and monitor national policy with respect to land and water development for agricultural, forestry, pastoral, and related activities. Finally, the Ministry of Livestock and Animal Husbandry – has as one of its responsibilities to develop and implement policy and programs with respect to animal production and animal health.

The use and protection of water resources in Niger are regulated by March 2, 1993 Law No. 93-014 defining the water regime. This law was revised on December 7, 1998, and the new Law No. 98-041 was adopted mainly to account for recommendations from international conferences on IWRM. The October 2, 1997 Decree No. 97-368/ PRN/MH/E was promulgated in order to reinforce the implementation of Law No. 98-041. The second Article of Law No. 98-041 stipulates that any usage of water, creation, and development of any hydraulic structure must be planned and developed in the context of the hydrological or hydrogeological basin in order to minimize any negative impact to the hydrological cycle, the quantity, and quality of the resource. In 2010, the Niger government adopted the ordinance no. 2010-09 on April 1, 2010, bearing the Water Code in Niger. The water code defines the main rules of water exploitation in the country. Niger was also able to pass a decree specifically targeting the groundwater sector and this follows two previous decrees passed in the same sector. These three decrees include:

a. Decree No. 2011-404/PRN/MH/E of August 31, 2011 aims at determining the nomenclature of developments, installations, works, and activities subject to declaration, authorization, and concession of water use;
b. Decree No. 67-143/PRN/MER of September 25, 1967 aims at setting the terms and conditions for the opening and closing dates of pumping stations located in pastoral zones.
c. Decree No. 61-254/MER-MAS of December 2, 1961 aims at determining the periods of use of pumping and grazing areas, pastures, bush fire control measures, veterinary sanitary measures, crops, and sanctions.

The rural sector is viewed as the main driving force of economic growth in Niger, even though extractive industries play an increasingly important role (Merrey and Sally, 2014). The Rural Development Strategy (SDR) provides a reference framework for all rural development interventions. In terms of assessing national policies for irrigation, the National Irrigation Development Strategy (SNDI/CER; Republic of Niger 2005) that is a sub-sectoral strategy in SDR sets an overall objective of improving the contribution of irrigated agriculture to agricultural GDP from 14% in 2001 to 28% in 2015. Furthermore, a clear pathway for the development of irrigation based on government disengagement in favor of farmers' organizations and the private sector was laid (Merrey and Sally, 2014). The SNDI/CER strategy faced a major obstacle due to the fact that it was not formally approved by the Nigerien government (Ehrnrooth et al., 2011). In addition, in December 2012, a small-scale irrigation strategy for Niger (SPIN) paper that covers all types of agricultural water management (AWM) except for Amenagements Hydro-Agricoles (AHAs) and large-scale commercial irrigated agriculture was also produced (Ministry of Agriculture, 2013).

In September 2011, the 3N (Nigériens Nourrissent les Nigériens) initiative with the objective to achieve food and nutritional security and sustainable development was created by the president of Niger. The 3N initiative is in line with other regional and continental agricultural policy frameworks such as the Comprehensive Africa Agriculture Development Plan (CAADP); the ECOWAS Common Agricultural Policy (ECOWAP); and the WAEMU Agricultural Policy (PAU). A High Commission including an operational team was set up to implement this initiative (Merrey and Sally, 2014).

Other key policy documents related to agricultural water according to Merrey and Sally (2014) include

a. The Strategy for Sustainable Development and Inclusive Growth adopted in 2011 articulate the overall vision of Niger's national development in the long term (horizon 2035).
b. The National Economic and Social Development 23 Plan (PDES) 2012–2015, adopted in August 2012, is the principal framework that guides interventions in support of the development agenda of the Government. The policy document is built on the objectives and progress in the implementation of the Accelerated Development and Poverty Reduction Strategy (SDARP) 2008–2012.

In 2017, the PANGIRE (National Action Plan for Integrated Water Resources Management) was adopted. The Plan defines the national framework for water resources management and is the operational tool for implementing the National Water Policy (MHA, 2017). PANGIRE also supports the better integration of the planned actions of various sectors, and the inter-sectoral strategies and programs on water. PANGIRE is planned from 2017 to 2030. The 3N initiative is consistent with the third pillar of PDES.

2.5.2 Groundwater supply and rural livelihood in Niger

Groundwater is an important water resource in Niger. It is used for both drinking and agricultural purposes. "Groundwater plays a significant role in agricultural growth and transformation, food security, and household livelihoods across rural Africa in general and Niger in particular" (UPGro, 2019). It serves as a source of water for more than 85% of the population during the dry season. It is used for watering animals, domestic water supply, small-scale irrigation, etc. Generally, groundwater is exploited mainly through wells and boreholes. Since many years in Niger, groundwater has been captured by using traditional methods such as hand-dug wells. In some cases, there is a direct flow of springs and groundwater into ponds. In rural areas, groundwater is abstracted from both unimproved hand-dug wells and improved hand-dug wells as well as drilled boreholes.

An estimated 85% of the abstracted water is used for drinking purposes, 10% for irrigation, and 5% for other purposes. At least two government databases collect information on water boreholes and hand-dug wells in the country, which includes the central region database and whole country database, although there is little groundwater source data for the north and east of Niger. According to the Ministry of Water, there are more than 24,000 wells and boreholes in the country. In another study by

UNICEF (2010), about 11,000 wells. Nonetheless, many of the wells did not have any information about the water level, geological log, and water quality.

2.5.2.1 Groundwater supply for irrigation

Niger is often hit by drought and famine, and this is mainly because the majority of the population depends on rainfed agriculture, which is highly impacted by climate variation and climate change. Irrigation is considered the best way to increase agricultural production and reduce vulnerability to climate change (Nazoumou et al., 2016). Even though over 90% of large private or government-owned irrigation schemes use surface water, the majority of small-scale irrigation undertaken on a limited scale uses groundwater abstracted from hand-dug wells and mainly applied by manual watering. Traditional bucket/calabash/watering cans play a dominant role (used by over 85% of the people) in irrigated agriculture in local communities (Dittoh et al., 2010). Shallow groundwater irrigation has significantly increased over the past decades, especially in the southwest and along the border with Nigeria. Other schemes are located in the Dallols of Niger (Cochand, 2007). The continuous increment in pump-based irrigation/irrigation calls for the need for better monitoring of groundwater. Figure 2.8 shows an irrigation system from a shallow aquifer.

After the 1970s and 1980s droughts, the country decided to make the development of irrigation farming one of the major axes of its food security policy. With the support of international donors, the state has heavily invested in large-scale irrigation, whose areas increased from 350 ha in 1934 to 13,000 ha in 2012 (CEIPI, 2011). The use of water for agriculture is neither monitored nor regulated, the users are diverse and dispersed thus making it difficult to monitor water usage and allocation (World Bank, 2000). According to the Ministry of Agriculture (2015), the potential irrigable land of

Figure 2.8 Irrigation from shallow groundwater (FAO, 2017).

Niger is more than 10 million hectares. About 5.7 million hectares are areas with water table level that ranges between 0 and 15 m depth and are suitable for small-scale irrigation. However, such small-scale irrigation schemes requires investment, which is often above the capacity of the local population and as a result not always implemented. Currently, only 93,000 ha are imperfectly exploited for small-scale irrigation (Ministry of Agriculture, 2015). In the desert zone, irrigation is developed in oases and valleys where the water table is superficial, e.g. in the valleys of the region of Agadez.

2.5.2.2 Groundwater supply for other uses

Livestock is an important sector in Niger and contributes significantly to the livelihood of the local population. Before, the activity was mainly practiced in the northern part, with pastoral vocation. Today, livestock farming is practiced everywhere, both in cities and rural areas. There are two types of livestock keeping that include nomadic and sedentary. The terminology "pastoral hydraulics" started around the 1950s and refers to the policy of multiplication of water points and modernization of animal watering (Baroin, 2003) through the construction of wells. Three types of wells were constructed, depending on the characteristics of the groundwater table: cemented wells (see Figure 2.9), cemented wells equipped with motorized pumping stations, and artesian wells. The objective was to intensify extensive livestock farming. The implementation of the policy to multiple water points in Niger contributed to the development of livestock breeding in the Sahel region. Small ponds fed with groundwater as shown in Figure 2.10, are also used for animal watering in the region.

2.5.3 Livelihood zoning in Niger

According to FEWS NET (2011), there are thirteen livelihood zones in Niger. The main objective of the FEWNSET study was to identify the highest priority areas for

Figure 2.9 Cemented well for animal watering (Ministry of Water and Environment, 2011).

Figure 2.10 Small pond fed by groundwater in the Dallol Maouri for animals watering (Cochand, 2007).

monitoring food security. The amount of annual rainfall received, and the inter-annual rainfall variability that increases moving toward the north contributed to determining the activities in the zone as well as in shaping the livelihoods map. The southern and southwestern parts of Niger are not intrinsically water-short but suffer from economic water scarcity (Molden et al., 2007). This means that access to water is limited by the availability of human, institutional, and financial capital. According to FEWS NET (2011), the moisture factor played an essential role in differentiating eight of the thirteen livelihood zones, including zones using groundwater. Zones using groundwater to enhance their livelihood include Northeast Oases Dates, Salt, and Trade; Aïr Massif Irrigated Gardening; Southern Irrigated Cash Crops; Dallols Seasonal Water-Course Irrigated Crops; Southeastern Natron Salt and Small Basin Irrigated Dates; and the Lake Chad Flood-retreat with Fishing.

In most cases, access to groundwater was used for irrigation or simply to enhance rainfed crop growth by protecting them against rainfall irregularities. Other factors used for mapping the livelihood zones include soil conditions, proximity to main market centers and cross-border trade, and special local resources such as salt deposits.

The Household Economy Analysis (HEA) methodology through a participative approach with representatives from each region was used for the study. The rural population themselves were not involved in the study. Due to the droughts of 1973–1974 and 1984–1985, and the food price hikes in 2005, the government initiated lasting concern about food security in Niger and triggered an early warning system to monitor shocks and to predict their effects. The aim for identifying and describing the trends and patterns in livelihoods was to serve as a starting point for early warning analysis and for planning for food security. Figure 2.11 shows the livelihood zone map of Niger. It should be noted that livestock is considered as a value asset in all thirteen zones.

Groundwater and livelihoods in Niger   45

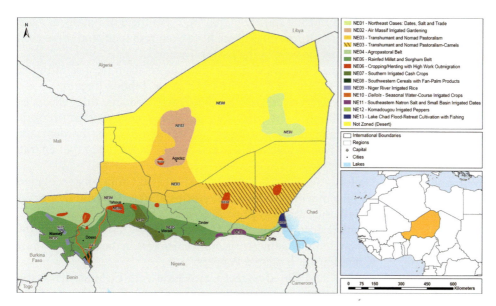

Figure 2.11 National livelihoods zones map of Niger (FEWS NET, 2011).

"Poverty is locally defined almost as much by lack of livestock as by lack of land and crop production capacity" (FEWS NET, 2011).

Of the thirteen zones, four (Zones 3–6) are considered high priority for monitoring food security because they are subjected to localized relief provision. Then, six (Zones 1, 2, 7, 10, 11, and 13) of the thirteen livelihood zones are considered as medium priority because they are not subjected to localized relief provision but are vulnerable to rainfall variations or climate shocks. Finally, three livelihood zones (Zones 8, 9, and 12) are of low priority for monitoring because they are of very low threat to any of the above-mentioned challenges.

The summarized description of the thirteen livelihood zones as presented below has been extracted from FEWS NET (2011).

Livelihood zone 1 – Northeast Oases: Dates, Salt and Trade: boundaries reshaped to reflect the position of oases.
Livelihood zone 2 – Aïr Massif Irrigated Gardening: shape extended to the north-west to show more accurately the coverage of the zone.
Livelihood zone 3 – Transhumant and Nomad Pastoralism: zone extended in the north-west to reflect more accurately the extent of seasonal pastures, especially at the western foot of the Aïr Massif. To the far east the boundary with the desert redrawn southwards to reflect the limit of usable pastures today.
Livelihood zone 4 – Agropastoral Belt: in the far west the boundary with the pastoral zone is placed a little further south to reflect the limits of cultivation today.
Livelihood zone 5 – Rainfed Millet and Sorghum Belt: far eastern limit now shortened to the beginning of the Komadougou zone; the former space taken over by the agropastoral zone.

46 Groundwater for Sustainable Livelihoods

Livelihood zone 6 – Cropping/Herding with High Work Outmigration: this is effectively the promotion to full zone status of what were identified originally as "subzones" of the rainfed agriculture zone. These have been reshaped, and three extra pieces have been added to the west in the pastoral areas.

Livelihood zone 7 – Southern Irrigated Cash Crops: a more accurate discontinuous shape and extension north into the lower Tarka Valley in Tahoua Region.

Livelihood zone 8 – Southwestern Cereals with Fan-Palm Products: this reflects especially the high rainfall, productive environment at the southwestern tip of the country, with its niche Palmyra palm products and its intensive cross-border trade.

Livelihood zone 9 – Niger River Irrigated Rice: more accurate placing of the separate schemes

Livelihood zone 10 – Dallols – Seasonal Water-Course Irrigated Crops: this recognizes a special ecological phenomenon allowing long strips of irrigated cultivation, formerly and very partially subsumed in Zone 7.

Livelihood zone 11 – Southeastern Natron Salt and Small Basin Irrigated Dates: this recognizes a limited but highly distinctive groundwater area in the arid south-east.

Livelihood Zone 12: Komadougou Irrigated Peppers.

Livelihood zone 13 – Lake Chad Flood-retreat Cultivation with Fishing: this recognizes a distinctive area of flood-retreat cultivation that was originally a mute appendix to Zone 12 – Komadougou Irrigated Peppers.

Without a lengthy description of the zones, agriculture stood out as an important livelihood strategy for many communities in Niger. The medium priority zones specifically mentioned the use of groundwater for irrigation. Although there was no mention of the quantity used, groundwater is a key factor for food security in the medium priority zones. The prevailing condition in the medium priority zones seats water intervention as an essential component to improving local livelihood as prescribed by the sustainable livelihood framework. The subsequent section seeks to demonstrate the contribution of water intervention in increasing the welfare and livelihood in Niger and specifically in the medium priority zones by referring to specific indicators from the World Bank and FAO databases.

2.5.4   Rural livelihoods: groundwater interventions and agriculture

Agriculture including raising livestock in most parts depends entirely on rainfall which is periodically affected by drought or a long period of no rain. The search for alternative water sources has been an important adaptation strategy for aridity and unreliable rainfall in the region. The medium priority zones consist of six regions with a total population of about 4 million inhabitants (see Table 2.4). Even though the economies of these regions are very different, the shared similarity is the use of groundwater for irrigation.

Rural population in Niger has been increasing over the years (FAO, 2020) (see Table 2.6). More than 50% of the population in Niger are located in rural areas. Even though the zoning according to FEWS NET (2011) makes no distinction between urban and rural, the livelihood activities help us understand the main source of income and types of livelihoods in rural communities for those zones. Table 2.5 shows that

Groundwater and livelihoods in Niger   47

Table 2.4 Population size of the selected livelihood zones

	Region	Population
Zone 1: North-east Oases: Dates, Salt and Trade	Agadez	17,080
Zone 2: Aïr Massif Irrigated Gardenin	Agadez	287,019
Zone 7: Southern Irrigated Cash Crops	Maradi, Tahoua, Zinder	2,249,710
Zone 10: Dallols – Seasonal Water-Course Irrigated Crops	Dosso, Tahoua, Tillaberi	1,241,122
Zone 11: Southeastern Natron Salt and Small Basin Irrigated Dates	Diffa, Zinder	187,664
Zone 13: Lake Chad Flood-Retreat Cultivation with Fishing	Diffa	91,989
	Total	4,074,584

Source: Adapted from FEWS NET (2011).

Table 2.5 Source of income and livelihood activities in the different zones

	Main income source	*Primary livelihood activities*	*Other activities and events*
Zone 1: North-east Oases: Dates, Salt and Trade	Date sales, salt sales, and livestock sales	Date trees, salt extraction and sale, gardening, camel sales	Wheat, work migration, and peak cereal purchase
Zone 2: Aïr Massif Irrigated Gardening	Crop sales, daily and agri. labor	Millet, wheat, maize – rainfed, maize – food-retreat, onions, onions – irrigated, garlic, potatoes, and agricultural labor	Goat milk production, goat sales, equipment rentals, credit cycles, peak cereal purchase, peak cereal prices
Zone 7: Southern Irrigated Cash Crops	Crop sales, agricultural labor, remittances	Onions, millet, maize, cowpeas sugar cane, tubers, sorghum, agricultural labor	Peanuts, livestock sales, milk peak production, work migration, peak cereal purchase, peak cereal prices
Zone 10: Dallols – Seasonal Water-Course Irrigated Crops	Cash-crop sales, livestock sales, petty trade	Millet manioc/sweet potato, sugar cane, mango sales, peak livestock sales	Off-season gardening, sorghum, egg sales, wild food gathering, labor migration, agri. input purchases, peak cereal purchase, credit cycles
Zone 11: Southeastern Natron Salt and Small Basin Irrigated Dates	Natron, dates remittances, daily labor, firewood and palm sales	Date sales, Natron extraction, maize – irrigated, millet, and livestock sales	Cowpeas, manioc, milk production, work migration, and peak cereal purchase
Zone 13: Lake Chad Flood-Retreat Cultivation with Fishing	Maize, cowpea sales, livestock sales, daily, and agri. labor	Maize, sorghum, cowpeas, livestock sales, milk production	Millet, agricultural labor, fishing, peak cereal purchase, peak cereal prices, peak fishing, and wood sales

Source: Adapted from FEWS NET (2011).

48 Groundwater for Sustainable Livelihoods

Table 2.6 Some general information about Niger

Indicators	2003–2007	2008–2012	2013–2017
Total population (1,000 inhab)	14,668	17,732	21,477
Rural population (1,000 inhab)	12,252	14,646	17,456
Gross Domestic Product (GDP) (Current US$)	4,291,363,339	6,942,209,595	7,142,951,342
Human Development Index (HDI) [highest = 1] (−)	0.257	0.3423	0.3483
Prevalence of undernourishment (3-year average) (%)	14.5	10.5	9.5
% of total country area cultivated (%)	11.84	12.63	13.34

Source: Adapted from FAO (2020a).

people in rural communities are involved in two sets of activities (Primary livelihood activities and other activities) for their livelihoods which is comprised mostly of agricultural activities and also the sales of agricultural products. Data from FAO AQUASTAT suggest that GDP and the value added from the agricultural sector in Niger are showing an increasing trend (see Table 2.6). This also corresponds to an increasing agricultural output (cereal and livestock production) trend as presented in Table 2.7. The increased values demonstrate a positive improvement in the local economy and livelihoods with a significant contribution coming from the agricultural sectors.

Table 2.6 shows an increasing trend in the Human Development Index and a decreasing trend in the prevalence of undernourishment. Even though there are no data from the medium priority zones to confirm these positive changes, it is important to note that both indicators are good for monitoring the change in the poverty rate. Further description of the zones is provided in Section 2.6. It could be assumed that the positive contribution from the agricultural sector led to the decreasing trend in the prevalence of undernourishment thus poverty rate. This can further be confirmed by the increasing trend in drinking water services, indicating an improvement in

Table 2.7 Progress outcome of the livelihood strategy

Indicators	2015	2016	2017	2018
Agriculture, forestry and fishing, value added (% of GDP)	36.3	38.8	39.7	39.2
Agriculture, forestry and fishing, value added (current US$)	2.6 million	2.9 million	3.2 million	3.6 million
Cereal production (metric tons)	5,464,683	5,852,529	5,899,777	NA
GDP (current US$)	7.2	7.5	8.1	9.2
Livestock production index (2004–2006 = 100)	102.7	103.9	NA	NA
People using at least basic drinking water services (% of population)	48.7	49.5	50.3	NA
People using at least basic drinking water services, rural (% of rural population)	41.6	42.6	43.6	NA

Source: Adapted from World Bank (2020).

socio-economy in the country. Eighty-five percent of the total groundwater exploited is used for drinking purpose. The increasing trend in drinking water services signifies an increasing exploitation of groundwater to cover for this increase since very little efforts have been made to extend water services to rural communities. It is easy for the local population to exploit groundwater and other sources which might be too far or of poor quality.

According to FEWS NET (2011), the medium priority zones are not subject to localized relief provision. However, the zones are not immune to serious rain failure or other shocks, which occasionally subject the poorer people to receive food and cash relief. More data are therefore needed to monitor groundwater interventions in enhancing rural livelihood for the medium priority zones.

Table 2.6 shows that the percentage of the total cultivated area is showing an increasing trend, which makes the argument for groundwater as a contributing factor for the improvement of livelihood very complex and requires more measured data for justification. Niger is a water-scarce country with rainfall concentrated within 4 months in a year, with increasing climate variability and climate change. About 85% of the population depends on groundwater as the only source of water during the dry season. Agricultural activities in the medium priority zone depend on groundwater as a coping strategy during a period of rainfall variation. Tables 2.8 and 2.9 show, respectively, information on water resources and on water used in the agricultural sector. According to data from FAO (2020a), total renewable groundwater resource has shown no increasing trend from 2003 to 2017. However, the amount of water withdrawn for agricultural purposes shows an increasing trend. This also corresponds to the amount of water withdrawn for irrigation in the country. There is the likelihood that the increasing trend in irrigation water has contributed to the increasing trend observed in agricultural output, e.g. cereal as presented in Table 2.7. According to FEWS NET (2011), in Zones 7, 9, 10, and 11 a certain amount of rainfed staples cultivation is buffered by the irrigation economy during rain failure thus reducing the vulnerability of the local population.

Water plays an important role in the agricultural sector of Niger because of the natural environmental conditions. Both surface and groundwater irrigation are common practices in the country because of their very large potential benefits (Giordano et al., 2012). According to Merrey and Sally (2014), there is an estimated 100,000 ha of the managed area, and 88,885 ha of exploited area is equipped for irrigation (see Table 2.10). Groundwater irrigation is more prominent in rural

Table 2.8 Water resource in Niger

Indicators	2003–2007	2008–2012	2013–2017
Long-term average annual precipitation in volume ($10^9\,m^3$/year)	191.3	191.3	191.3
Surface water produced internally ($10^9\,m^3$/year)	1	1	1
Groundwater produced internally ($10^9\,m^3$/year)	2.5	2.5	2.5
Groundwater: entering the country (total) ($10^9\,m^3$/year)	NA	NA	NA
Groundwater: leaving the country to other countries (total) ($10^9\,m^3$/year)	NA	NA	NA
Total renewable groundwater ($10^9\,m^3$/year)	2.5	2.5	2.5

Source: Adapted from FAO (2020a).

50 Groundwater for Sustainable Livelihoods

Table 2.9 Agriculture water in Niger

Indicator	1988–1992	2003–2007	2013–2017
Agricultural water withdrawal ($10^9 m^3$/year)	0.41	0.6565	1.536
Total water withdrawal ($10^9 m^3$/year)	0.5	0.9836	1.751
Irrigation water withdrawal ($10^9 m^3$/year)	NA	0.6565	1.366
Irrigation water requirement ($10^9 m^3$/year)	NA	NA	NA
Fresh groundwater withdrawal ($10^9 m^3$/year)	NA	NA	NA
Total freshwater withdrawal ($10^9 m^3$/year)	0.5	0.9836	1.751

Source: Adapted from FAO (2020a).

Table 2.10 Summary of agricultural water areas in Niger

Category of management	Managed area (ha)	Exploited area (ha)	Brief description
AHA (Amenagements Hydro-Agricoles) (National Office for Irrigation Systems – ONAHA)	13,850[a]	12,735[a]	Most of these schemes are for rice though, in some, high-value vegetable crops are grown in the dry season. Account for about 14% of the irrigated land
Commercial irrigation	<1,000	<1,000	Large-scale export-oriented commercial irrigated agriculture but this is not yet well developed in Niger
Small-scale private irrigation	16,150[a]	16,150[a]	Small-scale private irrigation has developed rapidly with little or no policy support or regulation. Most small-scale irrigation is based on groundwater
PCS (Périmètres de Contre-Saison)	70,000[a]	60,000[a]	These are dry season (October–May) irrigation systems collectively managed at the community watershed level. They include manual or mechanized irrigation from wells, streams, and ponds as well as flood recession agriculture.
Flood recession cropping area, non-equipped	12,000	10,000	These are areas that are cultivated after river floods and seasonal ponds recede, i.e. as the rainy season ends. This is common in the Lake Chad area and Niger River basins.
Total agricultural water-managed area[b]	112,000	98,885	
Collection of runoff water (water harvesting)	300,000	300,000	The collection of rainfall runoff through a variety of techniques (including pits, bunds, and dikes) thereby improving water availability for crop production.

Source: Adapted from World Bank (2009) cited in Merrey and Sally (2014).
[a] Estimates based on 2005 Stratégie nationale du développement de l'irrigation.
[b] Excludes commercial irrigation.

communities because they are mostly small-scale irrigation. In addition, the communities are often located farthest from main river courses to facilitate surface water irrigation. This study did not manage to establish a direct connection between groundwater development and the improvement of livelihood in Niger because of the lack of data in line with the different capitals of the sustainable livelihood framework. However, a connection on the use of groundwater to enhance livelihood was established in medium priority zones through the description of the farming activities as presented in the subsequent sections. Furthermore, there is a lot of information about the use of groundwater for irrigation purposes in Niger. The study manages to establish that there is an increasing trend between well-being and developing irrigation schemes.

2.6 DISCUSSIONS

In the livelihood zoning by FEWS NET (2011), the livelihood patterns clearly vary from one geographic area to another. The livelihood zone map has been useful as the first step for livelihoods-based analysis. In Niger, the degree to which rural population relies on agriculture be it crop farming or cattle farming is highly determined by the availability of water. The medium priority zones (see Figure 2.12) are all diverse in terms of specific livelihoods strategies, resource bases, and wealth distribution. The common factor in the zones has been the significant presence of irrigation and water resources used for off-season cultivation, diversifying production potential in a given year. The agricultural system is not entirely dependent on rainfed systems as a means of producing staple crops for consumption. Furthermore, the irrigation economy (employment, harvest, and payments in kind) provides a buffer in the face of poor

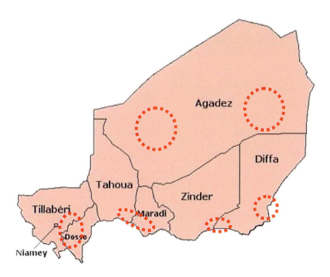

Figure 2.12 Map of Niger showing the medium priority zones (FEWS NET, 2014).

seasonal performance (Sisa, 2014). This section seeks to discuss the main characteristics of the medium priority zones and the contribution of groundwater as a coping strategy for the respective zones. Figure 2.12 shows the map of Niger with the location of the medium priority zones indicated with red circles.

Zone 1: Northeast Oases Date, Salt, and trade zone mostly around the Eastern Agadez is a net importer of food and famous for irrigation cultivation by constantly mitigating their food insecurity risks with water from oases. Beyond the immediate environs of oases lies pure desert. The zone is the least populated of all the livelihood zones with both poorer and wealthier households having land irrigated from wells. According to FEWS NET (2011), the wealthier have more land and tend to irrigate with motor pumps, whereas the poorer lift water with hand-pumps or using the "shaduf." Dates are the major crop exported from the zone. On average a date plantation may have a stand of 100 trees. Some amount of wheat and vegetables, animal rearing, and other fruits are common activities for home consumption or for the limited local market. Additional income is also generated from the sale of salt (largely found in Djado, Bilma, and Fachi Communes), local casual labor, and migrant work.

The key important productive assets in the zone include land, with wells for irrigation, motor pumps, non-motorized water pumps/"shaduf," date palms, vegetable gardens, livestock, and capital for trading. The availability of groundwater for irrigation in the zone has greatly supported agriculture activities. This is; however, not short of insecurity because of groundwater level fluctuation that depends on rainfall. Furthermore, the availability of irrigation has attracted migrants who have put pressure on land thus reducing plot sizes and increasing the vulnerability of the poor population. In addition, the zone is prone to food price spikes, disease of date palms, and periodic civil insecurity leading to market disruption. According to Merrey and Sally (2014), the use of technology (e.g. drip irrigation) to improve water productivity; easy access to groundwater depth as well as improving agricultural input would increase output for the markets thus increasing household income.

Zone 2: Aïr Massif Irrigated Gardenin comprises communities mostly around West/Central Agadez. The Aïr Massif community is a sparse rural population that also depends on irrigated cultivation by extracting water from near-surface aquifer through wells. Groundwater levels also depend on rainfall. Annual rainfall is less than 200 mm/a, which is not sufficient for cereal cultivation but supports the growth of grass for goat and sheep rearing. The most valuable product in the zone is onions, which are sold onto the national market as well as beyond in West Africa. Other crops out from the zone include millet, wheat, maize, Irish potatoes, garlic, and citrus fruit. Even though the zone is geographically isolated, it still depends on the national market for food and cash.

The most important productive assets include labor capacity, irrigable land, own well, small livestock, motor pumps, and camels for traction. According to FEWS NET (2011), cultivation is significantly enhanced by well ownership or the payment for water. The method of water lifting from the well visibly distinguishes the wealthy farmers from poor farmers. The common practice is the use of bucket or skin bags and rope via a pulley system powered by a camel. In other cases, wealthy people used motor pumps. Poorer farmers who cannot afford to own well pay for water with their labor (FEWS NET, 2011). Additional common shocks and hazards in the zone include flash floods damaging gardens, drought leading to the lower water table, food price hikes, civil insecurity disrupting markets, land pressure, etc. Technological enhancement to improve water productivity, e.g. drip irrigation, modern water lifting systems depending on the aquifer depth, has been recommended as a way forward to greatly improve farming input and output for the market.

Zone 7: Southern Irrigated Cash Crops comprises Southeast Tahoua, Southwest Maradi, and Southern Zinder (see Figure 2.11). The zone is regarded as one of the wealthiest livelihood zones within the medium priority zones (Sisa, 2014). The zone is a large basin area of land fed with water from a near-surface water table. The zone is famous for the cultivation of quality onion, which is sold in both local and international markets. Other products from the region include tobacco, watermelon, tomatoes, chili peppers, sugar cane, etc. "Due to intense production year-round, labor demand is also high, and a key pillar of poor household food security" (Sisa, 2014). Rainfall (above 500 mm/a) is therefore important in the zone since a large part of the population depends on rainfed agriculture which grows millet and sorghum for domestic consumption, as well as cash crops such as peanuts, cowpeas, and sorrel for income (FEWS NET, 2011).

The most important productive assets include household labor capacity, irrigated and rainfed land, hoes, machetes, small livestock, oxen, and plow cattle. Despite these assets, a large part of this densely populated zone owns extremely small plots of irrigable land and lacks the capital for inputs to maximize production (FEWS NET, 2011). On the other hand, they have the means to purchase food despite this insufficiency. Local employment reduces the need to migrate and thus provides income to sustain their livelihood. According to FEWS NET (2011), some of the common shocks and hazards include unusual bad insect attacks, rainfall failure/irregularities, flooding of irrigated fields, fall in cash-crop price, and a hike in prices. In all, one key adaption strategy to consider is the use of technology to improve water productivity as well as to explore the application of rainwater harvesting, other Sustainable Land and Watershed Management (SLWM) practices, and supplementary irrigation to increase productivity (Merrey and Sally, 2014).

Zone 10: Dallols – Seasonal Water-Course Irrigated Crops are located within the Niger River basin, Eastern Tillaberi, and Dosso regions. *Dallol* (in the local Djerma language) denotes a seasonal water course that retains sufficient water for cultivation through irrigation year-round. It is equivalent to "wadi" or "khor" in Arabic. The *dallols* are fossil riverbeds that were once tributaries to the Niger River and offer the local population off-season cultivation, in addition to the basic rainfed cropping. This also depends on the level of ground water that usually experienced lower annual rainfall and is at risk of rainfall irregularities. The *dallols* zone consist of a dense population by Niger rural standard that depends on crop production on small plot and mostly household labor for their livelihood. Common crops cultivated are millet, sorghum, rice, sugar cane, cassava, sweet potatoes, mangoes, citrus fruit, and vegetables. In addition, there is natron and white salt surface-mining in the Dallol Foga.

The most important productive assets in Zone 10 are the *Dallol* land, labor capacity, poultry, ox-plows and some tractors, cattle and small stock, capital for trade. The "households do not earn sufficient income from cereals and garden production" (Merrey and Sally, 2014). They depend on paid work, which often requires migration to urban centers and elsewhere in the region, including Nigeria, Benin, Ghana, and Ivory Coast" (Merrey and Sally, 2014). Richer farmers have more land and capacity to produce (especially sugar cane) and to maintain cattle. Common shocks and hazards include flooding, irregularities in rainfall, unusual serious attacks of insect pests and birds, and food price hikes. The improvement of water productivity through the use of better and efficient technologies such as drip irrigation and the use of rainwater harvesting technologies as well as other SLWM practices for rainfed crops would increase resilience in the agricultural sector.

Zone 11: Southeastern Natron Salt and Small Basin Irrigated Dates comprises Southwestern Diffa and Southeastern Zinder. It is a zone of moist depressions or small basins (cuvettes) that provide water for irrigation. The cuvettes provide a niche environment for date production which is a valuable crop for both local and international markets. Groundwater for irrigation poses some degree of food insecurity because the groundwater levels depend on rain. Common crop in the region includes maize, cassava, cabbages, and other vegetables. Little millet and cowpeas are cultivated on rainfed fields. Beyond the cuvettes are livestock farms mostly owned by wealthier households. The zone is also known for the availability of natron salt (hydrated sodium carbonate) which is extracted for commercial sale both for export and the local market. Dates and natron are the main source of income and are used for the purchase of food and other essentials.

The most important productive assets for the local population include the local natron pans, irrigated land, labor capacity, large and small livestock, and Ox-carts for rent. Annual rainfall is generally 300 mm or less, thus groundwater is needed as a booster to the farming sector allowing all year-round cultivation. Common shocks and hazards in the area include pest attacks on date-palm, rain failure, and market price fluctuation. Investment also in technologies to improve water production and efficiency through drip irrigation is highly recommended for building resilience in the agricultural sector.

Zone 13: Lake Chad Flood-Retreat Cultivation with fishing is a zone located at the border with Chad in the southeast Diffa region. The zone became available for flood-retreat cultivation as Lake Chad progressively reduced in size. The zone is sparsely populated with farmers who are mostly immigrants from Mali, Nigeria, Senegal, and Cameroon. The main staple grown is maize, followed by sorghum, while cowpeas are an important crop for home consumption and sale. The zone also has groundwater for irrigation, but the level is highly dependent on rainfall which occasionally fails in the overall drainage region feeding the lake. Cattle, camel, sheep, goats, fish, and poultry also bring useful cash for the poorest households, whereas wealthier farmers trade predominately with cattle.

The most important productive assets for the local population are the flood-retreat land fishing lines, nets, livestock, oxen and ox-plows, and stands of prosopis trees. The cultivation work in the dry season and the fishing resource attract seasonal migrant laborers. Common shocks and hazards in the region include low flood level of the lake by end of rains, unusual severity of crop pest, and unusual outbreak of animal disease. "Acute food insecurity is infrequent, but can occur if seasonal performance is insufficient to replenish the lake to levels that allow for adequate flood recession and irrigated cereals and cash crops" (Sisa, 2014). The long-term solution would require programs that improve the management of the Lake Chad basin, explore ways to enhance storage of water in flooded soil profiles, explore using shallow groundwater to irrigation after flood recedes, and explore the use of efficient irrigation technologies that would support a better farm input and output for the market (Merrey and Sally, 2014).

To wrap up, subsistence production at the household level accounts for up to 40% of rural household food consumption. The sale of cash crops provides additional income to the local population in Niger. Besides that, "casual agricultural labor is a primary income and livelihood source to many Nigeriens, who are consequently very vulnerable to several climate shocks and hazards" (Sisa, 2014). The pastoral sectors are less frequently acute food insecure, however, the sectors remain highly vulnerable to price hikes. Water is the most critical constraint on agricultural production. It is no surprise that irrigated land or water plays a major role in sustaining livelihood in a country already deficient in water resources as a result of the natural climatic conditions. Market purchase, therefore, remains the primary food access strategy for 8–9 months of the year, comprising nearly 100% of food needs year-round (Sisa, 2014).

Previous studies (e.g. Tillie et al., 2019) that analyze the microeconomic effects of small irrigation schemes using an agricultural household model show that an increase of 47,000 hectares or 44% of irrigated areas in the dry season in Niger bring significant benefits to Nigerien producer households. Average farm income increases by about 12% and the rural household income inequality would decrease by almost five points or about 9%. The extension of irrigated areas would also generate a large number of jobs, thus contributing to the reduction of the rural poverty rate by more than one point (from 52.4% to 50.8%). Furthermore, groundwater is the only water resource available for communities in desert and these communities represent more than 50% of the territory. Therefore, any livelihood strategy that seeks to improve food security and the well-being of the people in desert and water-restricted communities must consider the development of groundwater resources.

Investments that focus directly on reducing rural poverty (and increasing household food security and nutritional status) may lead to more rapid and more durable improvements in rural households' well-being. Such investment also contributes to GDP growth but may not be easy to measure in the early years because these involve small increment among many households (Merrey and Sally, 2014). This is an important consideration especially in situation where there is high rural poverty rate and food insecurity like in Niger. According to Merrey and Sally (2014), activities undertaken in many large irrigation schemes in Niger are not sustainable partly because they lack sense of ownership, the community is not involved in the design and implementation of project activities and finally, there is little or no freedom in the choice of crops to produce. In addition, there is a lack of essential maintenance of the scheme and no follow-up project or a sustainability plan for such schemes. This, therefore, makes sense to promote decentralized small-scale systems of irrigation that are mostly connected to groundwater supply for the sustainment of rural livelihood in the region.

On the other hand, such investment should consider the design of water supply systems that serve multiple purposes such as growing their crops, watering their livestock, drinking, laundry, bathing, etc. The performance of single-purpose schemes is undermine and sometimes leads to deterioration because the scheme is often modified by the local population. For example, if no facilities are available for domestic use, poor water quality reserve for irrigation will be used for this purpose. It is evident that the full involvement of the local community in the identification and prioritization of water needs and design of water supply schemes to meet their needs have always led to higher performance, greater sustainability, and improved outcomes in terms of people's well-being (Adank et al., 2012). It is also important to consider higher potential

56 Groundwater for Sustainable Livelihoods

investment that could make a major difference in terms of rural poverty reduction, improved food security, and economic growth. In this case, dry season irrigation from groundwater for high-value crops is strongly recommended because it targets the most vulnerable people (including women and pastoralists).

2.7 CONCLUSIONS

Water is a "binding constraint" to the development and growth of Niger's economy. The vast majority of the population is rural and overwhelmingly depends on agriculture (including livestock) for their livelihood. It is therefore unlikely that another sector in the country will develop at a sufficiently rapid rate to create job opportunities for the rapidly growing population. Many households in rural Niger have very little physical and financial capital. The key assets for these communities include a piece of land and their labor, which make small-scale farming a very lucrative livelihood strategy. Despite the deficiency in local data, the study examines how the increment of agricultural water could contribute to the well-being of the rural population, thus enhancing their livelihood. The study also shows that investing in physical capital, such as building new irrigation schemes can increase agricultural output, thus supporting also the enhancement of rural livelihoods.

Ensuring food security should be considered in the overall livelihood strategy especially because food loss is not only caused by the shortage of rain but also by the failure of having water at critical stages in the growing cycle of the crop. Access to groundwater for irrigation has enhanced rainfed agriculture in Niger. Even though a large part of livelihood zones presented in Figure 2.11 is mostly agriculture-based,there are no sufficient data to support if the mapped activities in the zones are the right strategies, especially with respect to the water interventions in those communities. The type and nature of intervention depend on the livelihood context. However, it is important to note that the application of water intervention options is specific and depends on the livelihood system. The decision to target the option of building a resilient community against risks and vulnerability or to target production and water values to enhance rural livelihoods will depend on the well-being – groundwater/water linkages or the poverty-groundwater/water linkages. The livelihood zoning map can help to determine future interventions. However, it is still important to conduct an area-specific livelihood analysis to identify the best intervention strategies.

Poverty and rural livelihood strategies should not be seen in isolation, but within the context of ongoing socio-economic changes. Agriculture is the core strategy, which is gradually transforming into more diversified forms of livelihood. The local population is engaged in two or three different types of agricultural activities which may include animal keeping, crop growing, food processing, commerce, and providing agricultural labor. This is important because it helps to diversify their risk and vulnerability. Rural economic development should find a balance between producing food, managing natural resources, promoting growth, and providing a livelihood base for the rural population for the wider good of sustainability. The potential for such interventions to enhance rural livelihood could be huge.

The potential for groundwater interventions in rural communities in Niger will largely be "development-driven" rather than "management-driven" considering current schemes use mostly traditional methods, which are labor intense and time-consuming. The introduction of modern methodologies will extend irrigation schemes in the countries thus increasing food production and saving time, which could be invested in other activities. Small-scale irrigation has been proven to increase household income as well as contribute to the reduction of poverty in rural communities. However, there is a need to better understand the aquifers in order to support and ensure sustainability among small-scale irrigation. In addition, the future design of rural electrification programs should include attention to supporting the electrification of pumps. The level of poverty is still high, the physical development of groundwater schemes will require a political-will as well as the intervention of the private sector for investment. Nonetheless, with the adoption of the National Action Plan for Integrated Water Resources Management, there is hope that groundwater exploitation will be better organized to promote the enhancement of livelihood strategies.

Rural people's constraints and opportunities should be recognized and addressed at all times. This study, therefore, recommends the implementation of more specific localized studies on groundwater and poverty linkages in selected poor regions, especially moving toward the north of the country, where dependence on groundwater is very high. In general, water availability is a major factor affecting their livelihoods, both in terms of basic services and in terms of building resilience and reducing vulnerability. Studies should also look at peri-urban livelihood systems especially because of the rapid expansion of urban areas to rural areas creating an interface between rural and urban areas. Peri-urban are strategic because they are market centers or link to markets and migration, which constitute part of the rural livelihood strategy. Understanding the dynamics and linkages to agriculture and groundwater management will be important for the designing of future poverty reduction programs.

Emerging economic development and the sensitive environmental context of Niger, including the impacts of climate change at both global and local levels, will shape and reshape rural livelihood strategies in the future. It is therefore very important to align data gathering around the different capitals of the sustainable livelihood framework because only with sufficient data and long-term monitoring will the right livelihood and adaptation strategies be established. Sufficient data also help for the proper evaluation of the groundwater intervention and its linkage to poverty reduction as well as with the enhancement of rural livelihood, which has been a major constraint for this study. In addition, livelihood zoning should complement the sustainable livelihood framework and vice versa. While livelihood zoning focuses on identifying areas with the same characteristic, a sustainable livelihood framework should focus on understanding the interaction between the different assets in the zone to establish sustainable connections. This helps to focus the limited investments so as to maximize the outcome in a more sustainable manner.

REFERENCES

Adank, M., van Koppen, B. & Smits, S. on behalf of the MUS Group. (2012) *Guidelines for planning and providing multiple use water services.* IRC International Water and Sanitation Centre and International Water Management Institute. Available from: http://www.musgroup.net [Accessed on 29th November 2020].

58 Groundwater for Sustainable Livelihoods

Baroin, C. (2003) L'hydraulique pastorale, un bienfait pour les éleveurs du Sahel? *Afrique contemporaine* 1 (205), 205–224.

BGS (British Geological Survey) Earthwise (2020) *Hydrogeology of Niger*, London. [Online] Available from: http://earthwise.bgs.ac.uk/index.php/Hydrogeology_of_Niger [Accessed on 27th October 2020].

Boubakar, A.H. (2010) Aquifères superficiels et Profonds et pollution urbaine en Afrique : cas de la communauté de Niamey (Niger). *Thèse de doctorat*, Université Abdou Moumouni de Niamey, Niger, p. 249.

Braune, E., Hollingworth, B., Xu, Y., Nel, M., Mahed, G. & Solomon, H. (2008) Protocol for the assessment of the status of sustainable utilization and management of groundwater resources - with special reference to Southern Africa. *WRC Report No.* TT 318/08, Water Research Commission, Pretoria, South Africa.

Burke, J.J., Sauveplane, C. & Moench, M. (1999) Groundwater management and socio-economic responses. *Natural Resources Forum*, 23, 303–313.

Calow, R.C., Macdonald, A.M., Nicol, A.L. & Robins, N.S. (2010) Ground water security and drought in Africa: Linking availability, access, and demand. *Ground Water*, 48, 246–256.

CEIPI (2011) *Projets et programmes de développement de l'irrigation au Niger (1960–2010): Eléments pour un bilan*. Centre d'Etudes et d'Information sur la Petite Irrigation.

Cochand, J. (2007) La petite irrigation privée dans le sud Niger: potentiels et contraintes d'une dynamique locale: Le cas du sud du Département de Gaya. *Mémoire*. Institut de Géographie, Université de Lausane.

Cook, S. & Gichuki, F. (2006) Analyzing water poverty: Water, agriculture and poverty in basins. *Basin Focal Project Working Paper No. 3*. CGIAR Challenge Program on Water and Food.

Dittoh, S., Akuriba, M.A., Issaka, B.Y. & Bhattarai, M. (2010). Sustainable micro irrigation systems for poverty alleviation in the Sahel: A case for micro public private partnerships? *Paper presented at the 3rd Annual African Association of Agricultural Economists (AAAE)*, Cape Town, South Africa.

Economic Community of West African States, Sahel and West Africa Club & Organization for Economic Co-operation and Development (2006). *The Ecologically Vulnerable Zones of the Sahelian Countries*. Atlas on Regional Integration in West Africa.

Ehrnrooth, A., Dambo, L. & Jauber, R. (2011) *Projets et programmes de développement de l'irrigation au Niger (1960–2010): Eléments pour un bilan*. Programme d'Appui au Développement de l'Irrigation Privée au Niger (PADIP). Centre d'Etudes et d'Information sur la Petite Irrigation (CEIPI), Niger.

FAO (2017) *40 ans de coopération entre la FAO et le NIGER*. [Online] Available from: http://www.fao.org/3/a-i7520f.pdf [Accessed 6th June 2020].

FAO (2020a) *Aquastat data*. [Online] Available from: http://www.fao.org/nr/water/aquastat/data/query/results.html [Accessed 4th June 2020].

FAO (2020b) FAOSTAT Database [Online] Available from: http://www.fao.org/faostat/en/#-data/QC [Accessed 4th June 2020].

Faurès, J.M. & Santini, G. (Eds.) (2008) *Water and the rural poor: Interventions for improving livelihoods in sub-Saharan Africa*. Rome: FAO and IFAD. [Online] Available from: www.fao.org/nr/water/docs/FAO_IFAD_rural-poor.pdf [Accessed 4th June 2020].

Favreau, G., Cappelaere, B., Massuel, S., Leblanc, M., Boucher, M., Boulain, N. & Leduc, C. (2009) Land clearing, climate variability, and water resources increase in semiarid southwest Niger: A review. *Water Resources Research*, 45, W00A16, 1–18.

FEG Consulting and Save the Children (2008) *The Practitioners' Guide to the Household Economy Approach*. Johannesburg: Regional Hunger and Vulnerability program.

FEWS NET (2011) *Livelihood Zoning "Plus" Activity in Niger*. Washington, DC: USAID.

FEWS NET (2014) *Niger Food Security Brief* [Online] Available from: https://fews.net/sites/default/files/documents/reports/Niger_Food_Security_Brief_Final.pdf [Accessed 31st March 2021].

FEWS NET (2020) *Niger* [Online] Available from: https://fews.net/west-africa/niger/key-message-update/september-2020 [Accessed 07th November 2020].

Giordano, M., de Fraiture, C., Weight, E., van der Bliek, J. (Eds.) (2012) Water for wealth and food security: supporting farmer-driven investments in agricultural water management. *Synthesis report of the AgWater Solutions Project.* Colombo, Sri Lanka: International Water Management Institute (IWMI).

Herrmann, S.M., Anyamba, A. & Tucker, C.J. (2005) Recent trends in vegetation dynamics in the African Sahel and their relationship to climate. *Global Environmental Change*, 15(4), 394–404.

Hesse, C., Anderson, S., Cotula, L., Skinner, J. & Toulmin, C. (2013). *Managing the Boom and Bust: Supporting Climate Resilient Livelihoods in the Sahel.* London: IIED Issue Paper. IIED.

Hussain, I. & Hanjra, M.A. (2004) Irrigation and poverty alleviation. Review of empirical evidence. *Irrigation and Drainage*, 53, 1–15.

Ickowicz, A., Ancey, V., Corniaux, C., Duteurtre, G., Poccard-Chappuis, R., Touré, I., Vall, E. & Wane, A. (2012) Crop–livestock production systems in the Sahel – increasing resilience for adaptation to climate change and preserving food security. *In Proceeding of FAO/OECD Workshop on "Building Resilience for Adaptation to Climate Change in the Agriculture sector.* 23–24 April 2012, Rome, FAO-OCDE. 40p.

International Federation of Red Cross (2020) *What is a Livelihood?* [Online] Available from: https://www.ifrc.org/en/what-we-do/disaster-management/from-crisis-to-recovery/what-is-a-livelihood/ [Accessed 21th May 2020].

International Fund for Agricultural Development (IFAD) 2010, Rural Poverty Report 2011 – New realities, new challenges: new opportunities. Rome, Italy. 322 pages ISBN 978-92-9072-200-7. [Online] Available from https://www.fao.org/fileadmin/user_upload/rome2007/docs/IFAD%20Rural%20Poverty%20Report%202011.pdf [Accessed 6th December 2021].

International Fund for Agricultural Development (IFAD) (2018) *Investing in Rural People in Niger.* Rome, Italy. [Online] Available from: https://www.ifad.org/documents/38714170/40843602/Enabling+poor+rural+people+to+overcome+poverty+in+Niger_e.pdf/8fb842f1-2a21-4de1-8017-7877282ccee1 [Accessed 5th June 2020].

Islam, T. & Ryan, J. (2015) *Hazard Mitigation in Emergency Management.* Waltham, MA: Butterworth-Heinemann. ISBN 978-0-12-420134-7.

Khanal, P.R., Santini, G. & Merrey, D.J. (2014) *Water and the Rural Poor Interventions for Improving Livelihoods in Asia.* Bangkok: Food and Agriculture Organization of the Nation Regional Office for Asia and The Pacific. ISBN 978-92-5-108263-8.

Kimba, H. (2009) Politique nationale de l'eau et de l'assainissement du Niger: Approche de gestion intégrée des ressources en eau à l'échelle locale, régionale, nationale et internationale. *Haut-commissariat à l'Aménagement de la Valée du Niger (HCAN).* Cabinet du Premier Ministre, République du Niger.

MacDonald, A.M., Bonsor, H.C., Dochartaigh, B.É.Ó. & Taylor, R.G. (2012) Quantitative maps of groundwater resources in Africa. *Environmental Research Letters*, 7, 024009.

MacDonald, A.M., Davies, J., Calow, R.C. & Chilton, J.P. (2005) *Developing Groundwater: A Guide for Rural Water Supply.* Rugby: ITDG Publishing.

Merrey, D.J. & Sally, H. (2014) Improving access to water for agriculture and livestock in Niger: A preliminary analysis of investment options for the Millennium Challenge Corporation. Submitted to Millennium Challenge Corporation (MCC) in Response to the Technical Directive Task Order (TO) against MBO Consultancy #MCC-10–0016-BPA, Call 13-CL-0074. Preliminary Background Research and Analysis to Inform the Development of an Irrigation Project in Niger. [Online] Available from: https://www.researchgate.net/publication/259911951_Improving_Access_to_Water_for_Agriculture_and_Livestock_in_Niger_A_Preliminary_Analysis_of_Investment_Options_for_the_Millennium_Challenge_

Corporation_Submitted_to_Millennium_Challenge_Corporation_MCC [Accessed 27th October 2020].

MHA (Ministère de l'Hydraulique et de l'Assainissement) (2017) *Plan d'Action National de Gestion Intégrée des Ressources en Eau*, PANGIRE Niger. République du Niger.

Ministry of Agriculture (2013) *Strategie de la petite irrigation au Niger (SPIN)*. Partie 1, Version préfinale. March. Republique du Niger.

Ministry of Agriculture (2015) Evaluation du potentiel en terre irrigable du Niger. Direction Générale du Génie Rural - Ministère de l'Agriculture, Niamey.

Ministry of Water and Environment (2011) *Guide National d'Animation en Hydraulique Pastorale*. Niger: Niamey. 31p.

Molden, D., Frenken, K., Barker, B., de Fraiture, C., Mati, B., Svendsen, M., Sadoff, C., and Max Finlayson, C. (2007) Trends in water and agricultural development. In *Comprehensive Assessment of Water Management in Agriculture* (D. Molden, ed.). Water for food, Water for life: A comprehensive assessment of water management in agriculture. Chapter 2. London: Earthscan and Colombo: IWMI.

Nazoumou, Y., Favreau, G., Adamou, M.M. & Maïnassara, I. (2016) La petite irrigation par les eaux souterraines, une solution durable contre la pauvreté et les crises alimentaires au Niger? *Cahiers Agricultures, EDP Sciences*, [Online] 25(1), 15003. Available from Doi 10.1051/cagri/2016005 [Accessed 27th October 2020].

Paterson, T., Pound, B. & Ziaee, A.Q. (2013) Landmines and livelihoods in Afghanistan: Evaluating the benefits of mine action. *Journal of Peace building & Development*, [Online] 8(2), 73–90. Available from: doi:10.1080/15423166.2013.814969 [Accessed 27th October 2020].

Pender, J., Jagger, P., Nkonya, E. & Sse-runkuuma, D. (2004). Development pathways and land management in Uganda. *World Development*, 32(5), 767–792.

Robins, N. S., Davies, J., Farr, J. L., & Calow, R. C. (2006). The challenging role of hydrogeology in semi-arid Southern and Eastern Africa. *Hydrogeology Journal*, 14, 1483–1492.

Sahel and West Africa Club/OECD (2010) *Security Implications of Climate Change in the Sahel Region: Policy Considerations*. [Online] Available from: https://www.oecd.org/swac/publications/47234320.pdf [Accessed 29th March 2021].

Sisa, M. (2014) *Niger Food Security Brief*. Washington, DC: FEWS NET.

Sissoko, K., van Keulen, H., Verhagen, J., Tekken, V. & Battaglini, A. (2011) Agriculture, livelihoods and climate change in the West African Sahel. *Regional Environmental Change*, [Online] 11, 119–125. Available from: https://doi.org/10.1007/s10113-010-0164-y [Accessed 27th October 2020].

Stephen, D. Jr (2014) Land degradation and agriculture in the Sahel of Africa: causes, impacts and recommendations. *Journal of Agricultural Science and Applications*, [Online] 03(03), 67–73. Available from: https://doi.org/10.14511/jasa.2014.030303 [Accessed 27th October 2020].

SWAC-OECD, ECOWAS (2008) Livestock and regional market in the Sahel and West Africa. Potentials and challenges. *Study carried out within ECOWAS Commission and SWAC/OECD partnership on the future of livestock in the Sahel and West Africa*.

Tillie, P., Louhichi, K. & Gomez-Y-Paloma, S. (2019) Impacts ex-ante de la Petite Irrigation au Niger: Analyse des effets micro-économiques à l'aide d'un modèle de ménage agricole, *EUR 29836 FR*, Publications Office of the European Union, Luxembourg, ISBN 978-92-76-09722-8. JRC115744.

UN Office for the Coordination of Humanitarian Affairs (2011) *Horn of Africa Drought Crisis*: Situation Report No. 11, 25 August 2011.

UNICEF (2010) *Etude de faisabilité des forages manuels: identifiaction des zones potentiellement favorables*. Republique du Niger Ministere de l'Eau, de l'Environnement et de la Lutte Contre Le Desertification.

United Nations Environment Programme (2011) *Climate Change, Conflict and Migration in the Sahel*. Geneva: United Nations Environment Programme. ISBN: 978-92-807-3198-9.

UPGro (2017) *Groundwater and Poverty in Sub-Saharan Africa.* St. Gallen: UPGro Working Paper, Skat Foundation.

World Bank (2000) *Niger - Towards Water Resource Management (English).* Washington, DC: World Bank Group. [Online] Available from: http://documents.worldbank.org/curated/en/821031468758341721/Niger-Towards-water-resource-management [Accessed 27th October 2020].

World Bank (2007) *Investment in Agricultural Water for Poverty Reduction and Economic Growth in Sub-Saharan Africa.* Synthesis Report. A collaborative programme of AfDB, FAO, IFAD, IWMI and the World Bank.

World Bank (2009) *Agricultural Land (Percentage of Land Area).* [Online]. Available from: http://data.worldbank.org/indicator/AG.LND.AGRI.ZS [Accessed 1st May 2011].

World Bank (2014) *Enhancing Knowledge of Groundwater Usage in the Sahel:* Brief - April 30, 2014. [Online]. Available from: https://www.worldbank.org/en/region/afr/brief/enhancing-knowledge-of-groundwater-usage-in-the-sahel [Accessed 21st May 2020].

World Bank (2020). *World Bank databank.* [Online] Available from: https://databank.worldbank.org/reports.aspx?source=2&country=NER [Accessed 6th June 2020].

World Water Assessment Programme (WWAP) (2006) *Water, a Shared Responsibility. The United Nations World Water Development Report 2.* World Water Assessment Programme. Paris: UNESCO, and New York: Berghahn Books.

Chapter 3

Groundwater, informal abstraction, and peri-urban dwellers in the Techiman Municipality of Ghana

L. Kwoyiga
University for Development Studies, Tamale, Ghana

CONTENTS

3.1 Introduction 63
    3.1.1 Peri-urbanism in Africa 64
    3.1.2 Peri-urban water supply in Africa 65
3.2 Methods, data and study area 66
3.3 Results and discussion 67
    3.3.1 Water supply in the Techiman Municipality 67
    3.3.2 Extent of dependence 69
    3.3.3 Drivers of informal water supplies 70
    3.3.4 Challenges 71
        3.3.4.1 Operators 71
        3.3.4.2 Customers 72
    3.3.5 Discussion 72
    3.3.6 Lessons learned from Techiman Municipality 73
3.4 Conclusion 74
References 74

3.1 INTRODUCTION

Urban and peri-urban areas continue to witness growth and expansion as "virtually all of the expected growth in the world population will be concentrated in the urban areas of the less developed regions" (Heilig, 2012). Nevertheless, the spate of growth of these urban areas outstrips the supply of utilities such as water by conventional distribution networks. In Sub-Saharan Africa, it is noted that "in most small urban centers, there is little or no public provision" (UN-HABITAT, 2003). The situation in Ghana is no different with few of the urban population having access to piped water in their dwelling, plot, or yard (Adank and Tuffour, 2013). This is particularly an issue in peri-urban areas in the country.

Thus, to address the water supply challenges in many peri-urban areas, including those in Ghana, groundwater abstraction has become a reliable option. This is because, groundwater is often locally available, more reliant especially during droughts, and provides better quality water than surface water (Food and Agriculture Organization of the United Nations, 2016). For example in the Techiman Municipality of Ghana, some peri-urban areas with limited domestic water supply services rely on

DOI: 10.1201/9781003024101-3

64 Groundwater for Sustainable Livelihoods

mechanized boreholes for their domestic water supply (Kortatsi and Quansah, 2004). Groundwater is also informally and privately exploited by individual enterprises. Thus, Kariuki and Schwartz (2005) described the activities of the enterprises as "gap fillers," "pioneers," or "sub-concessionaire," while Baker (2009) suggested that factors, such as low-level coverage, absence of utility services, high user demand, favorable hydrogeological conditions among others account for their emergence. Furthermore, Healy et al. (2020) documented the important role of private enterprises in promoting and expanding urban self-supply of water in Lagos, Nigeria. Yet, informal water providers often face some challenges such as lack of financial support, lack of autonomy in tariff setting, excessive government interference, land insecurity, lack of recognition, and public perception of poor or inferior services (Baker, 2009). On the part of customers of these service providers, Ahlers et al. (2013) alleged that "once a provider has captured the market, the customer is locked in due to the infrastructure used and the high connection fees and has little (if at all) room to switch providers". Despite the importance of informal water suppliers and their activities in peri-urban areas, they remain under-investigated (Peloso and Morinville, 2014).

The goal of this chapter was to examine the contribution of informal groundwater abstraction for domestic water supply in peri-urban areas of the Techiman Municipality in Ghana. Specifically, it focuses on the following: (a) the context and drivers of groundwater abstraction; (b) extent of dependence on informal groundwater abstraction; (c) the challenges of informal groundwater supply services; and (d) lessons learned from informal groundwater suppliers.

3.1.1 Peri-urbanism in Africa

Even though peri-urbanism has been studied in planning, economics, sociology, and others, Karg et al. (2019) recognized the lack of a standard definition of what constitutes "peri-urban." While it is often tempting to define peri-urban areas as an interface between rural and urban areas, Willis (2007) cautioned that categorizing a place as peri-urban defies any singular spatial measure. Therefore, to describe it, Iaquinta and Drescher (2002) noted it as a place comprising a heterogeneous mosaic of land use, with highly diverse, dynamic socio-economic structures. Webster (2002) reflected the dynamic nature of peri-urban areas noting "land that can be characterized as peri-urban shifts over time as cities, and the transition zone itself, expand outward." Considering peri-urbanism in Africa, Karg et al. (2019) associated it with features of fuzzy boundaries, unclear remits, high dynamic in character, non-linear development, and weak institutions.

In Ghana, Ocloo (2011) noted that discussions about peri-urban areas relate to settlements that fall within 40 km from the urban area. Looking at the Techiman Municipality, the characteristics resonate with Willis' (2007) admission of the difficulty in categorizing regions and sub-regions as either urban, rural, or peri-urban. As a result, water supply in the municipality typifies what Adank and Tuffour (2013) described as areas where water services exhibit features of both urban areas (high volume, high-quality water services provided at people's doorsteps) and rural areas (lower volume, low-quality water services provided at some distance from people homes through point sources).

3.1.2 Peri-urban water supply in Africa

Water supply in (peri) urban areas in African countries remains largely the responsibility of formal governments. Nonetheless, growth and expansion of (peri) urban areas with a corresponding increase in demand outstrip the capacity of formal water suppliers. Therefore, informal water suppliers have emerged providing alternative water supply services alongside formal water suppliers. Thus, one finds such terms as "formal" which means having a statutory or legal recognition, owned by the state or an incorporated private entity, and frequently uses modern technology and network while "informal" means all other modes without legal status (Misra, 2014). Several terms have been assigned to informal water suppliers, these include small-scale water providers, informal operators, and many others (Peloso and Morinville, 2014). Their nature, scale, and operations vary and are determined by factors such as water resources availability, topography, population distribution, and technology (Kariuki and Schwartz, 2005). Suppliers finance, develop and manage the delivery of small-scale services to their clients (Baker, 2009).

Ghana is witnessing a phenomenon where cities are spreading beyond the urban fringes. Yet, few urban dwellers have a 24-hour supply of water and many depend on water rationing. The situation of water supply is more challenging in peri-urban areas as these places tend to receive water supply just once a week or no supply at all (Ministry of Water Resources, Works and Housing, 2007). A study by the Public Utilities and Regulatory Commission (PURC) concluded that the majority of urban dwellers are unserved or under-served directly by Ghana Water Company Limited (GWCL) except through informal services (Ministry of Water Resources, Works and Housing, 2011). "In Accra, for example, it has been estimated that only approximately 25% of residents enjoy a 24-hour water supply. About 30% have an average of 12 hours of service every day for five days a week. Another 35% have service for two days each week while the remaining residents on the outskirts of Accra are completely without access to piped water supplies." (WaterAid, 2005).

Table 3.1 shows the gap in water supply and water demand for Ghana, showing that for half the regions coverage is less than 50%. Undoubtedly, it is evident that the urban water demand gap is one that may not immediately be filled by formal water supply

Table 3.1 Supply and demand gap in Ghana

Regions	Supply 2010 (m^3/day)	Demand (m^3/day)	Gap (m^3/day)	Coverage (%)
Upper West	1,257	10,800	9,543	11.6
Brong Ahafo	14,088	49,257	35,169	28.6
Upper East	5,851	17,983	12,132	32.5
Eastern	21,125	62,245	41,120	33.9
Volta	17,101	45,064	27,963	37.9
Central	45,562	78,288	32,726	58.2
Northern	33,644	49,665	16,021	67.7
Ashanti	110,345	162,829	52,484	67.8
Greater Accra	406,045	557,549	151,504	72.8
Western	32,932	42,846	9,914	76.9
Total	687,950	1,076,526	388,576	63.9

Source: Ministry of Water Resources, Works and Housing (2010).

66 Groundwater for Sustainable Livelihoods

systems in the country. The Techiman Municipality is within the Bono East Region (originally part of the Brong Ahafo Region until 2018) has a gap of >70% (Table 3.1) which helps to explain the rationale for the emerging activities of informal water suppliers, particularly since 64.5% of the municipality population lives in the urban areas.

3.2 METHODS, DATA AND STUDY AREA

This chapter is based on a qualitative study that was conducted between April and October 2019. Data were obtained through in-depth interviews, informal discussions, personal conversations, observations, and a desk review resulting in the generation of both primary and secondary data. The individual interviews involved a total of 362 respondents: 11 independent informal water suppliers, 350 customers, and 1 official of the Techiman Municipal Assembly. The interviews took place in the homes of users and at the sites of the boreholes using transcription of English and were conducted in both English and the local language of the people. Interviews were semi-structured and with open-ended questionnaires. Two research assistants, who were familiar with the study area and had interacted with some of the users and operators conducted the interviews. The study covered some peri-urban areas of the municipality noted for groundwater abstraction for domestic purposes.

According to the Ghana Statistical Service (2012), the Techiman Municipality is situated in the Bono East Region (Figure 3.1) and shares boundaries with districts/

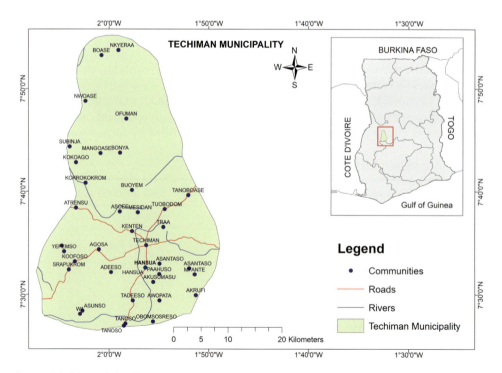

Figure 3.1 Map of the Techiman Municipality in Ghana.

municipalities such as Techiman North, Wenchi, and Nkronza Municipalities and Offinso-North. The Techiman Municipality has a land surface area of around $649\,km^2$. Almost two-thirds (64.5%) of the population live in urban areas. During the 2010 Population and Housing Census, the Municipality had 34,137 households. Of these households, 69.0% of them are found in urban areas. With the average household being 4.3 persons, the urban average household is 4.0. The peri-urban areas studied in the Municipality were Abosso, Fante New Town, Brigade, James Town, and Hansua.

The topography of the Municipality is low lying and gently undulating, with the lowest topography of 305 m found in the Southwestern part of the Techiman Municipality. The Techiman Municipality is drained by the rivers Tano, Subin, Kar, Brewa, Taifi, Kyini, and Fia. The Tano River is dammed at Tanoso to provide pipe-borne water to the people in Techiman and for irrigational purposes. The climate of the Techiman Municipality is characterized by semi-equatorial and tropical conventional or savannah climates associated with moderate to heavy rainfall. The mean annual rainfall ranges between 1,260 and 1,660 mm. The dry season period starts from November to March within which the highest monthly temperature of about 30°C is recorded. The relative humidity is usually high throughout the year. The vegetation zones of the area are the guinea-savannah woodland located in the Northwest, the semi-deciduous zone in the South, and the transitional zone which stretches from the Southeast and West up to the North of the Municipality.

Hydrogeological studies by Kortatsi and Quansah (2004) have indicated that the Techiman Municipality is underlain by sandstones of Upper Voltaian age. The weathered rocks thus vary from 20 to 50 m in depth. The yields of standard boreholes here range from less than 0.2 to about $29\,m^3/h$ with a median value of $8.5\,m^3/h$. Adequate groundwater, therefore, exists here which is exploited to augment municipal water supply. In terms of quality, apart from pH and total hardness, all major chemical constituents are within the limits of WHO (1993) guidelines limit for drinking water. Bacteriological results show zero fecal coliforms of all boreholes (Kortatsi and Quansah, 2004).

3.3 RESULTS AND DISCUSSION

3.3.1 Water supply in the Techiman Municipality

Water supply in the Techiman municipality is traced to multiple sources such as surface water, groundwater, rainwater harvesting among others. Nonetheless, groundwater abstraction in the municipality has been an age-long practice. Results of the study showed that 86% of the respondents in the study area rely on groundwater as a source of water supply in the municipality. As mentioned in Section 3.2, the 2010 Population and Housing Census revealed that there were 34,137 households in the Techiman Municipality. From the interview results, one mechanized borehole/water facility supplied water to an average of approximately twenty individual households. The reason being that there are fewer borehole operators at the moment but the demand for domestic water supply is high. This implies that there may be pressure on the existing boreholes as the population grows. Moreover, the results indicated that about 1% of household income is spent on water while almost every member of the household (99%) knew where and when to get their water.

Responses from operators showed that informal groundwater abstraction has evolved through personal initiatives in the urban fringe. Figure 3.2 shows various aspects of private wells in the study area, peri-urban dwellers here who could afford the cost of mechanized boreholes have drilled boreholes in their houses. From the interviews, all operators developed their own mechanized boreholes which cost up to US$2500 (GHS 12,000) depending on the depth. Water was used within the dwelling and as a utility through a system of piped connections.

From Figure 3.2, electricity (a) is used to pump water into an elevated storage tank (b). The water from the tank through the pipes is chlorinated and filtered (c) before distribution. A platform (d and e) is created where taps are mounted, permitting direct point source services. With a meter (f and g), customers whose homes were connected through pipes were billed monthly. From the interactions, all metered customers (who were connecting through pipes) for the first time paid a non-refundable connection fee of about US$76 (GHS 400). This amount varies depending on the informal water supplier and it is reviewed over time for new customers. Customers also paid for the

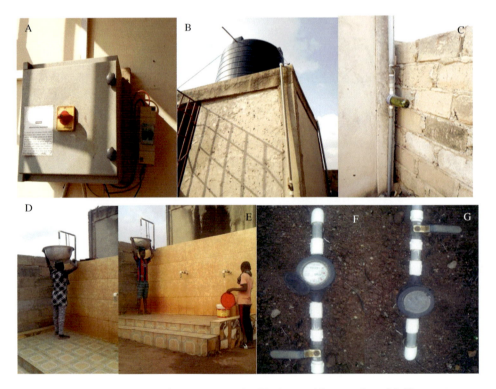

Figure 3.2 Groundwater supply system in the Techiman Municipality. (a) Electricity supply and meter; (b) elevated storage tank; (c) water filter; (d) and (e) platform and collection point; (f) and (g) individual meters for household connections. (Author.)

services of a plumber and purchased pipes themselves or gave money to the operator to secure these.

From the interviews, every person dwelling in the study area has access to the services of water suppliers without any discrimination provided the person can pay the costs. Metered customers have no challenges with distance to access water supply as water is in their dwellings. However, non-metered customers who use bicycles and motorbikes (35%) spent at most 10 minutes to return to their dwelling units. On the other hand, non-metered customers who carry water manually (on their heads) spent between 5 and 30 minutes to reach the water sources depending on the distance between their homes and the water source. All non-metered customers spent about a minute queuing.

From the interviews, it was revealed that all metered customers, enjoyed more than 15 L of water per day per person because they can afford and access water easily. In contrast, non-metered customers, used less than 15 L of water per person per day due to issues of distance and affordability. To further highlight the situation, the processes involved in groundwater abstraction, distribution, and use are illustrated in Figure 3.3.

3.3.2 Extent of dependence

Although the results indicated multiple sources of water for domestic purposes as shown in Figure 3.4, out of the 350 water users/respondents interviewed, the results revealed that 86% of them patronized the services of informal groundwater suppliers. Factors such as availability, uninterrupted supply, quality, convenience, and easy access, determined the level of use or patronage of these water sources.

The study revealed that the farther the location from the heart of the Techiman Municipality, the poorer the coverage of the formal water network by GWCL. By location, Abosso and Fante New Town lie closer to the municipality, so there was some coverage of formal conventional network supply of the GWCL. However, there was no regular supply of water in such areas. Thus, users here tended to patronize other sources of water especially the services of informal groundwater suppliers. It is also

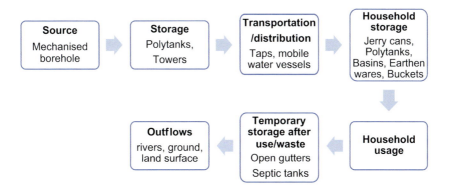

Figure 3.3 Processes involved in water service delivery. (Adapted from Farajalla et al. (2017).)

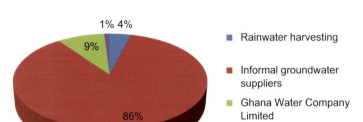

Figure 3.4 Source of water for domestic purposes in the Techiman Municipality.

important to state that 70% of the dwellers in these communities particularly the poor could not afford the connection fees of both GWCL and informal groundwater suppliers hence relying on point-of-source services by informal water suppliers.

3.3.3 Drivers of informal water supplies

The emergence of informal groundwater abstraction in study areas of Techiman Municipality was attributed to the slow pace of the Techiman Municipality to extend utility networks and the desire of informal suppliers to complement utility services. Although there was some formal water coverage in some parts of Abosso and Fante Newtown, places such as Hansua and Jamestown are without any coverage/connections. It has therefore become necessary for dwellers in Hansua and Jamestown to rely solely on informal suppliers for their domestic water supply.

Also, due to poor services (unreliable flow of water through taps) rendered by the GWCL, customers of formal/municipal suppliers sometimes had to rely on informal suppliers for water supply. This, they did by fetching from the tap of informal water suppliers whenever the GWCL failed to supply them with water, as many could not afford double connection fees from both the GWCL and the informal suppliers. There were no water vendors (trucks) whose source of water is from the boreholes. Thus, customers access water directly from boreholes themselves. The hydrogeological conditions of the Techiman Municipality show the potentials of groundwater. As such, people who could afford the cost of drilling, have sunk boreholes in their dwelling units for varied purposes.

Furthermore, typical of small-scale businesses is the condition of easy entry and exit. Profiles of the operators showed that the initial cost of drilling a borehole with the additional installation costs was easily borne by them. Most of the operators had major income sources as they were into professions such as teaching, nursing, lecturing, business, and farming. A cocoa farmer was able to raise enough income to drill a borehole, which he thought would generate supplementary income for his family. Other operators were able to raise income from savings from their formal jobs. None of these operators took a loan from the bank purposely to drill the borehole. As established earlier from the interviews, the primary aim of exploiting groundwater informally by the operators was to provide water for their dwelling units but not to sell it. This had influenced their decisions not to seek loans to finance such activity as they were not initially conceived as profit-driven ventures. Furthermore, there were no bureaucratic processes in the form of registration that could frustrate informal suppliers. There was no cost associated with obtaining a license. Operators were not responsible to any formal agency as their activities did not require any license covering abstraction and operation of the enterprise.

The passage of the Drilling and Groundwater Development Regulations, 2006 (L.I.1827) has also simplified the activities of private drilling companies. These companies only needed to register the boreholes with the Municipal Assembly, which is easy to do. The companies did not have to travel to Accra or other distant places to register their activities. Once they had obtained the license and registered with the Assembly, they were ready to operate. Some of them did not even have specific site locations (offices) yet through informal channels of communication such as social networks, it was easy for water suppliers to contact them for their services.

Technological advancement coupled with the availability of informal groundwater-related skilled workers is boosting informal suppliers' activities. Borehole materials, spare parts, and storage facilities are available and affordable. Private plumbers who are skillful are easily contacted and their services are affordable. This, therefore, made it easy to start and maintain a mechanized borehole business.

A cheap source of labor with basic operational skills and experiences is an important factor that is enabling the activities of informal groundwater suppliers to thrive. From the interviews, informal suppliers mostly depended on family members to support them run their businesses. Usually, husbands who were better off drilled the boreholes and handed them over to their wives to operate as a family business. In other instances, women owners depended on other family members to assist them to operate the business.

3.3.4　Challenges

Informal groundwater abstraction and supply are associated with some challenges on the part of both the suppliers and the customers as described in the following sections.

3.3.4.1　Operators

- From the interviews, the cost of operation was considered high especially in terms of the purchase of electricity. There had been a frequent increase in electricity tariffs in the country which operators complained as having an impact on their cost of production. More money was needed to purchase the same amount of credit that could pump sufficient water for customers. However, the prices of water had remained unchanged. This resulted in little returns for informal suppliers after they had purchased electricity for operations.
- Similar to the above-mentioned point, electricity/power outages were affecting the activities of informal groundwater suppliers. Ghana has become a country

characterized by erratic outages, especially in the dry season. There is often load shedding and this affects the pumping of water into the tanks for distribution.

- Besides, some metered customers were not committed to payment deadlines as they delayed in paying bills and this affected the purchase of electricity and other materials for the business. In association with this, operators admitted to the manipulation of water meters by users which resulted in wrong figures being recorded. By this act, operators are underpaid.
- Misunderstanding emanating from wrong interpretations of figures read on the meters (the basis of pricing) was also a problem that suppliers encountered. Some customers, especially the less educated, assumed that they were supposed to pay fixed prices monthly regardless of their level of consumption. This sometimes created some misunderstandings between customers and informal suppliers. It also delayed payment hence affecting the operations of the water supply systems.

3.3.4.2 Customers

- From the interviews, low pumping pressure usually affected the normal flow of water. According to the metered customers, sometimes, water from their taps dripped. In fact, customers whose houses were situated in hilly areas or uphill faced this challenge.
- Moreover, prices of water were determined only by informal suppliers with little room for negotiations. Customers had lamented about the lack of involvement by them in determining prices charged. The interviews showed that only operators decided on the amount they will charge per unit of consumption and for buckets, basins, jerry cans, or gallons at a point in time.
- Sometimes informal suppliers were absent from home due to engagement in other activities like attending church services, markets, other jobs, funerals without making arrangements for continuous operation and this affected the activities of customers who come to fetch directly.

3.3.5 Discussion

As Foster and Vairavamoorthy (2013) noted of the vital role that groundwater plays in the evolution and development of cities and urban settlements, the role of groundwater in meeting domestic water supply needs of some peri-urban areas of the Techiman Municipality is evident in this study. The importance of groundwater, especially as a source of potable water to human beings (Food and Agriculture Organization of the United Nations, 2016) is reflected in this study. Noteworthy, however, instead of formal authorities exploring groundwater to meet domestic water needs as studied elsewhere (Foster and Vairavamoorthy, 2013), in the studied municipality, groundwater is rather exploited by informal private individuals. Conditions such as favorable hydrogeology have helped facilitate the use of groundwater abstraction in the Techiman Municipality.

The study results further confirmed the argument by Njiru (2003) that the inability of formal water suppliers to meet the water demands of urban areas has resulted in the emergence of informal water suppliers. Similar to Solo's (2003) findings, the case of the study areas of Techiman Municipality was also attributed to the slow pace of the Techiman Municipal Assembly to extend formal water supply network, hence the desire of informal suppliers to complement utility services. The roles that these informal groundwater suppliers play in the municipality is similar to what Kariuki and Schwartz (2005) described as either "gap fillers," "pioneers," or "sub-concessionaire."

Furthermore, even though there are water vendors using trucks in the study areas, their source of water is not from groundwater suppliers. Groundwater users directly access water from the boreholes themselves. It is interesting to note that all the study areas are within the ambit of Tanoso area where the Tano River is dammed to provide potable water for the municipality. Yet, but for the exploitation of groundwater through informal means, such dwellers would have been without water since there are no formal/municipal connections in the areas. Therefore, the study confirms that informal water suppliers provide services that improve the living conditions of people living in deprived areas and offer services that are comparable to the ones offered by main operators despite the fact that they do not enjoy subsidies (Cave and Blanc, 2012). In line with Blanc and Botton (2012), the suppliers provide water to low-income households in difficult-to-reach urban areas and are efficient.

As noted by Njiru (2003) and others that such operators are often constrained in terms of social discrimination, financial resources, unfair competition and unclear regulation, informal groundwater suppliers in the municipality are also confronted with some challenges. However, operators in Techiman appear to be able to immediately address them. Customers in the study areas also face some challenges similar to what Ahlers et al. (2013) mentioned that due to the required infrastructure and high connection fees, customers are compelled to patronize the services of a particular operator even if such services are unsatisfactory.

The study again revealed that in terms of customer satisfaction, although informal groundwater suppliers are without any external subsidies, they are outperforming larger formal providers in meeting the demand for household, lending support to the findings of Schaub-Jones (2008).

3.3.6   Lessons learned from Techiman Municipality

- Increased population growth in peri-urban areas has not been accompanied by a corresponding increase in municipal water supply in the Techiman Municipality, which has pushed urban dwellers to search for alternative ways of meeting domestic water supply – including informal groundwater abstraction.
- Noted in the past for exploiting groundwater through manually constructed wells, coupled with better technology in recent times that enables the easy drilling and mechanization of boreholes, the study revealed that places like the Techiman Municipality will continue to witness informal groundwater abstraction even in places where there is coverage of municipal water supply systems.
- Informal groundwater suppliers operate a system that allows for the abstraction of groundwater, treatment, and distribution to customers. Operators bear the cost of their operational activities alone without subsidy, yet their charges are affordable to customers. There is also a regular supply of water to customers. This has increased access to the domestic water supply.

74 Groundwater for Sustainable Livelihoods

- Operators are mostly sole proprietors who also recover their costs fully, have no unaccounted for water, and are financially stable. Their activities are largely informal but both operators and customers relate in a mutually beneficial manner.

Rather than stifling the growth and activities of informal groundwater exploiters in peri-urban areas, the Municipal Assembly could promote training programs about basic water supply regulations, treatment, customer services, and operational activities. This can be achieved through the registration of the operators by the Municipal Assembly. Coupled with this, detailed mapping of the groundwater source points should be done to enable the Assembly to monitor well siting and withdrawal of groundwater.

3.4 CONCLUSION

The study set out to interrogate the contexts and contributions of groundwater in addressing the domestic water needs of some peri-urban areas in the Techiman Municipality of Ghana using mainly qualitative research methods. The case of the Techiman Municipality revealed that rapid peri-urbanization without corresponding increase in coverage of municipal water supply has rendered some peri-urban areas with limited or no access to the formal water supply. However, groundwater exists in these areas and is being exploited to meet the gap between demand and formal supply. This study highlighted the importance of simple affordable technology for both individual household supply and for informal groundwater suppliers who use both household connections and collect and carry services, which were used by the majority of the population, and were generally affordable.

It is recommended that the municipal water suppliers collaborate with and assist the informal suppliers. This can be done through formal registration of the enterprises, and broadening intervention from monitoring the activities of these suppliers especially regarding the quality to training and sharing of information on ways of enhancing quality services delivery.

REFERENCES

Adank, M., & Tuffour, B. (2013). *Management Models for the Provision of Small Town and Peri-Urban Water Services in Ghana: TPP Synthesis Report.* TPP Project/Resource Centre Network Ghana. https://www.ircwash.org/sites/default/files/tpp_synthesis_report_final_medium_res_min_size.pdf [Accessed 23rd April 2019].

Ahlers, R., Perez Güida, V., Rusca, M., & Schwartz, K. (2013). Unleashing entrepreneurs or controlling unruly providers? The formalisation of small-scale water providers in greater Maputo, Mozambique. *The Journal of Development Studies, 49*(4), 470–482. https://doi.org/1 0.1080/00220388.2012.713467 [Accessed 5th April 2019].

Baker, J. L. (Ed.) (2009). Opportunities and challenges for small scale private service providers in electricity and water supply: Evidence from Bangladesh, Cambodia, Kenya, and the Philippines. *World Bank*, Washington, DC. Download (ppiaf.org) [Accessed 11th June 2019].

Cave, J., & Blanc, A. (2012). Review of the international literature on drinking water distribution SSWPs. In A. Blanc & S. Botton (Eds.), *Water Services and the Private Sector in Developing Countries.* Citeseer. http://citeseerx.ist.psu.edu/viewdoc/download?-doi=10.1.1.363.4158&rep=rep1&type=pdf [Accessed 25th July 2019].

Farajalla, N., et al. (2017). *The Role of Informal Systems in Urban Sustainability and Resilience: A Review*. Issam Fares Institute for Public Policy and International Affairs office at the American University of Beirut. https://www.aub.edu.lb/ifi/Documents/publications/research_reports/2016-2017/20170706_informal_systems.pdf [Accessed 17th August 2019].

Food and Agriculture Organization of the United Nations. (2016). *Global Framework for Action to Achieve the Vision on Groundwater Governance*. http://www.fao.org/3/a-i5705e.pdf [Accessed 19th September 2019].

Foster, S., & Vairavamoorthy, K. (2013). *Urban Groundwater – Policies and Institutions for Integrated Management* (Perspectives Paper) [Technical Report]. Global Water Partnership. https://www.researchgate.net/publication/280491584_Urban_groundwater_-_policies_and_institutions [Accessed 9th April 2019].

Ghana Statistical Service. (2012). *2010 Population and Housing Census: Summary Report of Final Results*. Government of Ghana, Accra.

Healy, A., Upton, K., Capstick, S., Bristow, G., Tijani, M., MacDonald, A., Goni, I., Bukar, Y., Whitmarsh, L., Theis, S., Danert, K., & Allan, S. (2020). Domestic groundwater abstraction in Lagos, Nigeria: A disjuncture in the science-policy-practice interface? *Environmental Research Letters*, *15*(4), 045006 https://doi.org/10.1088/1748-9326/ab746.3 [Accessed 12th May 2020].

Heilig, G. K. (2012). *World Urbanization Prospects: The 2011 Revision*. United Nations, Department of Economic and Social Affairs (DESA), Population Division, Population Estimates and Projections Section. https://www.un.org/en/development/desa/population/publications/pdf/urbanization/WUP2011_Report.pdf [13th October 2019].

Iaquinta, D., & Drescher, A. (2002). Food security in cities–A new challenge to development. *WIT Transactions on Ecology and the Environment*, *54* https://www.witpress.com/elibrary/wit-transactions-on-ecology-and-the-environment/54/738 [Accessed 12th June 2019].

Karg, H., Hologa, R., Schlesinger, J., Drescher, A., Kranjac-Berisavljevic, G., & Glaser, R. (2019). Classifying and mapping Periurban areas of rapidly growing medium-sized Sub-Saharan African cities: a multi-method approach applied to Tamale, Ghana. *Land*, *8*(3), 40 https://doi.org/10.3390/land8030040 [Accessed 11th May 2019].

Kariuki, M., & Schwartz, J. (2005). *Small-Scale Private Service Providers of Water Supply and Electricity: A Review of Incidence, Structure, Pricing, and Operating Characteristics*. The World Bank. https://elibrary.worldbank.org/doi/pdf/10.1596/1813-9450-3727 [Accessed 3rd October 2019].

Kortatsi, B., & Quansah, J. (2004). Assessment of groundwater potential in the Sunyani and Techiman areas of Ghana for urban water supply. *West African Journal of Applied Ecology*, *5*(1) https://www.ajol.info/index.php/wajae/article/view/45592 [Accessed 3rd May 2019].

Ministry of Water Resources, Works and Housing. (2007). *National Water Policy*. Ministry of Water Resources, Works and Housing, Accra. https://www.gwcl.com.gh/national_water_policy.pdf [Accessed 18th October 2019].

Ministry of Water Resources, Works and Housing. (2010). *Water and Sanitation Sector Performance Report*. Ministry of Water Resources, Works and Housing, Accra.

Ministry of Water Resources, Works and Housing. (2011). *Improvement of Water Sector Performance Management Framework. National Rainwater Harvesting Strategy Final Report*. Ministry of Water Resources, Works and Housing, Accra.

Misra, K. (2014). From formal-informal to emergent formalisation: fluidities in the production of urban waterscapes. *Water Alternatives*, *7*(1), 20. http://www.water-alternatives.org/index.php/volume7/v7issue1/231-a7-1-2/file [Accessed 30th August 2019].

Njiru, C. (2003). Improving water services: utility-small water enterprise partnerships. *29th WEDC International Conference, towards the Millennium Development Goals*, Abuja Nigeria. Njiru.pmd (ircwash.org) [Accessed 27th September 2019].

Ocloo, K. A. (2011). *Harnessing Local Potentials for Peri-Urban Water Supply in Ghana: Prospects and Challenges.* PhD Thesis. Dissertation.pdf (tu-dortmund.de) [Accessed 21st August 2019].

Peloso, M., & Morinville, C. (2014). 'Chasing for Water': everyday practices of water access in peri-urban Ashaiman, Ghana. *Water Alternatives, 7*(1). https://www.water-alternatives.org/index.php/alldoc/articles/vol7/v7issue1/237-a7-1-8/file [Accessed 25th May 2019].

Schaub-Jones, D. (2008). Harnessing entrepreneurship in the water sector: expanding water services through independent network operators. *Waterlines, 27*(4), 270–288 https://www.ircwash.org/sites/default/files/Schaub-Jones-2008-Harnessing.pdf [Accessed 10th April 2019].

Solo, T. M. (2003). *Independent Water Entrepreneurs in Latin America-The Other Private Sector in Water Services.* The World Bank. http://documents.worldbank.org/curated/en/786541468012040884/pdf/multi0page.pdf [Accessed 21st September 2019].

UN-HABITAT. (2003). *Water and Sanitation in the World's Cities: Local Action for Global Goals.* Earthscan. https://unhabitat.org/water-and-sanitation-in-the-worlds-cities-local-action-for-global-goals [Accessed 7th October 2019].

WaterAid. (2005). *WaterAid—National Water Sector Assessment, Ghana.* WaterAid. https://www.ircwash.org/sites/default/files/WaterAid-2005-Ghana.pdf [Accessed 30th October 2019].

Webster, D. (May, 2002). *On the Edge: Shaping the Future of Peri-Urban East Asia.* (Working Paper number 53) Shorenstein Asia/Pacific Research Center. https://aparc.fsi.stanford.edu/publications/on_the_edge_shaping_the_future_of_periurban_east_asia [Accessed 20th June 2019].

World Health Organization (1993). Guidelines for drinking-water quality. World Health Organization, Geneva

Willis, A.-M. (2007). From peri-urban to unknown territory. *Design Philosophy Papers, 5*(2), 79–90. https://doi.org/10.2752/144871307X13966292017432 [Accessed 23rd June 2019].

Chapter 4

Urban development and intensive groundwater use in African coastal areas

The case of Lomé urban area in Togo

R. Barry
GEOTOP, Université du Québec A Montréal (UQAM)

F. Barbecot
GEOTOP, Université du Québec A Montréal (UQAM)

M. Rodriguez
Université Laval

A. Djongon
GEOTOP, Université du Québec A Montréal (UQAM)
Université Paris Saclay

W. Akakpo
Ministère de l'Eau et de l'Hydraulique villageoise

CONTENTS

4.1 Introduction...78
4.2 General settings of African coastal urban areas 78
    4.2.1 Importance and specificities of African coastal areas...........................78
    4.2.2 Benefits of groundwater exploitation..79
4.3 Case study of Lomé urban area 80
    4.3.1 Context 80
        4.3.1.1 Uncontrolled urban development 80
        4.3.1.2 Togolese coastal sedimentary basin.................................81
    4.3.2 Groundwater contribution to urban growth 84
        4.3.2.1 The area covered by water utilities.................................84
        4.3.2.2 Outside the area covered by water utilities 84
        4.3.2.3 Household water access 84
        4.3.2.4 An urban economy based on groundwater.......................86
    4.3.3 The challenges of groundwater management in coastal urban areas 87
        4.3.3.1 Impacts of urbanisation on groundwater 87

DOI: 10.1201/9781003024101-4

78 Groundwater for Sustainable Livelihoods

        4.3.3.2   The climate threat...90
        4.3.3.3   Management challenges..90
4.4   Conclusion...91
References...92

4.1 INTRODUCTION

In Africa, due to a combination of historical (colonisation), social (migrations), and economic (trade and commerce) factors, the most dynamic cities are generally located on the coasts, and this phenomenon is increasing. Many elements govern the establishment of urban environments and their evolution; and the availability of water resources is an indispensable condition for the settlement of populations and the development of cities. However, due to its hidden nature, groundwater is often ignored among these other factors, yet often serves as a reliable water supply for both drinking and industry. It also enables the development of economic activities in the form of raw materials or as a major component of wealth creation chains. Groundwater contributes to the improvement of the living conditions of city dwellers and to the dynamism of urban environments, in a context of limited resources. However, the intense demand linked to population growth, in the context of climate change, generates many management problems that threaten the sustainability of groundwater and therefore the sustainability of life in these urban environments. This poses new management challenges in developing a sustainable approach that integrates all components and stakeholders. We examine the role of groundwater in the development of African cities, through its contribution to the well-being of urban populations, using the city of Lomé in Togo as an example.

4.2 GENERAL SETTINGS OF AFRICAN COASTAL URBAN AREAS

More than half of the world's population, now lives in cities and this figure is expected to rise to 75% by 2050 (UN, 2018). Africa is projected to have the fastest urban growth rate in the world – by 2050, cities in Africa will be home to an additional 950 million people (OECD, 2020). This evolution is not uniform on the continent because populations are concentrated mainly on the coasts (Maghreb, Gulf of Guinea, South-East Africa) and secondarily in the large valleys (Nile, Inner Niger Delta), as well as on the high plateaus (East Africa, Ethiopia, Great Lakes). The high population density in the coastal regions can be explained by historical, economic, and migratory reasons. But also because of a lesser-known advantage offered by coastal environments: the availability of fresh groundwater.

4.2.1 Importance and specificities of African coastal areas

The importance of coastal cities in Africa can be explained first by historical reasons: the colonial penetration was primarily through the coast. The English, French, Germans, Spanish, and Portuguese first settled on the coast. They built commercial cities, fortified them, and gave them administrative statutes. And it was from these cities that they set out to conquer the interior. It is because of this historical heritage that even today, the majority of the political and economic capitals of twenty-four of the fifty-four African states are located on the coast.

    The second factor that has contributed to the dynamism of coastal cities in Africa is economic. Indeed, most of the African economies are based on the export of raw materials and the import of finished goods, where coastal cities became primary transit points for goods between the continent and the rest of the world (Wall et al., 2018). Therefore, businesses and industries are generally concentrated on the coast.

In addition to economic and historical factors, migration has strengthened Africa's coastal cities (Mercandalli and Losch, 2017; Valsson and Ulfarsson, 2012). At the national level, it is mainly the rural exodus, which refers to the sustainable movement of people from rural areas to cities. On a continental scale, the demographics of coastal cities are fuelled by migrants from landlocked countries (Mercandalli and Losch, 2017).

This strong dynamic has consequences with particularities related to the African context. On the African coasts, cities develop along the coast and inland, but the habitat is more dispersed and less compact than in Asia and Europe (Nlend et al., 2018; Xu et al., 2019). The low density is explained by the lower construction height, compared to other continents, due to spontaneous and unplanned expansion. This results in rapid urban sprawl with many challenges to be met. Over the next 15 years, the ten fastest-growing cities in the world will be in Africa (UN, 2018), but infrastructure and industrial development has not kept pace with urban population growth (60% of urban dwellers in sub-Saharan Africa live in slums and only 25% have access to safe water) (Pinault, 2019). In general, these inadequacies are often accompanied by poverty, land degradation, and lack of basic services (Martine and Marshall, 2007), such as drinking water supply, sanitation, and waste management.

4.2.2 Benefits of groundwater exploitation

This gap between the expansion of cities and the delay in infrastructure construction is pronounced in the case of water access services. Indeed, these infrastructures have a high cost for low-income countries but for social reasons are not considered as ordinary market goods. At the same time, lack of water is a threat to the very existence of a city, and in many regions groundwater is used to fill these gaps. Following the example of megacities such as Lagos (Healy et al., 2020) and Nairobi (Oiro et al., 2020), up to 75% of the population of Africa uses groundwater as the main drinking water source (AfDB, 2000) especially in urban areas (Foster et al., 2008; Healy et al., 2020; Oiro et al., 2020). Indeed, one-third of Africa's urban population is likely to use groundwater for their domestic use (Chávez García Silva et al., 2020).

Groundwater is generally considered a "safe source" of drinking water because it is abstracted with a low microbial load with little need for treatment before drinking. This resource is generally potable or has a low treatment cost. In quantitative terms, the volume of groundwater was estimated to be 0.66 million km^3 for Africa as a whole, more than hundred times the annual renewable freshwater resources and twenty times the freshwater stored in African lakes (MacDonald et al., 2012). A perennial source of water, it provides a buffer in times of drought that can be developed for local use (Masiyandima and Giordano, 2007), and shelters users from seasonal variations. Furthermore, twelve of Africa's forty main transboundary aquifers are located in coastal areas (Altchenko and Villholth, 2013). Therefore, African coastal areas present a significant groundwater potential.

The conjunction between sites where groundwater is available and where demand is high makes groundwater one of the major contributors to the well-being of coastal urban populations in Africa as illustrated in the following case study.

4.3 CASE STUDY OF LOMÉ URBAN AREA

4.3.1 Context

4.3.1.1 Uncontrolled urban development

With a population of fewer than 2 million inhabitants (DGSCN-Togo, 2011), Lomé in the Republic of Togo is not the size of Africa's major megacities, but its rapid development makes it an interesting case study. The capital of this small West African country (56,000 km^2) presents, on a smaller scale, the main trends observed in many African cities.

Lomé city showed rapid planned urbanisation during the colonial period and the first decades of independence: 4.9% per year from 1924 to 1958 and 7.3% per year between 1960 and 1970 (Nyassogbo, 1984). The city grew outward from the historic centre, first rapidly and then at a slower rate (Figure 4.1). This second phase was not planned, due to the deterioration of the socio-political situation.

Currently, Lomé sustains a third of the national population, and there are several reasons for this. It is the political capital of the country, and therefore the seat of national institutions. This centralised governance creates a natural convergence towards the capital. Lomé is also the economic capital of the country, which concentrates almost half of the country's activities. It, therefore, attracts many people from rural areas looking for jobs. For a long time, the city expanded due to rapid population growth, but since the 1970s urban sprawl has outpaced population growth, and thus density is declining (Figure 4.2).

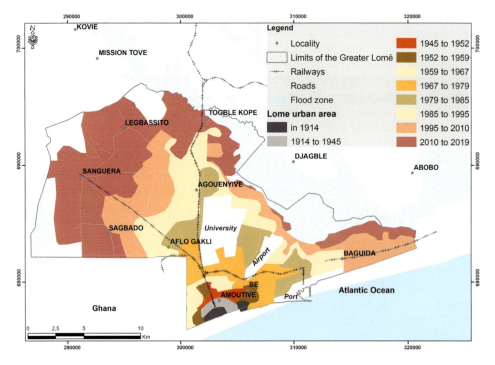

Figure 4.1 The stages of Lomé's growth from 1914 to 2010. (Adapted from Guézéré (2013).)

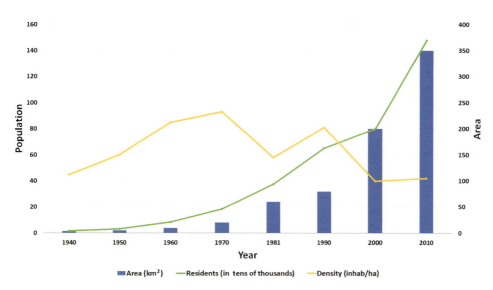

Figure 4.2 Surface and population evolution of the agglomeration of Lomé, from 1940 to 2010. (Data from Guézéré (2013).)

Lomé is also home to the headquarters of many sub-regional institutions such as the West African Development Bank, the Economic Community of West African States (ECOWAS) Bank of Investment and Development, the pan-African bank ECOBANK, etc., which makes it attractive. Togo is a member of the ECOWAS, with free movement of goods and people, and ease of settlement for all West African nationals. Lomé is, therefore, a destination for many migrants, particularly from Sahelian countries. This sub-regional dimension of the city is accentuated as it is home to the only deep-water port in West Africa. This makes it an important point of entry and exit of goods not only for Togo but also for the Sahelian countries, in a North-South direction. In the East-West direction, the city is also located halfway along the Abidjan-Lagos coastal corridor, making its port a suitable logistics base for the movement of goods in the sub-region.

4.3.1.2 Togolese coastal sedimentary basin

The city of Lomé is located in the Togolese coastal sedimentary basin (Figure 4.3). This is the Togolese portion of the Keta transboundary basin, which extends from Ghana in the West to Nigeria in the East (Akouvi et al., 2008; Gnazou et al., 2017). These are sedimentary deposits that lie on the Precambrian basement. The people of Lomé have under their feet an aquifer system consisting of the unconfined Continental Terminal aquifer, the confined Palaeocene aquifer, and the semi-confined Maastrichtian aquifer (Akouvi et al., 2008; Gnazou et al., 2017; Slansky, 1962) (Figures 4.3–4.5). The general direction of the flow is from northeast to southwest discharging into the Atlantic Ocean and the basin's renewable groundwater resources are estimated at 1.7 billion m^3 per year (MEAHV, 2010).

The coastal sedimentary basin is the southern part of the Lake Togo basin with surface water resources estimated at an average of 0.63 billion m^3 per year, and

82 Groundwater for Sustainable Livelihoods

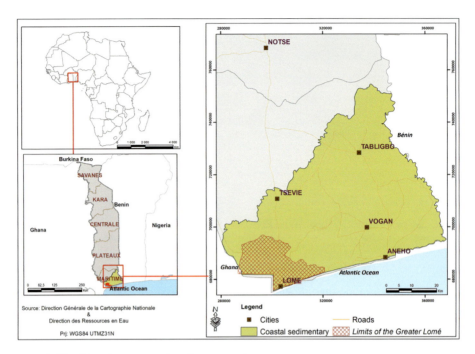

Figure 4.3 The Togolese coastal sedimentary basin.

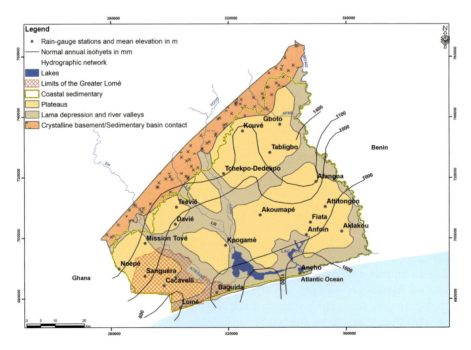

Figure 4.4 Geomorphological units of the coastal sedimentary basin in Togo and limits of Lomé. (Adapted from Akouvi et al. (2008).)

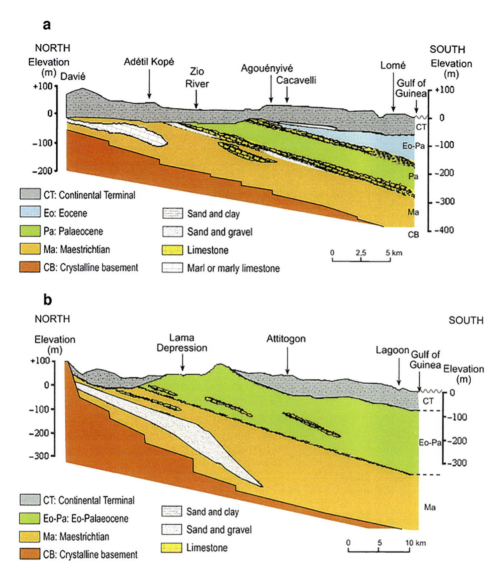

Figure 4.5 Transverse N–S geological cross-sections, as reported in Figure 4.4 (Akouvi et al., 2008).

annual precipitations of 7.62 billion (MEAHV, 2010). Therefore for a specific water demand of 80 L/day and per inhabitant and the basin's population of about 3 million inhabitants, including 1.5 million in the city of Lomé (DGSCN-Togo, 2011), there is theoretically enough water (i.e. 87.6 million m^3 per year) to meet their needs, and cover the demand for other socio-economic sectors. But achieving this objective requires significant technical, human, and financial resources that the country does not yet have.

4.3.2 Groundwater contribution to urban growth

A city's dynamism is characterised by its ability to attract and retain residents. And the ability to meet the water needs of the population is a prerequisite. The region of Lomé contains a lagoon system that is connected to Lake Togo (which is the lake where the river Zio flows, north of the city). But to satisfy the city's water demand, the brackish water of the lagoons is not fit for drinking and the resources from the River Zio would be too expensive to treat and transfer to the city. That's why in the case of the city of Lomé, groundwater is fully relied upon to meet the city's needs, but in different ways depending on the location and uses.

4.3.2.1 The area covered by water utilities

The first means by which the city of Lomé obtains its water supply is from Togolese Water Company ("Togolaise des Eaux," TdE). TdE supplies water to the city of Lomé, entirely from twenty-nine boreholes. It is the approved distributor for households, administrations, shops, and industries. Water is taken from the aquifers of the Continental Terminal, Maastrichtian, and Palaeocene and discharged to the Cacavelli production site. The production capacity that has increased over time is currently around 40,200 m^3/day. However, the demand for water is not fully met by TdE, and the service is frequently interrupted.

Due to the unreliability in the water supplied by TdE, in all parts of the city, many residents have installed private wells and boreholes to meet their continuous water needs despite concerns on local contamination of shallow groundwater. The same applies to buildings housing major national and international institutions, industries, companies, hospitals, educational institutions, embassies, etc. By 2030, the National Action Plan for the Water and Sanitation Sector aim to achieve optimal coverage of 85% in the capital by choosing to rely on groundwater (MEAHV, 2011).

4.3.2.2 Outside the area covered by water utilities

The services of the TdE are absent in a large part of the Lomé agglomeration (Figure 4.6), and the population rely on drilling wells or boreholes to access groundwater. In general, for households, it is the Continental Terminal aquifer that is used. Small networks are therefore created between nearby houses, with boreholes that supply the neighbourhoods for a fee. The aquifer of the Palaeocene and Maastrichtian are generally exploited by industries. Recently, the authorities have been drilling deep boreholes to supply the population through small networks on the outskirts of the city (see the area not covered by TdE – Figure 4.6). Groundwater complements the supply of water services in the peripheral areas of the city, pending the extension of the TdE network.

4.3.2.3 Household water access

In Lomé groundwater resources supply almost all the water requirements of industries and households. The daily production of TdE (40,200 m^3/day) is three times less

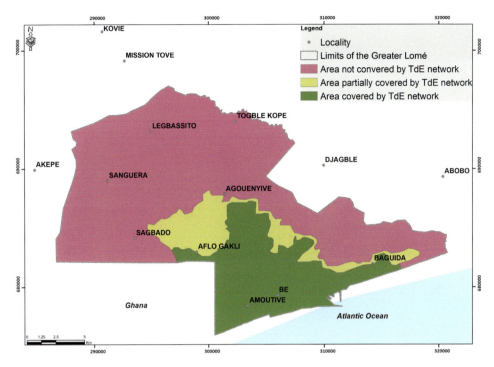

Figure 4.6 Situation of water supply by the TdE in the city of Lomé. (Adapted from Somadjago et al. (2017).)

than the needs (i.e. 120,000 m³/day, based on 80 L/day/person for 1.5 million people). Indeed, the majority of households use a private borehole/pump well (43.3%), an outdoor public tap (21.4%), or the tap in the dwelling or concession (14.6%) as their main source of drinking water supply (Figure 4.7). There are several reasons for the strong preference for wells: not all of the city is served by the TdE (and the service is not available 24 hours a day), as well as the lower price of well water compared to other sources (Figure 4.7), given that the minimum monthly salary is 60 US$ (or 2 US$ per day).

Because of this diversified supply, and because users take less time to access water points in the capital (Figure 4.8), the proportion of households whose basic needs are met is 37.8% in Lomé and 38.7% in other urban areas, compared to 24.9% in rural areas (N'Guissan, 2015). Groundwater is therefore accessible at a short distance and at a modest price, making life easier for the residents of Lomé. It should be noted that due to a lack of data, it is not yet possible to assess the quality of water from the different sources.

Theoretically, even with an annual population growth of almost 3% in Lomé and in the basin as a whole (DGSCN-Togo, 2011), the renewable groundwater potential should largely cover the domestic needs of the population.

86 Groundwater for Sustainable Livelihoods

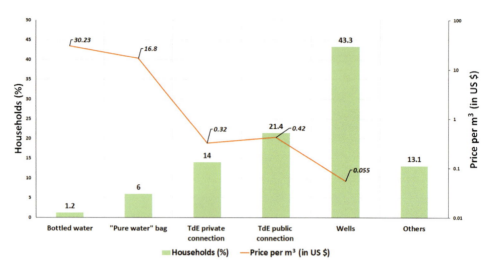

Figure 4.7 Households distribution (in %) by the water supply mode and selling price. (Water supply data from Quibb-Togo 2015 (N'Guissan, 2015).)

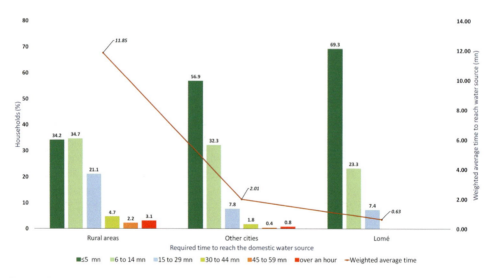

Figure 4.8 Percentage of households by required time to reach household water source. (Data from Quibb-Togo 2015 (N'Guissan, 2015).)

4.3.2.4 An urban economy based on groundwater

In the Togolese capital, groundwater has not only enabled the city to expand by meeting domestic needs, but it has also contributed significantly to its economic development, both in the formal and informal economies.

4.3.2.4.I FORMAL ECONOMY

Groundwater provides most of the water used in industrial processes. Almost all the companies based in the Lomé-free industrial zone have boreholes. The largest samplers identified by the Ministry of Water are the agri-food industries (BB brewery, Fan Milk, Nioto oil mills, etc.), cement plants (Cimtogo), and chemical industries. Groundwater is also used in companies' firefighting systems and cooling systems, including thermal power plants operated by the national electricity company.

Over the past 20 years or so, a new sachet water industry known as "Pure water" has emerged and prospered. The groundwater packaged in bags (water sachets) is sold at 25 FCFA (less than 0.05 US $) for 250 mL. Due to this low cost, it is a product that most people can easily afford and purchase. In several developing countries establishing a bagged water plant can easily be a very profitable business with low startup costs and the profits range from $10,000 to over $100,000 per month (waterbusiness.com, 2019). In 2017, according to the Water Resources Directorate of Togo, thirty-three of the fifty-five legally authorised pure water companies were located in or around Lomé.

4.3.2.4.2 INFORMAL ACTIVITIES

As the access to drinking water services is not universal, water sales activities have developed. Thus, houses with a connection to the TdE network, or a borehole, sell water to their neighbours who do not have it. A 75-L jerrycan can be sold for 25 FCFA (0.042 US$), while the cubic metre according to the billing bands of TdE, varies from 190 FCFA (0.32 US$) to 500 FCFA (0.84 US$). This can provide significant income for a household. A study carried out by Somadjago et al. (2017) in the canton of Aflao-Sagbado (northwest of Lomé) has shown that: (a) only 9% of the households are connected to the TdE network (b) of all the concessions with boreholes, 85% sell water (c) trading in borehole water is a lucrative activity that has generated about 20,000 jobs (d) water vendors make average profits ranging from 20,000 F CFA (US$33) to more than 40,000 F CFA (US$66) per month (the minimum monthly salary is 35,000 FCFA, i.e. 60 US$).

Groundwater is also used for irrigation in urban and peri-urban market gardening. Food crops and horticultural products, particularly in the barrier beach (Kanda et al., 2009), play a crucial role in the supply of vegetables to the city of Lomé.

4.3.3 The challenges of groundwater management in coastal urban areas

4.3.3.I *Impacts of urbanisation on groundwater*

Urbanisation modifies underlying groundwater systems resulting in hydrological, geotechnical, socio-economic, or water quality effects that can compromise the sustainability of the resource (Tellam et al., 2006). Although specific studies on the impacts of urbanisation on groundwater have not been carried out in the basin, there are signs that vigilance is required. Indeed, the piezometric monitoring carried out by the Water Resources Directorate (which resumed in early 2010 after a 20-year interruption due to the socio-political crisis that affected the country), reveals some interesting information (see Figure 4.9).

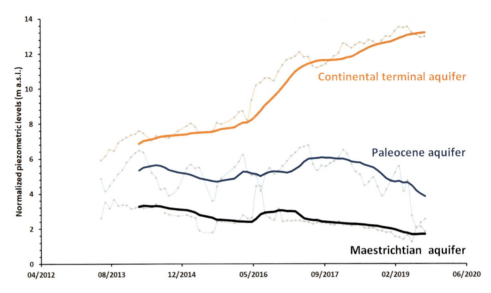

Figure 4.9 Groundwater levels evolution in stations located in the city of Lomé. Campus Nord for the continental terminal aquifer, Campus Sud for the Palaeocene aquifer, and Apessito for the Maastrichtian aquifer.

Since the 1970s, we observed decreases in the level of the Maastrichtian semi-confined aquifer, and the Palaeocene confined aquifer water tables (DHE, 1984). We observed decreases in the level of the Maastrichtian semi-confined aquifer, and the Palaeocene confined aquifer water tables. The decrease is less for the Maastrichtian than the Palaeocene because of its significant depth. In 5 years, the water level decreased by about 1.5 m in the Maastrichtian and 2 m in the Palaeocene (see Figure 4.9). The water from these aquifers is taken by TdE and industries that have very high demand.

The recharge area of the Maastrichtian (semi-confined) is located in the north of the basin (in sparsely populated areas), through the Crystalline Basement and Continental Terminal (Akouvi et al., 2008). The recharge of Palaeocene is mainly through the Continental Terminal. These factors contribute to these aquifers' qualitative preservation, although the withdrawals endanger the aquifers.

The continental terminal unconfined aquifer shows a different response. Groundwater levels are almost constant over time in rural areas (Figure 4.10), but in urban areas, we observe a general upward trend and strong intra-annual variations. Studies must be carried out to understand the mechanisms at work. It can nevertheless be stated that the increase in piezometry is not necessarily linked to that of rainfall. Otherwise, it would also have been observed in the rural areas. The most likely hypothesis to explain this phenomenon would be that of the developments in the city of Lomé, and the basins built in the same period, to collect runoff water and avoid flooding. Indeed, the developments may contribute to a localised increase in the recharge in urban areas (Lerner, 2002).

In qualitative terms, the various water supply most used (well, drilling, public network) in urban environments of some African countries are vulnerable to physical, chemical, and microbiological pollution (Hounsounou et al., 2017). In Lomé, Soncy et al. (2015) showed non-compliance of well water (related to faecal contamination

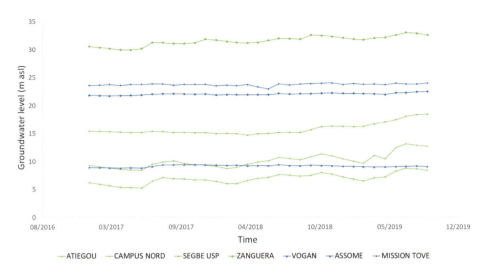

Figure 4.10 Continental terminal aquifer groundwater level evolution in stations located in the city of Lomé (Green) and in rural areas (Blue).

germs in 65% of cases and in 70% of cases related to *Escherichia coli*), the contamination of the drilling's waters (in 53.54% cases by mesophilic aerobic flora, in 26.77% cases by total coliforms, and in 2.03% cases by faecal streptococci), the presence of indicator organisms of faecal contamination in the analysed waters exposes consumers to the risk of gastroenteritis.

They suggest that measures should be taken for monitoring and disinfecting borehole water before use. In 2017, during the project to reduce the environmental and health risks linked to the activity of borehole water vendors in Greater Lomé (PRRESAF carried out by EAA: Eau et Assainissement pour l'Afrique) (Gnazou, 2016), it was determined that, out of nearly 15,000 boreholes surveyed in the city, about 30% contained bacterial contamination and nearly half had nitrate levels above 50 mg/L (i.e. above WHO recommendations, Figure 4.11).

This anthropogenic pollution affects the entire aquifer system, but the Continental Terminal unconfined aquifer is also impacted by a marine intrusion due to increased pumping (MEAHV, 2010). The pumping also causes the release of salts contained in clayey deposits, linked to paleo-intrusions of seawater (Akouvi, 2001; Akouvi et al., 2008). The salinity of the waters generally increases from the north towards the south. Electrical conductivity increases from 315 µS/cm in the town of Tabligbo in the northeastern part of the basin to 1,100 µS/cm in Abobo in the southern part. High electrical conductivities were found in the city of Lomé in the order of 2,640 µS/cm at the Campus Nord site and 2,380 µS/cm in Atiegou, while they are lower in the periphery of the capital, 409 µS/cm in Zanguera in the northwestern suburbs and 173 µS/cm in Mission Tové a little further north in rural areas. This pattern reflects a change in the recharge quality of the Continental Terminal unconfined aquifer issuing from anthropogenic activities in the Togolese capital.

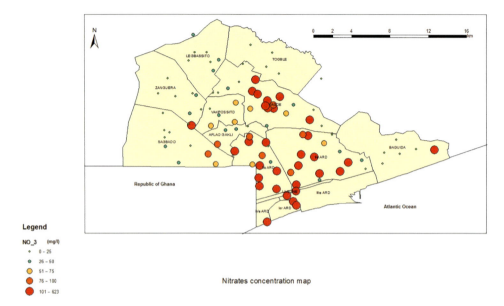

Figure 4.11 Map of nitrate concentrations in the terminal continental aquifer. (Data from the PRRESAF/2017 Project (Gnazou, 2016).)

4.3.3.2 The climate threat

As in all areas, climate change impacts groundwater and causes upheaval. However, hydrological variability (including in groundwater dynamics) in Africa is higher than in any other region of the world (Taylor et al., 2009), which could indicate a greater vulnerability. In the event of a possible decrease in groundwater resources, the consequences would be serious for the population.

On the other hand, the IPCC (2019) underlines that the impacts on coastal cities in Africa as a result of sea-level rise with the consequences of coastal retreat and the aggravation of seawater intrusion. This phenomenon is observed in a spectacular way on the Togolese coast under the combined effect of anthropogenic actions and climate change, with a retreat of between 5 and 10 m per year (Blivi, 1993; Ozer et al., 2017).

4.3.3.3 Management challenges

The Continental Terminal aquifer is polluted by bacterial pollution linked to the lack of a collective sewage system and by chemical discharges linked to economic activities. The Palaeocene aquifer is declining due to withdrawals for drinking water supply and by industries. Seawater intrusion is increasing due to pumping and the rising sea level. There is therefore a real management problem.

The sustainable development of groundwater in general and in urban coastal contexts, in particular, requires sustainable management choices. The task is complex because it involves several stakeholders who do not necessarily interact. Already it will be necessary to be able to provide water and sanitation services throughout the city. And certain management requirements need to be followed: first, the reservation of good quality water (deeper) for sensitive uses and shallow water (of poor quality) for non-sensitive uses, second, the definition of protection zones for priority control of contaminant loading, and third, the planning of wastewater treatment/disposal.

The first challenge is to raise awareness of the preciousness of groundwater and the threats to it. Acceptability is a fundamental concept in sustainable groundwater management. Only through information and awareness raising can the necessary support be generated for the sustainable management of the resource.

The second challenge is that of controlling urban expansion through planning that allows the establishment of water and sanitation infrastructure before the population settles. This approach must consider elements that minimise the impact of the different components of the urban environment of the water cycle, and the control of marine influences through adapted and rigorously implemented standards and regulations, as well as adequate development.

The third challenge is that of taking into account the modifications induced by urbanisation on hydrogeological dynamics, with urban hydrogeology which is in full expansion (Schirmer et al., 2013; Tubau et al., 2017).

4.4 CONCLUSION

Throughout the world, urban dynamics have accelerated in recent decades. Forecasts suggest an accentuation of this trend in the future. In Africa, for historical, economic, and social reasons, coastal cities will be at the forefront of these changes. But unlike other continents, urban expansion in Africa is being followed by a decrease in population density, because there is no development of cities in the vertical direction. For the urban perimeter to expand, access to water is needed to meet the needs of the population.

The case study of the city of Lomé illustrates how in the context of limited resources, where the use of surface water would have been very expensive, groundwater makes it possible to satisfy the needs of the population through a variety of approaches. The demand of domestic users is met by the official distribution network, when it is available, and by direct withdrawals to supplement the supply of the official network or to replace it when absent. Groundwater supports economic activities in both the formal and informal sectors. The Togolese coastal aquifers, therefore, play a major role in the urban development of the Togolese capital. This role is similar to most of the continent's coastal cities, according to numerous studies in West, Central, East, and North Africa. This role will be strengthened as a substantial increase in groundwater abstraction is expected throughout sub-Saharan Africa as part of the achievement of UN SDGs 2 and 6 (Taylor et al., 2019).

The priority for Lomé is to protect deep aquifers and prioritise their use for drinking water. These actions must be part of a broader framework of integrated management, which carefully plans the city development in such a way as to limit its impact on the sustainability of the coastal water resources as much as possible. The challenge will be to develop a framework that can simultaneously reconcile physical and socio-economic dynamics. The success of such an undertaking depends on the mobilisation of all the stakeholders, starting with information and awareness raising.

REFERENCES

AfDB. (2000). The Africa water vision for 2025: Equitable and sustainable use of water for socioeconomic development. Retrieved from https://www.ircwash.org/resources/africa-water-vision-2025-equitable-and-sustainable-use-water-socioeconomic-development

Akouvi, A. (2001). Etude géochimique et hydrogéologique des eaux souterraines d'un bassin sedimentaire cotier en zone tropicale. *Thèse de doctorat- Université Paris VI.*

Akouvi, A., et al. (2008). The sedimentary coastal basin of Togo: example of a multilayered aquifer still influenced by a palaeo-seawater intrusion. *Hydrogeology Journal, 16*(3), 419–436. https://doi.org/10.1007/s10040-007-0246-1.

Altchenko, Y., & Villholth, K. G. (2013). Transboundary aquifer mapping and management in Africa: a harmonised approach. *Hydrogeology Journal, 21*(7), 1497–1517. https://doi.org/10.1007/s10040-013-1002-3.

Blivi, A. (1993). Morphology and current dynamics of the coast of Togo. *Geo-Eco-Trop (Belgium).*

Chávez García Silva, R., et al. (2020). Estimating domestic self-supply groundwater use in urban continental Africa. *Environmental Research Letters, 15*(10). https://doi.org/10.1088/1748-9326/ab9af9.

DGSCN-Togo. (2011). *Quatrième Recensement Général de la Population et de l'Habitat 2010.* Ministère Auprès du Président de la République, Chargé de la Planification, du Développement et de l'Aménagement.

DHE. (1984). *Alimentation en eau de Lomé: Ressources en eau souterraine, synthèse des données hydrogéologiques.* Lomé- Togo: Direction de l'Hydraulique et de l'Energie.

Foster, S. S., et al. (2008). Groundwater in Sub-Saharan Africa: a strategic overview of developmental issues. In: S.M.A. Adelana and A. M. MacDonald (eds.) *Applied Groundwater Studies in Africa,* CRC Press, Leiden, the Netherlands (IAH selected papers in Hydrogeology, 13), pp. 19–34.

Gnazou, M. D. T. (2016). *Cartographie hydrochimique et bacteriologique.* Retrieved from Rapport partiel du Projet de Réduction des Risques Environnementaux et Sanitaires liés à l'Activité des Vendeurs d'Eau de Forage dans le Grand-Lomé (PRRESAF).

Gnazou, M. D. T., et al. (2017). Multilayered aquifer modeling in the coastal sedimentary basin of Togo. *Journal of African Earth Sciences, 125*, 42–58. https://doi.org/10.1016/j.jafrearsci.2016.10.008.

Guézéré, A. (2013). Deux roues motorisées et étalement urbain à Lomé, quel lien avec la théorie des «trois âges» de la ville? *Norois. Environnement, aménagement, société* (226), 41–62.

Healy, A., et al. (2020). Domestic groundwater abstraction in Lagos, Nigeria: a disjuncture in the science-policy-practice interface? *Environmental Research Letters, 15*(4). https://doi.org/10.1088/1748-9326/ab7463.

Hounsounou, E. O., et al. (2017). Pollution des eaux à usages domestiques dans les milieux urbains défavorisés des pays en développement: Synthèse bibliographique. *International Journal of Biological and Chemical Sciences, 10*(5). https://doi.org/10.4314/ijbcs.v10i5.35.

IPCC. (2019). Global Warming of 1.5C An IPCC Special Report on the Impacts of Global Warming of 1.5C Above Pre-Industrial Levels and Related Global Greenhouse Gas Emission Pathways, in the Context of Strengthening the Global Response to the Threat of Climate Change. *Sustainable Development, and Efforts to Eradicate Poverty.* Retrieved from https://www.ipcc.ch/sr15/ [Accessed, May 2019].

Kanda, M., et al. (2009). Le maraîchage périurbain à Lomé: pratiques culturales, risques sanitaires et dynamiques spatiales. *Cahiers Agricultures, 18*(4), 356–363 (351).

Lerner, D. N. (2002). Identifying and quantifying urban recharge: a review. *Hydrogeology Journal, 10*(1), 143–152. https://doi.org/10.1007/s10040-001-0177-1.

MacDonald, A. M., et al. (2012). Quantitative maps of groundwater resources in Africa. *Environmental Research Letters, 7*(2). https://doi.org/10.1088/1748-9326/7/2/024009.

Martine, G., & Marshall, A. (2007). State of world population 2007: unleashing the potential of urban growth. In *State of World Population 2007: Unleashing the Potential of Urban Growth*: UNFPA.

Masiyandima, M., & Giordano, M. (2007). Sub-Saharan Africa: opportunistic exploitation. In *The Agricultural Groundwater Revolution: Opportunities and Threats to Development* (pp. 79–99).

MEAHV. (2010). *Plan d'Actions National de Gestion Intégrée des Ressources en Eau -Togo*. Lomé: Gouvernement du Togo.

MEAHV. (2011). Plan d'Action National pour le Secteur de l'Eau et de l'Assainissement (PAN-SEA) du Togo. Lomé-Togo.

Mercandalli, S., & Losch, B. (2017). Rural Africa in motion. Dynamics and drivers of migration South of the Sahara.

N'Guissan, K. (2015). *Questionnaire des indicateurs de base du bien-être 2015 (QUIBB 2015): rapport final*. Retrieved from https://www.africabib.org/rec.php?RID=412438305 [Accessed May 2019].

Nlend, B., et al. (2018). The impact of urban development on aquifers in large coastal cities of West Africa: Present status and future challenges. *Land Use Policy, 75*, 352–363. https://doi.org/10.1016/j.landusepol.2018.03.007.

Nyassogbo, K. (1984). L'urbanisation et son évolution au Togo. *Les cahiers d'outre-mer, 37*(146), 135–158.

OECD/SWAC (2020). Africa's Urbanisation Dynamics 2020: Africapolis, Mapping a New Urban Geography, OECD Publishing, Paris. https://doi.org/10.1787/69e2a9a9-en.

Oiro, S., et al. (2020). Depletion of groundwater resources under rapid urbanisation in Africa: recent and future trends in the Nairobi Aquifer System, Kenya. *Hydrogeology Journal*. https://doi.org/10.1007/s10040-020-02236-5.

Ozer, P., et al. (2017). Evolution récente du trait de côte dans le golfe du Bénin. Exemples du Togo et du Bénin. *Geo-Eco-Trop: Revue Internationale de Géologie, de Géographie et d'Écologie Tropicales, 41*(3) 529–541. http://hdl.handle.net/2268/217276.

Pinault, N. (2019). Rapid urbanization presents new problems for Africa. *Voice Of America*, May 20th 2019. https://www.voanews.com/a/rapid-urbanization-presents-new-problems-for-africa/4925127.html.

Schirmer, M., et al. (2013). Current research in urban hydrogeology – A review. *Advances in Water Resources, 51*, 280–291. https://doi.org/10.1016/j.advwatres.2012.06.015.

Slansky, M. (1962). *Contribution à l'étude géologique du bassin sédimentaire côtier du Dahomey et du Togo*. Mémoires du BRGM, 11, BRGM, Paris.

Somadjago, M., et al. (2017). Etalement urbain et dynamique du commerce de l'eau à aflao-sagbado, un canton périphérique de Lomé (Togo). *Notes scientifiques, homme et société, Université de Lomé*.

Soncy, K., et al. (2015). Évaluation de la qualité bactériologique des eaux de puits et de forage à Lomé, Togo. *Journal of Applied Biosciences, 91*(1). https://doi.org/10.4314/jab.v91i1.6.

Taylor, R. G., et al. (2009). Groundwater and climate in Africa—a review. *Hydrological Sciences Journal, 54*(4), 655–664. https://doi.org/10.1623/hysj.54.4.655.

Taylor, R. G., et al. (2019). Topical collection: determining groundwater sustainability from long-term piezometry in Sub-Saharan Africa. *Hydrogeology Journal*. https://doi.org/10.1007/s10040-019-01946-9.

Tellam, J. H., et al. (2006). Towards management and sustainable development of urban groundwater systems. In: J.H. Tellam et al. (eds.), *Urban Groundwater Management and Sustainability*. NATO Science Series (IV: Earth and Environmental Sciences), vol 74. Springer, Dordrecht. https://doi.org/10.1007/1-4020-5175-1_1.

Tubau, I., et al. (2017). Quantification of groundwater recharge in urban environments. *Science of the Total Environment, 592*, 391–402. doi:10.1016/j.scitotenv.2017.03.118.

UN. (2018). World urbanization prospects: The 2018 revision of world urbanization prospects. *United Nations*. https://www.un.org/development/desa/publications/2018-revision-of-world-urbanization-prospects.html.

Valsson, T., & Ulfarsson, G. F. (2012). Megapatterns of global settlement: typology and drivers in a warming world. *Futures, 44*(1), 91–104.

Wall, R., et al. (2018). The state of African cities 2018: the geography of African investment. United Nations Environment Programme, United Nations.

waterbusiness.com. (2019). Bagged water, water bags, water pouches, water sachets. *Water Business USA*. San Marcos, CA. https://www.waterbusiness.com/business-opportunities/pouch-bagged-water/ [Accessed, May 2019] .

Xu, G., et al. (2019). Urban expansion and form changes across African cities with a global outlook: spatiotemporal analysis of urban land densities. *Journal of Cleaner Production, 224*, 802–810. https://doi.org/10.1016/j.jclepro.2019.03.276.

Chapter 5

Contribution of groundwater towards urban household water security

N. Mujere
University of Zimbabwe

CONTENTS

5.1 Introduction...95
5.2 Occurrence of groundwater in Budiriro 5B...96
5.3 Municipal water supply in Harare...96
5.4 Groundwater and household water security in Budiriro 5B.............................97
    5.4.1 Boreholes and wells 97
5.5 Summary 99
References...99

5.1 INTRODUCTION

Availability of water in acceptable quantity and quality is critical for the socio-economic well-being of people. Almost 68.5% of the fresh water on Earth is found in ice sheets, icecaps and glaciers. Just about 31.2% is found in ground water, and only 0.3% occurs as surface water in lakes, ponds, bogs, rivers, streams and swamps (Boberg, 2005).

Access to adequate quantities of acceptable quality water, in ways that safeguards livelihoods and ecosystems, and makes societies resilient to water-related hazards, underpins all the UN seventeen sustainable development goals (SDGs). Specifically, Goal 6 which calls for ensuring availability and sustainable management of water and sanitation for all people by the year 2030 (Zimbabwe Peace Project, 2019). The World Health Organization (WHO)'s water and sanitation component of Vision 2030 also aims to ensure increased accessibility to safe water and sanitation to all people in rural and urban areas beyond the 2015 levels. The organization stipulates that ideally every person should access between 50 and 100 L of water per day to ensure the most basic needs are met and the outbreak of disease is prevented (WHO and DFID, 2009). At the African continental level, the N'gor Declaration by water ministers also highlights the importance of accessibility to clean and affordable water (Zimbabwe Peace Project, 2019).

The right to water is also echoed in Zimbabwe's Constitution Section 77, which states that every person has a right to safe, clean and potable water. In addition, Zimbabwe's National Transitional Programme (NTP) regards access to water as a springboard towards achieving an upper-middle income economy by 2030 (GoZ, 2018). Thus, the country needs to take reasonable legislative and other measures, within the limits of available water resources to achieve the progressive realisation of this right.

DOI: 10.1201/9781003024101-5

Water is an enabler for the country to improve its industry, mining, transport, agriculture, tourism and power.

Nevertheless, socio-economic development especially in arid and semi-arid areas is hampered by short periods of rainfall, prolonged dry spells, frequent droughts and high rates of surface water evaporation. Thus, groundwater is critical in arid and semi-arid areas as surface water resources are scarce and rainfall is erratic. It is free from evaporation and is ubiquitous. It is a vital natural capital for the consistent and economic provision of potable water supply for both rural and urban environments. However, the importance and contribution of groundwater resources towards improving water security are often underestimated. Water security means the reliable availability of an acceptable quantity and quality of water for health, livelihoods and production, coupled with an acceptable level of water-related risks.

5.2 OCCURRENCE OF GROUNDWATER IN BUDIRIRO 5B

Groundwater occurs everywhere beneath the unsaturated zone where open spaces between sedimentary materials or in fractured rocks are filled with water whose pressure is greater than atmospheric pressure. These water-bearing formations which yield and transmit water in usable quantities are called aquifers. The capacity of water-bearing formations to store water is chiefly determined by the nature of the voids or pores. Groundwater is recharged from infiltration and is discharged via evapotranspiration, river baseflow, springs, inter-aquifer transfers and pumping from wells and boreholes (Holman, 2006). Through discharge, water stored underground eventually makes its way back to the surface streams, lakes or oceans.

In Budiriro 5B, groundwater occurs in crystalline rocks belonging to the Harare greenstone belt. The rocks are of intrusive igneous and metamorphic origin comprising granite and gneiss rocks of the Precambrian Basement Complex. Basement rocks have very low permeability and are devoid of primary porosity. Thus, groundwater occurs where fracturing, jointing, gneissocity planes and weathering have created adequate secondary porosity and permeability. Consequently, the aquifer has high groundwater development potential and supports yields of 1.2–2.9 L/s or 50–100 m^3/day. The average borehole depth is in the range 30–50 m and the water level is typically from 5 m to greater than 20 m below ground level. Transmissivity is 1–10 m^2/day and specific capacity varies between 2 and 50 m^3/day/m (Owen, 2000).

5.3 MUNICIPAL WATER SUPPLY IN HARARE

Harare is the capital city of Zimbabwe and has a population of almost 2 million people (Zimstat, 2012). The city needs 1,200 ML/day of potable water. Raw water from Lake Chivero (built in 1952) with a storage capacity of 247,181 ML and Lake Manyame (built in 1975) with a storage capacity of 480,236 ML is treated at Morton Jaffray water treatment works with a capacity of 614 ML/day. The treatment works were commissioned in 1954 and had a capacity of 160 mL/day. Its capacity was extended to 220 and 614 ML/day in 1976 and 1994, respectively (Nhapi, 2008). After treatment at the plant, water is pumped to a storage reservoir from which it is distributed to consumers. The waterworks pumps 350–380 ML/day depending on seasons. Leakages along the 6,000 km long pipes are estimated to be 100 ML/day (Harare City Council engineer, Personal Communication, 2018).

Municipal water shortages are a result of low reservoir water levels, leakages and lack of chemicals. Thus, the city council often embark on rationing of water. Lack of water treatment chemicals forced Morton Jaffray to halt operation on 23 September 2019, leaving over one million people without running water. The reason for the shutdown was due to shortage of foreign currency to import water treatment chemicals. Harare City council needs at least US$2.7 million a month for water chemicals, against a monthly revenue of about US$0.19 million (Harare City Council engineer, Personal Communication, 2018).

5.4 GROUNDWATER AND HOUSEHOLD WATER SECURITY IN BUDIRIRO 5B

Budiriro 5B is a low-income residential suburb in western Harare. It has 5,200 residential stands comprising 3,200 units developed by a local bank and 2,000 units developed by a group of private housing cooperatives. Stand areas vary from 200 to 1,200 m^2. The suburb was opened in 2012 and had a population of 15,580 people (Zimstat, 2012). It has been experiencing municipal water shortages since commissioning in 2012. Water supply is erratic and some areas can go for 6 months without municipal water. In 2019 and 2020, most residents received no municipal piped water from May to November.

During field visits and household interviews, it was observed that less than 10% of the household stands (especially those on low gradient) receive municipal water at most three days in a week. Water is received on Tuesdays, Thursdays and Fridays around 10 o'clock in the evening and supply stops very early in the following morning. Stands on high ground lack municipal water during most of the dry season, from May to October. This is due to low water pressure, and hence, pipes first run dry as compared to low elevation stands.

During times of municipal water shortage, residents collect water for domestic purposes from wells at their stands and wells dug in wetlands along seasonal streams. It was also observed that the Harare City council sunk ten boreholes in the area but only four are functional while six are dysfunctional due to lack of maintenance. Three boreholes were abandoned due to failure to get water after drilling for more than 40 m.

5.4.1 Boreholes and wells

Almost 90% of the households have dug shallow wells in their backyards to reduce water scarcity. Only 2% of households have drilled boreholes, while 8% have no underground water source at their premises. Residents who have resorted to drilling boreholes to access clean and safe water face some power supply challenges. Due to frequent load shedding extending for 18 hours daily, the boreholes often fail to pump enough water into overhead storage tanks which in turn supply water to premises by gravity. Alternatively, households who have resorted to using diesel, petrol and solar power complain of high costs of fuel, pumps, generators, solar panels, inverters and batteries because they are imported.

98 Groundwater for Sustainable Livelihoods

Water vending is also an issue in the area. During periods of water scarcity, some residents rely on individuals and privately-owned companies who bring and sell water at cost of US$50 per 5,000 L. Also, stand owners with boreholes charge an average of US$1 for 100 L of water.

Residents use groundwater for drinking, washing clothes, food preparation, personal and household hygiene among other things (Table 5.1).

Women and children (especially young girls) bear the brunt of the water crisis. They spend most of their time looking for water. School children spend long early morning hours in water queues resulting in them getting to school tired and late.

Most (95%) residential wells run dry during the dry season. Also, the wells often fail to provide safe water due to contamination from burst sewer pipes and waste dumps. This has exposed people to health hazards as some residents use untreated water from unprotected wells for drinking.

There are almost fifty unprotected open wells dug in riverine wetlands outside the residential boundary in Budiriro 5B. The wells are in open spaces or green belts along rivers and have depths varying from 1 to 4 m. Wetland wells rarely dry up, although their water storages reach very low levels during peak withdrawal periods. Drying of residential wells during the dry seasons result in over-exploitation of wetland wells. Fetching water in wetlands starts as early as 3 o'clock in the morning and ends late at midnight. Residents collect water from the wells using tins tied to ropes which empty water in containers of varying sizes, from 2 L to 5,000 L. Some residents have to endure long distances of more than 2 km to get to wells and boreholes. Thus, negatively impacting on the productive time of most of the residents, especially those who are entrepreneurial.

Wetland water collectors include children, youths, adults, and the aged. Women form a greater (72%) proportion of the water collectors. Most (33%) people who fetch water are women aged between 36 and 60 years. Table 5.2 shows age–sex variations of wetland water collectors.

Some people have resorted to fence off wells they dug in the wetlands and plant vegetables around wells. To access water from wells in the fenced garden, those fetching water are required to irrigate vegetables as payment of water.

Table 5.1 Uses of groundwater in Budiriro 5B

Groundwater use	Percentage (%) of households
Domestic needs (e.g., drinking, washing, laundry and cooking)	100
Watering backyard vegetable gardens	90
Watering lawn and flowers	65
Building	55
Moulding bricks	37
Selling	5
Watering orchards	20
Watering pets (e.g., dogs and cats)	85
Poultry production	25
Others (e.g., religious activities like baptisms)	5

Groundwater and household water security   99

Table 5.2 Gendered differentiation in collecting water from wetland wells

Gender	Percentage (%)
Young boys (less than 15 years)	6
Young girls (less than 15 years)	9
Youth males (15–35 years)	12
Youth females (15–35 years)	27
Adult men (above 36–60 years)	5
Adult women (36–60 years)	33
Older men (above 60 years)	3
Older women (above 60 years)	5
Total	100

5.5 SUMMARY

This chapter has shown that municipal water supply is unreliable in Budiriro 5B high-density suburb in Harare. Thus, residents resort to groundwater as a coping strategy in times of water scarcity. Borehole water is commonly used for drinking, cooking and bathing, whereas open wells outside residential premises are the most common source of water for other uses which rarely require good quality water. Based on the research findings, the Harare City Council needs to invest in rehabilitating the water supply infrastructure as is too old and inadequate for the rapidly growing city. Galvanized water pipes can be replaced with poly pipes which do not rust. Also, drilling of community boreholes is a noble option to redress the prevailing water scarcity.

REFERENCES

Boberg, J. (2005) Freshwater availability. In: Arlington, V.A. and Pittsburgh, P.A. (eds.) *Liquid Assets: How Demographic Changes and Water Management Policies Affect Freshwater Resources.* RAND Corporation, Santa Monica, CA, 15–28.
Government of Zimbabwe (GoZ) (2018) *Towards an Upper-Middle Income Economy by 2030: New Dispensation Core Values.* GoZ, Harare.
Holman, J.P. (2006) Climate change impacts on groundwater recharge-uncertainty shortcomings, and the way forward. *Hydrogeology Journal*, 14, 635–645.
Nhapi, I. (2008) Inventory of water management in Harare, Zimbabwe. *Water and Environment Journal*, 22, 54–63.
Owen, R.J.S. (2000) *Conceptual Models for the evolution of groundwater flow paths in shallow aquifers in Zimbabwe.* DPhil thesis, University of Zimbabwe, Zimbabwe.
WHO and DFID (2009) *Summary and Policy Implications Vision 2030: The Resilience of Water Supply and Sanitation in the Face of Climate Change.* WHO, UK.
Zimbabwe Peace Project (2019) *Water Crisis Fact Sheet.* Number 2. Zimbabwe Peace Project, Harare.
Zimbabwe National Statistics Agency, Zimstat (2012) *Population Census 2012.* Zimstat, Harare.

Chapter 6

Sustainable and resilient exploitation of small alluvial aquifers in the Brazilian semi-arid region

The experience of Sumé

J.C. Rêgo and J.P. Albuquerque
Federal University of Campina Grande

J.D. Pontes Filho
Federal University of Campina Grande
Federal University of Ceará

B.B. Tsuyuguchi, T.J. Souza, and C.O. Galvao
Federal University of Campina Grande

CONTENTS

6.1 Introduction 101
6.2 The Sumé alluvial aquifer 103
6.3 The alluvial aquifer role in resilience to drought 106
6.4 Appropriate technologies for increasing water availability 108
    6.4.1 Underground dams 108
    6.4.2 "Duck bill" well 111
6.5 Integrated use of reservoir and aquifer 114
6.6 Community organization and engagement 114
6.7 Remaining challenges 115
    6.7.1 Appropriation of modeling tools 115
    6.7.2 Managed aquifer recharge 115
    6.7.3 Groundwater salinity 115
    6.7.4 Governance 116
6.8 Summary 117
Acknowledgments 118
References 118

6.1 INTRODUCTION

The Brazilian semi-arid (BSA) region experiences low rainfall and high seasonality, with 80% of rainfall concentrated in 4 months, intense interannual variability, causing frequent and prolonged droughts. Such features, associated with high temperatures and evapotranspiration, contribute to low surface water availability and the

DOI: 10.1201/9781003024101-6

intermittence of rivers. Moreover, the hydrogeology in the region is characterized mainly by a crystalline-rock aquifer system of very low productivity, with smaller areas comprising sedimentary aquifers. All these conditions expose the region to great drought vulnerability, especially with the failures on governing water resources. Also, with increasing desertification risk in the area due to land use processes and climate change, vulnerability is expected to increase (Vieira et al., 2015).

Most of the BSA (approximately 90%) is part of the Brazilian northeast, which is the region in the country where poverty among agricultural households is the highest (Soares et al., 2016). Surface reservoirs construction policy was greatly encouraged by governments in the region, initially by the federal government and followed by the states. Ambitious irrigation projects followed such a policy, supported by the idea that reservoirs would allow sustainable water use for irrigation and wealth generation. However, unregulated construction and increasing demands exposed the harsh reality of the limited water availability, and many of these projects collapsed.

Considering this scenario, the small alluvial deposits formed along the riverbeds over the crystalline substrate, despite their small size, are an important source of water resources. They have been increasingly exploited in the annual dry periods and during drought occurrences, for domestic water supply and mainly for irrigation (Tsuyuguchi et al., 2020; Braga, 2016; Burte et al., 2005; Mackay et al., 2005; Rêgo, 2012). As streams in the region are mostly intermittent, streamflow occurs only during the rainy season and the recharge of these alluvial aquifers is sporadic. Due to their size, during these episodes, the aquifers can be almost fully recharged and then be used throughout the dry spells or droughts. The short residence time and easiness of recharge collaborate to good water quality in these aquifers. On the other hand, the shallow depth and high permeability of the aquifers increase their vulnerability to pollution (Andrade et al., 2008; Blackburn et al., 2002; Salgado et al., 2018).

Aquifers with similar hydrogeological settings in arid and semi-arid regions worldwide have been referred to by the terms wadi, sand rivers, small alluvial aquifers, and alluvial strip aquifers (Al-Shaibani, 2008; Bazuhair & Wood, 1996; Love et al., 2007; Moyce et al., 2006; Sarma & Xu, 2017; Sorman & Abdulrazzak, 1993; Taher et al., 2012). The Arabic term "wadi," which is used to define valleys of ephemeral rivers exploited in arid regions, has become well known due to their intense investigation and exploitation, especially in the north of Africa and in the Middle East (e.g. Missimer et al., 2012, 2015). The aquifers reported in the scientific literature are mostly of tens or hundreds of meters depth, with some shallower aquifers. The aquifers in BSA are particularly small, with average depths ranging from 0.5 to 20 m (Burte et al., 2005, 2011, 2009; Montenegro & Montenegro, 2006; Rêgo, 2012).

The use of these shallow aquifers has sustained traditional communities spread over the BSA region for centuries, especially due to the particular climatic conditions. In addition, the easiness of access to this water source, due to the small depth, and of excavation, due to unconsolidated rock formation, favored its use. Nevertheless, they have been largely neglected by public policies designed to deal with these natural conditions, mainly because they constitute a small and "invisible" resource. There is a lack of studies on these shallow aquifers' functioning, and their potential is considered relatively small to sustain larger irrigation projects (Tsuyuguchi et al., 2020): 144×10^8 m³/year is the estimated water availability of alluvium in Northeast Brazil (Sá & Diniz, 2012).

One of the irrigated perimeters (IPs) created as part of a public policy for semi-arid region development is Sumé, which was considered a major success story between 1970 and 1987 for allowing tomato and corn production for agribusiness supplied by a surface reservoir. When the project collapsed in 1987, due to reservoir incapacity to supply both the growing city and the IP, many families gave up rural activities and migrated to big cities (Mendonça, 2010). Some of the remaining families began to use the small alluvial aquifer downstream the surface reservoir as the source of water.

It has been now more than 30 years since the interruption of water supply for irrigation by the reservoir, and these families continue to develop agriculture focused on subsistence and local distribution and commercialization. This chapter presents this successful case of groundwater exploitation from a small alluvial aquifer that has been sustainably supplying an IP after its supply by surface reservoir collapsed and farmers had to adapt to the new existing hydrological conditions. Our main objective was to advertise the importance of developing neglected small alluvial aquifers in BSA, and call attention to the challenges and opportunities on the use of such aquifers.

6.2 THE SUMÉ ALLUVIAL AQUIFER

The alluvial deposits show very large variations in the grain size distributions. This is due to the differences in time and space in the course of the deposition process. The deposited sediments consist mostly of fine to coarse sand; however, they can be distributed over all grain classes, from fine grain or clay to coarse gravel (Albuquerque & Rêgo, 1987). The recharge of these aquifers results mostly from the infiltration of part of the streamflow volume over the riverbed, which results mainly from the tributary streams and rainfall over the basin.

The geological and geomorphological characteristics of the alluvial sedimentary soil layers determine their properties as aquifers. Usually, the coarser sediments are deposited in the middle of the riverbed and the finer ones on the riverside. This condition forms the so-called alluvial terraces, which have even smaller depths and concentrate the finer sediments carried by the extreme floods. Closer to the river beds, the sand or gravel layers are more concentrated, despite also being heterogeneously embedded with clay and silt layers.

Direct groundwater recharge, through infiltration of the rainwater, plays a secondary role in alluvial aquifers in the BSA region due to their small size. The indirect recharge, through infiltration of runoff over the riverbed is the most significant. After the short and intense rains, which characterize the rainy season, there is also extensive surface runoff, which increases the amount of infiltrated water. The effective infiltration rate depends not only on the soil properties and soil water content, but also on the intensity, duration and extent of the flooding over the aquifer system areas. The annual groundwater recharge resulting from such processes can often take up the total storage capacity of the alluvial due to its small size. The stored water should be used by the end of the dry season, usually 9 months, as the aquifer gets almost empty by this time. Due to their small dimensions, such alluvial aquifers function as a reservoir capable of being completely filled even in less rainy years.

The Sumé alluvial aquifer is located in the State of Paraíba, 12 km along the Sucuru River and downstream of the Sumé Reservoir, with storage capacity of 45 hm^3 (Figure 6.1). Sumé IP was supplied by this reservoir, and farmers experienced high productivity and profits. However, the reservoir water availability was significantly

104 Groundwater for Sustainable Livelihoods

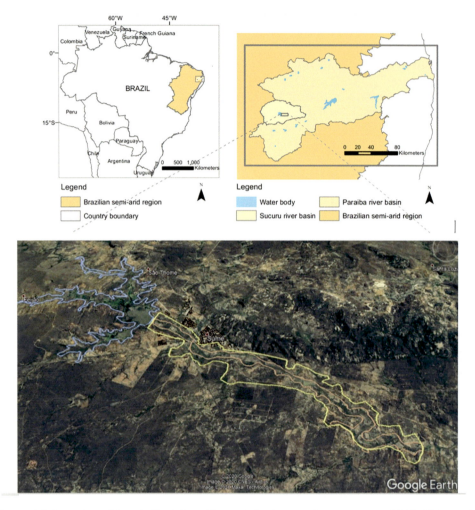

Figure 6.1 Location of Sumé alluvial aquifer and the irrigated perimeter in 2018. (Adapted from Tsuyuguchi et al. (2020).)

reduced, due to streamflow interception by other reservoirs constructed over the upstream catchment. Moreover, the growth of Sumé city and surrounding areas caused an increase in urban water demand, resulting in the interruption of water supply for irrigation in 1987.

Such interruption in water supply by the surface reservoir has eventually led farmers and the government to start using the alluvial aquifer to maintain their activities. As mentioned, the aquifer is easily recharged especially through eventual runoff over the intermittent river and can be exploited along the dry period. The easiness of drilling/excavating dug wells and the proximity to the crop areas facilitated the exploitation. Although this water source is indeed more limited, it should be highlighted that it was previously disregarded and suddenly became the only available resource.

Alluvial aquifers have also been used in other IP projects, such as Eng. Arcoverde, also in Paraíba, and of Custódia, in the State of Pernambuco, among others.

Along the 12 km length of the Sumé IP, the alluvial aquifer occupies an area of 351 ha, with a width of 50–500 m and a depth of 0.5–15 m (Vieira, 2002; Schimmelpfennig et al., 2018) (Figures 6.1 and 6.2). Its storage capacity is estimated at 1,700,000 m^3, and its lithology and shape present relatively high variability, resulting in differences in groundwater availability in the aquifer (Tsuyuguchi et al., 2020). During the year, the water table presents significant variation because of the rapid recharge during the short rainfall period combined with high permeability and abstractions (Figure 6.3).

Figure 6.2 View of Sumé alluvial aquifer.

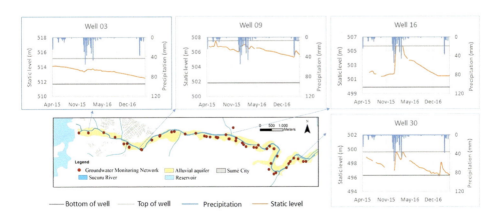

Figure 6.3 Water levels along the aquifer in response to precipitation (Schimmelpfennig et al., 2018).

6.3 THE ALLUVIAL AQUIFER ROLE IN RESILIENCE TO DROUGHT

Alluvial aquifers can increase smallholder farmers' income by allowing them to grow high-value crops such as vegetables and fruits during the dry season or droughts when the lands would be unproductive. As a small source of water, their use by farmers is much more controlled and efficient than when using canals during the initial years of the IP. The inefficient furrow irrigation system was changed to the use of microsprinkler and drip irrigation supported by Paraíba State initiatives (Silva, 2006).

Determination of when and how much to irrigate and how to reduce evaporation losses have been incorporated in everyday irrigation practices by the farmers to ensure the best use of the resource. Furthermore, as it is usually annually renewable, its use allows the sustainable production of perennial crops that have higher added value. The use of groundwater allowed farmers at Sumé to develop subsistence agriculture, and also generate income through selling the production into the local market.

As previously described, the Sumé IP was considered a successful case due to the high production of corn and tomato. Before the interruption of water supply from the reservoir, there were forty-seven active farms, with areas varying between 8 and 14 ha, with a total irrigated area of 273 ha (Mendonça, 2010). With the end of the water supply by the surface reservoir in 1987, there was a sudden change in the irrigated area. Tomato production was nearly ended and corn production, which previously reached up to 120 ha, fell to the 40–60 ha range as shown in Figure 6.4. After switching the water supply source, the farmers adapted to the new reality, making the production more stable with the use of the alluvial aquifer. In the period of 1998–2003, a severe drought significantly affected crops production. The increase in resilience to drought was observed during the following severe drought, from 2012 to 2018, when there was comparatively a lower impact over the cultivated areas.

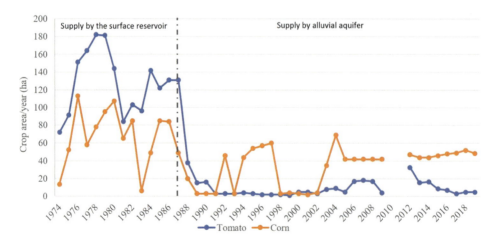

Figure 6.4 Cultivated areas of corn and tomato at Sumé IP. Detail to year 1987 when the project collapsed. (DNOCS (2012–2019) and Mendonça (2010).)

A plan for the development of the IP was formulated and a previous diagnosis informed an irrigated area of 59.85 ha through the exploitation of the alluvial aquifer (MIN, 2007). This diagnosis generated an assessment of the annual production in the IP of Sumé in US$ 738,035.00, from which 94% were commercialized. Dantas Neto et al. (2013) analyzing the crop yield of tomatoes and pepper in two areas inside of a lot identified production of 30,000 and 7,000 kg/ha, respectively, in 2009. They estimated that their production in an area of 2 ha would generate an income of US$ 47.510,00 per year, showing that the crop irrigation is viable to small farmers using groundwater and can generate wealth.

After 2010, the production was diversified, including fruits, grains, vegetables and forage, to feed the livestock (mainly sheep, goat, cattle and pigs). The cultivated areas and those used by livestock have strong variability along the years and seasons, depending on climate, market and family labor force availability. Besides subsistence, trading and commerce at the local market, farmers have signed contracts with food industries, with cases of farmers receiving awards for productivity performance (Souza, 2013). Table 6.1 shows the crop areas cultivated between 2012 and 2019, even during the drought (besides tomato's and corn's areas that were presented in Figure 6.4).

Resilience to drought was very much increased during the last 10 years, as demonstrated before, due to an increase in irrigation efficiency and crop diversification. For example, during the period of 2015–2018, when the region was facing a severe drought initiated in 2012, of thirty-six identified in the area, twenty-eight wells were exploited. The usual pumping rates vary approximately from 10 to 100 m³/day. Such exploitation supported the cultivated area in the IP, between 8 and 43 ha, with seasonal variation.

Current agricultural and livestock explorations are shown in Figure 6.5. Although, in this figure, farms are represented with only one main crop category, in most of them different types of crops are cultivated and diverse domestic animals are raised. Figure 6.6 shows some of the cultivated areas and livestock. Moreover, livestock production has been maintained with low seasonal variation in the total number of animals of around 1,400. Also, data of fish culture in the IP from 2018 indicate that 22,500 kg of fish were produced in the year.

Table 6.1 Main crop areas cultivated from 2012 to 2019

Crop area (ha)	Year							
	2012	2013	2014	2015	2016	2017	2018	2019
Onion	1.0	-	-	0.4	1.0	-	0.5	-
Coconut	1.7	0.5	0.5	-	-	-	-	-
Bean	44.0	44.0	44.0	44.0	44.0	44.0	49.0	44.0
Passion fruit	0.5	-	0.6	0.6	1.5	1.5	-	-
Watermelon	2.3	2.3	2.5	-	1.0	-	-	-
Bell pepper	-	1.8	1.0	1.5	3.5	2.0	4.5	1.0

Source: DNOCS (2012–2019).

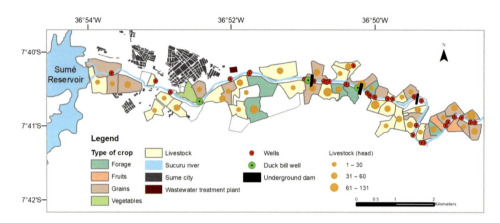

Figure 6.5 Agricultural and livestock activities during the period of 2016–2018 at Sumé IP, location of underground dams, wells and "duck bill" wells.

6.4 APPROPRIATE TECHNOLOGIES FOR INCREASING WATER AVAILABILITY

The long tradition of using the shallow alluvial aquifers in BSA region was improved in the Sumé IP by the introduction of appropriate technologies designed to increase aquifer potential of exploitability, such as (a) underground dam, which increases groundwater retention and (b) the "duck bill" well, which improves well's production.

6.4.1 Underground dams

Underground dams are structures used in alluvial aquifers throughout the BSA (Cirilo et al., 2017; Costa et al., 1998). A ditch is opened into the ground up to the bedrock along a cross section of the river, the wall is covered with an impermeable material and then the ditch is filled back with the material removed (Figure 6.7). These dams are used to retain groundwater flow and improve extraction by an upstream well. In 2018, underground dams had their importance recognized, as a finalist in the prize Sustainable Development Objectives Brazil 2018. With this, the underground dam technology became part of the "bank of best practices", which will be a reference in the implementation and dissemination of Agenda 2030 for Sustainable Development, in Brazil.

The underground dam has become a major appropriate technology of several public policy programs, as it is a practice of relatively low operational cost, simplicity and functional replicability, serving a significant number of households. Thousands of them have been constructed through government programs, such as the Program One Land and Two Waters (P1+2), and were proven to support the reduction of poverty and hunger in rural areas (Campelo, 2013). Thus, these technologies perform an important role in the decrease of water shortage and in the guarantee of food security for rural communities.

Figure 6.6 Farm activities at Sumé IP in 2018.

The predominant axial flow direction of groundwater is a result of the geological formation of shallow, narrow and elongated alluvial aquifers. This condition favors the use of underground dams built transversely to the main flow direction. The dam can fully cover the cross section of the aquifer but this condition is not ideal as no downstream groundwater flow would occur. The main objective is the retardation of the flow, and consequent water table elevation, increasing aquifer capabilities to supply users' demands (Alves et al., 2018).

Figure 6.7 Construction of groundwater dam at alluvial aquifer in BSA region (Cirilo, 2008).

In the Sumé IP, three underground dams were built partially covering the river cross section (Figures 6.5 and 6.8). The advantages of using these underground dams for groundwater exploitation efficiency in the area were verified by Vieira (2002) and Alves et al. (2018) through the simulation of scenarios using numerical methods (Figure 6.9). In the scenarios in which the underground dam was inserted, the possible areas to be irrigated were always larger than in the same recharging conditions but without the existence of the dam.

Figure 6.8 View of an aquifer reach with an underground dam at Sumé.

Small alluvial aquifers in Brazil   111

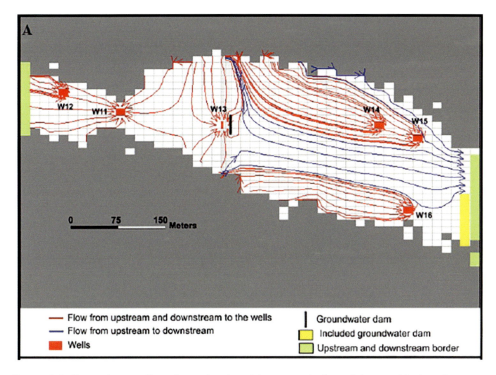

Figure 6.9 Groundwater flow lines simulated by numerical models considering the presence of an underground dam at Sumé Irrigated Perimeter (Alves et al., 2018).

6.4.2 "Duck bill" well

Excavated wells of low depth and large diameter are frequently used for exploitation of shallow water table aquifers, such as the small alluvium. The construction of large wells with walls entirely sealed or made of concrete rings (Figure 6.10) is quite common in the region. However, operation of these wells is similar to a reservoir that dries up rapidly. Since there is no filter section, the inflow rates into the well (only through the bottom of the aquifer) are usually lower than the low pumping rates in practice. Then, after its emptying, it is necessary to wait for its recharge to restart pumping. As a consequence, these wells have been built from ancient times with diameters up to 5 m to have large volumes inside, what elevates costs of construction.

With this in mind, alternatives for allowing radial flow into the well and continuous exploitation have been developed (Albuquerque & Rêgo, 1990; Braga, 2016). The well called "duck bill" was designed and constructed in 1997 along the Sumé IP, considering the local characteristics of the aquifer and social aspects to increase the well efficiency and lower costs of construction (Rêgo et al., 2014), as shown in Figures 6.11 and 6.12. The name "duck bill" refers to the type of brick whose format is trapezoidal (similar to a duck bill) and that is produced from the extraction of soil with higher

112 Groundwater for Sustainable Livelihoods

Figure 6.10 Well constructed with concrete rings at Sumé IP.

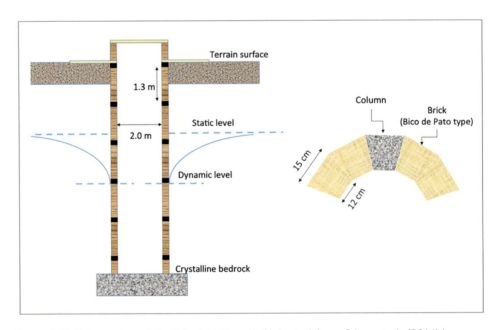

Figure 6.11 Schematics of the "duck bill" well. (Adapted from Rêgo et al. (2014).)

Small alluvial aquifers in Brazil 113

Figure 6.12 "Duck bill" well construction at Sumé IP in 1997. (Adapted from Rêgo (2012).)

clay content present in the region. Their dimensions were chosen to allow the bricks to form an interlocked structure of 2 m-diameter, avoiding the use of mortar sealing the walls of well. Thus, lateral flow of water is allowed along the entire length of the well walls. To improve stability, the well design has structural elements every 1.2 m, as well as three vertical columns, to support the structure.

In order to define the construction sites, data from about 100 boreholes were used, which identified the depth and constitution of the soil. Sites with greater depth and greater percentage of sand were chosen to allow the greatest potential for water exploitation. Uninterruptedly exploitation rates of around 9 L/s have been measured for hours and the structure has resisted for more than 20 years. The owner of a duck bill well has been maintaining irrigation throughout the year, even during the long droughts, and heading the group to bring projects and investments to the perimeter. The use of such enhancement in the design of wells, conjunctly with other appropriate technologies such as the underground dams, can improve conditions for sharing wells among farms, which can be helpful for better planning and implementation of different crop strategies (Rêgo, 2012; Tsuyuguchi et al., 2020).

6.5 INTEGRATED USE OF RESERVOIR AND AQUIFER

Since the Sumé IP started to use groundwater from the alluvial aquifer, sometimes, when the Sumé Reservoir presented higher stored volumes, it was possible to release water to recharge the alluvium during the dry season. Such operations are very important to implement an integrated management of surface and groundwater resources (Billib et al., 1991; Burte et al., 2009). During the recent great drought (2012–2017) in the region, reservoir releases to the alluvium were canceled, to preserve the surface water for supplying the city of Sumé.

Recently, in 2020, a new water allocation plan for the Sumé Reservoir was developed with participation of government, water users and civil society, and the River Basin Committee, under the coordination of the National Water Agency (ANA). It was accorded that a volume of 2.1 hm³ per year would be targeted for use by the IP and for water release onto the Sucuru riverbed (Sumé alluvial aquifer). Currently, the IP will benefit from this agreement through the recharge of the aquifer and the following exploitation from wells.

6.6 COMMUNITY ORGANIZATION AND ENGAGEMENT

The Sumé IP farmers are organized in a cooperative association. It was created when the IP started operating in the 1970s, as part of the management model established by the National Department against Droughts (DNOCS), the government body responsible for implementing the IPs in the BSA. The cooperative was designed to be the entity responsible for governing the production dynamics and the financial resources of the IP (i.e., intermediating the payments by enterprises buying the crop production and the credits granted). Thus, all negotiations were accorded through a cooperative with democratic values. Given the long drought and the interruption of the water supply by the reservoir, the members found it difficult to deal with the maintenance costs, so that they voted for ending the cooperative in 1991. However, in 1994, with the growing evidences that the IP would be again viable by using the alluvial aquifer resources, the farmers decided to reopen the cooperative, as they realized that working through it (as previously) would be important to booster the agricultural activities (Monte, 2013).

As described above, the exploitation of the alluvial aquifer increased resilience of local farmers, improving the livelihood conditions and the engagement of the community in the aquifer governance. The decision of reopening the cooperative is a meaningful signal of this, as it is currently composed only by the own farmers. It should be highlighted that during the high-production-years of the perimeter, DNOCS represented a major actor in the guarantee of provision for farmers, and its role has been very reduced since then. Meanwhile, farmers have demonstrated resilience on actions such as building wells and underground dams, using other methods of irrigation, requesting allocation of water from the surface reservoir to recharge the aquifer, participating as active members of the river basin committee, among others (Tsuyuguchi et al., 2020). Therefore, the alluvial aquifer was an essential resource to keep farmers in the rural area, and such a fact is recognized by the group.

6.7 REMAINING CHALLENGES

6.7.1 Appropriation of modeling tools

Groundwater flow modeling has been applied in the last 20 years to increase knowledge concerning water availability in the alluvial aquifer of the IP of Sumé. These numerical model simulations are important to understand the likelihood of improving crop production in the area and, therefore, farmers' resilience. Vieira (2002), Tsuyuguchi et al. (2017) and Alves et al. (2018) modeled portions of the aquifer, estimated its storage capacity and analyzed the impact of underground dams and wells' pumping rates over groundwater flow and exploitation. They demonstrated that the technologies could improve the water supply for irrigation. Nóbrega et al. (2008) developed an integrated tool to estimate recharge in the alluvium based on seasonal forecasts of rainfall produced by atmospheric models, which is important considering the high interannual climate variability. Silva (2016) used multi-criteria decision analysis to investigate the integration of surface, groundwater and wastewater for irrigation considering technical, economic and environmental criteria in order to optimize the water use. If these results and tools are appropriately adopted by the farmers, they can use them to support decision-making concerning the implementation of strategies to better harness the groundwater resource. This can be developed through the increase of the water volume exploited and through improvements on crop and livestock planning. This planning includes several practices, such as the selection of species and definition of the area to be cultivated and the animals to be raised, which allow to reduce risks and mitigate impacts of the droughts on production.

6.7.2 Managed aquifer recharge

Deposition of sewage from neighboring cities or communities onto dry riverbeds generates an unmanaged aquifer recharge in the alluvial aquifers all over the BSA, and also in Sumé. There is an urgent need to develop strategies for appropriate reuse of sewage and management of aquifer recharge. Recently, an appropriate Managed Aquifer Recharge (MAR) strategy for such alluvial aquifers was formulated, using monitored data of Sumé alluvial aquifer (Pontes Filho, 2018). Sewage presence in groundwater was detected by chloride analyses and it remained high in the urban zone, decreasing in terms of values and variation along the aquifer. Despite this apparent natural Soil-Aquifer Treatment, salts remain in groundwater, demonstrating the need to recharge the aquifer with a lower concentration source to avoid salinization hazards. The strategy was formulated as measures considering the joint utilization of sewage and surface water resources with the aim of promoting water reuse, reducing evaporation losses and, consequently, increasing water availability.

6.7.3 Groundwater salinity

The extensive use of groundwater for small-scale agriculture is a common activity in alluvial aquifers such as Sumé. However, these aquifers are susceptible to the salinization process, which can be influenced by hydroclimatic, geological, land use and occupation aspects and by recharge of treated and untreated wastewater. Recent research based on secondary data, fieldwork and surveys regarding land use and occupation in the IP, including extensive monitoring of electrical conductivity, showed that most of the solutes in the groundwater are originated from wastewater disposal, followed by the ones from agricultural activities. In order to deal with such problem in Sumé, we recommend the adoption of water and soil management strategies, which should consider the saline residues from irrigated agriculture and livestock, as well as the wastewater disposal from the neighboring city. These strategies are: replacing current less-tolerant with appropriate salt-tolerant crops, mixing groundwater and surface water, sanitary protection measures for the wells, and agricultural soils drainage. They can help in the crop planning by farmers to increase production. Since salinity levels vary seasonally and annually, according to the amount of recharge by waters with lower salinity, well-informed seasonal crop planning is a necessary step to be taken in the IP.

6.7.4 Governance

Tsuyuguchi et al. (2020) analyzed the IP of Sumé under the perspective of a social-ecological system, aiming to identify problems and opportunities to improve governance of the aquifer (Figure 6.13). The authors identified a reasonable aquifer characterization, the existence of technologies that are appropriate for the region, participation of the community on water management and assistance for family farming. However, issues on monitoring the resource, on sharing knowledge and on integrated water management at local level considering the aquifer as a management unity have limited the exploitation efficiency and the equitable distribution of water among the users. Therefore, it was suggested to empower community through changes

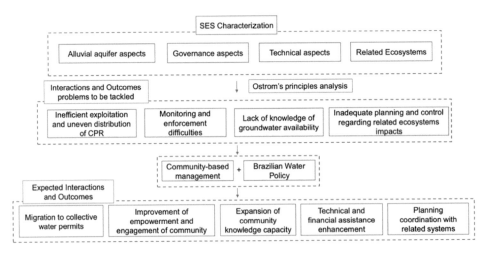

Figure 6.13 Opportunities for improving governance of Sumé alluvial aquifer (Tsuyuguchi et al., 2020).

in the water permits schemes and improving dialogue between all actors involved in the process.

6.8 SUMMARY

This chapter highlights the importance of small alluvial aquifers to the BSA region. Although having sustained traditional communities for centuries, alluvial aquifers have been largely neglected as a meaningful resource to cope with drought and water scarcity issues. This chapter recognizes the strategic role of those small alluvial aquifers in maintaining rural communities active, promoting the sustainable use of water for food production that is marketed locally and generating income, promoting social and economic development.

The Sumé case study illustrates this importance as it comprises an IP that has been supplied by the alluvial aquifer for more than 30 years. It is a real success story of sustainable and long-term use of groundwater to mitigate the harmful effects of drought, especially poverty. Farmers had to adapt to sustainably exploit the small aquifer to meet their needs. The change from the inefficient furrow irrigation system to the use of micro-sprinkler and drip irrigation, allied to crops diversification, made their production more resilient to droughts and more sustainable. Appropriate technologies have also been adopted to improve exploitation performance. Integrated management of surface and groundwater resources is stimulated and water release to recharge the alluvium during the dry season has been used. Despite the adverse conditions and abrupt changes in water availability, farmers have demonstrated resilience and organized in a cooperative.

The success of Sumé IP and other alluvial aquifers over the State of Paraíba (e.g., Albuquerque, 1984; Rêgo, 2012) brought attention to these important water resources. In 2006, the Water Resources Master Plan of the State of Paraíba recognized these aquifers as a regional aquifer system and launched a 20-year program for developing studies and guidelines for rational exploitation of these resources in the state (Paraíba, 2006).

Our analyses show that the use of the small alluvial aquifer in Sumé IP meets the Sustainable Development Goals of Agenda 2030, established by the United Nations. Among the goals, we emphasize six of them, which are described as follow: End hunger, achieve food security and improved nutrition and promote sustainable agriculture (Goal 2); Ensure availability and sustainable management of water and sanitation for all (Goal 6); Build resilient infrastructure, promote inclusive and sustainable industrialization and foster innovation (Goal 9); Ensure sustainable consumption and production patterns (Goal 12); Take urgent action to combat climate change and its impacts (Goal 13); and Protect, restore and promote sustainable use of terrestrial ecosystems, sustainably manage forests, combat desertification, and halt and reverse land degradation and halt biodiversity loss (Goal 15). In general, the achievement of these goals can be seen in the Sumé through increase of water availability, food security and diversified crops to guarantee family income. Also, the adoption of appropriate technologies and more efficient irrigation methods for saving water and reducing degradation of soil, demonstrates innovation for improving ecosystems' sustainability.

We expect that our findings call attention to the opportunities on the strategic use of the small alluvial aquifers spread over BSA region and other small and neglected aquifers around the world.

ACKNOWLEDGMENTS

The research reported in this chapter was funded by the Brazilian agencies CAPES (Finance Code 001), CNPq (grants 550059/2014-8 and 310789/2016-8), FINEP (grant 01.13.0340.00) and FAPESQ/PB, through the R&D Project "BRAMAR – Strategies and Technologies for Water Scarcity Mitigation in Northeast of Brazil: Water Reuse, Managed Aquifer Recharge and Integrated Water Resources Management." Funding from INCT Climate Change (CNPq and FAPESP) is also highly appreciated. The participation and support of the local community were essential to the development of this chapter.

REFERENCES

Al-Shaibani, A. M. (2008) Hydrogeology and hydrochemistry of a shallow alluvial aquifer, western Saudi Arabia. *Hydrogeology Journal*. [Online] 16(1), 155–165. Available from: https://doi.org/10.1007/s10040-007-0220-y [Accessed 20th July 2020].

Albuquerque, J. do P. T. (1984) *Os Recursos de Água Subterrânea do Trópico Semi-Árido do Estado da Paraíba*. Master thesis. Universidade Federal da Paraíba. Campina Grande, Paraíba.

Albuquerque, J. do P. T., & Rêgo, J. C. (1987) Estudos de avaliação e utilização de um aqüífero aluvial do semi-árido paraibano. In *VII Simpósio Brasileiro de Hidrologia e Recursos Hídricos*. Salvador, Bahia.

Albuquerque, J. do P. T., & Rêgo, J. C. (1990) Estruturas de Captação de Aqüíferos Aluviais Ocorrentes na Região Semi-Árida do Estado da Paraíba. In *VI Congresso Brasileiro de Águas Subterrâneas*. Porto Alegre. [Online]. Available from: https://aguassubterraneas.abas.org/asubterraneas/article/view/24547 [Accessed 12th July 2020].

Alves, E. J. C., Rêgo, J. C., Galvão, C. de O., & Vieira, J. B. de A. (2018) Limits and conditions for the exploitation of alluvial aquifers in the Brazilian semi-arid region. *Brazilian Journal of Water Resources*. [Online] 23(34), 1–8. Available from: https://doi.org/10.1590/2318-0331.0318160049 [Accessed 15th July 2020].

Andrade, E. M. de, Palácio, H. A. Q., Souza, I. H., de Oliveira Leão, R. A., & Guerreiro, M. J. (2008) Land use effects in groundwater composition of an alluvial aquifer (Trussu River, Brazil) by multivariate techniques. *Environmental Research*. [Online] 106(2), 170–177. Available from: https://doi.org/10.1016/j.envres.2007.10.008 [Accessed 15th June 2020].

Bazuhair, A. S., & Wood, W. W. (1996) Chloride mass-balance method for estimating ground water recharge in arid areas: examples from western Saudi Arabia. *Journal of Hydrology*. [Online] 186, 153–159. Available from: https://doi.org/10.1016/S0022-1694(96)03028-4 [Accessed 15th July 2020].

Billib, M., Boochs, P. T., & Rêgo, J. C. (1991) Management of an artificial underground reservoir for irrigation. In *Third Chinese-German Symposium on Hydrology and Coastal Engineering*. Nanjing.

Blackburn, D. M., Montenegro, A. A. A., & Montenegro, S. M. G. (2002) Recarga de aquífero aluvial a partir da agricultura irrigada e suas implicações na qualidade da água subterrânea em Pesqueira - PE. In *XII Congresso Brasileiro de Águas Subterrâneas* (pp. 1–9) Florianópolis, SC. Available from: https://aguassubterraneas.abas.org/asubterraneas/article/viewFile/22712/14913 [Accessed 20th July 2020].

Braga, R. A. P. (2016) *Águas de Areias*. [Online] Recife, Clã. Available from: http://www.aguas-donordeste.org.br/website/wp-content/uploads/2018/04/Livro-%C3%81guas-de-Areias.pdf [Accessed 15th July 2020].

Burte, J. D. P., Coudrain, A., Frischkorn, H., Chaffaut, I., & Kosuth, P. (2005) Human impacts on components of hydrological balance in an alluvial aquifer in the semiarid Northeast, Brazil. *Hydrological Sciences Journal* [Online] 50(1), 37–41. Available from: https://doi.org/10.1623/hysj.50.1.95.56337 [Accessed 20th July 2020].

Burte, J. D. P., Coudrain, A., & Marlet, S. (2011) Use of water from small alluvial aquifers for irrigation in semi-arid regions. *Revista Ciencia Agronomica*. [Online] 42, 635–643. Available from: https://doi.org/10.1590/S1806-66902011000300009 [Accessed 12th July 2020].

Burte, J. D. P., Jamin, J. Y., Coudrain, A., Frischkorn, H., & Martins, E. S. (2009) Simulations of multipurpose water availability in a semi-arid catchment under different management strategies. *Agricultural Water Management*. [Online] 96(8), 1181–1190. Available from: https://doi.org/10.1016/j.agwat.2009.03.013 [Accessed 15th July 2020].

Campelo, D. A. (2013) Public policies for Brazilian family farming in a semiarid climate: from combat to coexistence. *Revista Brasileira de Pós-Graduação* [Online] 10(21), 851–873. Available from: https://doi.org/10.21713/2358-2332.2013.v10.587 [Accessed 12th August 2020].

Cirilo, J. A. (2008) Public water resources policy for the semi-arid region. *Estudos Avançados* [Online] 22(63), 61–82. Available from: http://dx.doi.org/10.1590/S0103-40142008000200005 [Accessed 10th August 2020].

Cirilo, J. A., Montenegro, S. M. G. L., & Campos, J. N. B. (2017) The issue of water in the brazilian semi-arid region. In C. de Mattos Bicudo & J. Galizia Tundisi (Eds.), *Waters of Brazil: Strategic Analysis*. Cham: Springer International Publishing. https://doi.org/10.1007/978-3-319-41372-3_5

Costa, W. D., Cirilo, J. A., Pontes, M., Maia, A. Z., & Sobrinho, O. P. (1998) Barragem subterrânea: uma forma eficiente de conviver com a seca. In *X Congresso Brasileiro de Águas Subterrâneas*. São Paulo: Brazil [Online]. Available from: https://aguassubterraneas.abas.org/asubterraneas/article/view/22277/14620 [Accessed 11th July 2020].

Dantas Neto, J., Lima, V. L. A. de, Silva, P. F. da, Santos, C. S. dos, & Silva, L. F. D. (2013) Unidades produtivas da agricultura familiar no Perímetro Irrigado de Sumé. *Enciclopédia Biosfera* [Online] 9(16), 2060–2070. Available from: http://www.conhecer.org.br/enciclop/2013a/agrarias/UNIDADES%20PRODUTIVAS.pdf [Accessed 11th July 2020].

DNOCS (2012–2019) Controle do plano de explotaçao agrícola. Departamento Nacional de Obras Contra as Secas. CAMIS monthly reports.

Love, D., Owen, R., Uhlenbrook, S., van der Zaag, P., & Moyce, W. (2007) The lower Mzingwane alluvial aquifer: managed releases, groundwater - surface water interactions and the challenge of salinity, (iv), 5. Available from: https://www.gov.uk/research-for-development-outputs/alluvial-aquifer-indicators-for-small-scale-irrigation-in-north-east-brazil [Accessed 11th July 2020].

Mackay, R., Montenegro, A. A. de A., Montenegro, S. M. G. L., & Wonderen, J. Van. (2005) Alluvial aquifer indicators for small-scale irrigation in Northeast Brazil. *Symposium of International Association of Hydrological Sciences-IAHS* [Online] (1), 1–5. Available from: https://doi.org/10.1007/s13398-014-0173-7.2 [Accessed 11th July 2020].

Mendonça, J. R. N. (2010) *Do Oásis à Miragem: uma análise da trajetória do Perímetro Irrigado de Sumé - PB no contexto das políticas de desenvolvimento para o Nordeste (From the oasis to the mirage: An analysis of the tale of the Sumé Irrigated Perimeter in the context of develop)*. Master thesis. Universidade Federal de Campina Grande. Campina Grande, Brazil. Available from: http://dspace.sti.ufcg.edu.br:8080/jspui/handle/riufcg/2356 [Accessed 13th July 2020].

MIN. (2007) *Elaboração de diagnóstico e plano de desenvolvimento do perímetro irrigado (Ellaboration of Dianosis and Development Plan of the Irrigated Perimeter)* (Vol. II). Brasilia, Brazil.

Missimer, T. M., Drewes, J. E., Amy, G., Maliva, R. G., & Keller, S. (2012) Restoration of wadi aquifers by artificial recharge with treated waste water. *Ground Water.* [Online] 50(4), 514–527. Available from: https://doi.org/10.1111/j.1745-6584.2012.00941.x [Accessed 13th July 2020].

Missimer, T. M., Guo, W., Maliva, R. G., Rosas, J., & Jadoon, K. Z. (2015) Enhancement of wadi recharge using dams coupled with aquifer storage and recovery wells. *Environmental Earth Sciences* [Online] 73(12), 7723–7731. Available from: https://doi.org/10.1007/s12665-014-3410-7 [Accessed 13th July 2020].

Monte, H. C. (2013) *Terceiro setor/cooperativismo: Desenvolvimento local, um estudo de caso da CAMIS.* Undergraduate thesis. Universidade Federal de Campina Grande. Sumé, Brazil.

Montenegro, A. A. de A., & Montenegro, S. M. G. L. (2006) Variabilidade espacial de classes de textura, salinidade e condutividade hidráulica de solos em planície aluvial. *Revista Brasileira de Engenharia Agrícola e Ambiental* [Online] 10(1), 30–37. Available from: https://doi.org/10.1590/S1415-43662006000100005 [Accessed 21th July 2020].

Moyce, W., Mangeya, P., Owen, R., & Love, D. (2006) Alluvial aquifers in the Mzingwane catchment: their distribution, properties, current usage and potential expansion. *Physics and Chemistry of the Earth* [Online] 31, 988–994. Available from: https://doi.org/10.1016/j.pce.2006.08.013 [Accessed 21th July 2020].

Nóbrega, R. L. B., Costa, I. Y. de L. G. da, Oliveira, K. F. de, Pinheiro, K. M., Almeida, E. M. M., Pn, J. C., & Galvão, C. de O. (2008) Uso da ferramenta SegHidro para previsão de recarga e níveis de água em aluviões no semi-árido brasileiro. In *IX Simpósio de Recursos Hídricos do Nordeste* (pp. 1–10). Salvador, Bahia.

Paraíba. Plano estadual de recursos hídricos do estado da Paraíba (Paraiba state water resources plan) (2006) *Secretaria de Estado da Ciência e Tecnologia e do Meio Ambiente, Agência Executiva de Gestão de Águas do Estado da Paraíba.* Brasília: Consórcio TC/BR - Concremat.

Pontes Filho, J. D. de A. (2018) *Da recarga não gerenciada à recarga gerenciada: estratégia para aquífero aluvial no Semiárido Brasileiro.* Master Thesis. Universidade Federal de Campina Grande. Campina Grande, Brazil.

Rêgo, J., Albuquerque, J. do P. T., & Pereira, I. J. (2014) Poço Bico de Pato (Duck bill well). In D. A. Furtado, J. G. de V. Baracuhy, P. R. M. Francisco, S. F. Neto, & V. A. de Sousa (Eds.), *Tecnologias Adaptadas para o desenvolvimento sustentável do semiárido brasileiro (Adapted Technologies for the sustainable development of the Brazilian semiarid region)* (pp. 250–255). Campina Grande: EPGRAF.

Rêgo, J. C. (2012) *Bewirtschaftung kleiner alluvialer Grundwasservorkommen im semiariden Nordosten Brasiliens (Management of small alluvial groundwater resources in the semi-arid northeast of Brazil).* PhD Thesis. Gottfried Wilhelm Leibniz Universität. Hanover, Germany.

Sá, J., & Diniz, J. A. O. (2012) Aproveitamento Das Aluviões Do Semiárido Do Nordeste. In *XVII Congresso Brasileiro de Águas Subterrâneas e XVIII Encontro Nacional de Perfuradores de Poços* (Vol. 0, pp. 1–4). Bonito, Brazil.

Salgado, J. P., Coura, M. A., Barbosa, D. L., Feitosa, P. H. C., Meira, M. A., & Rêgo, J. C. (2018). Influence of sewage disposal on the water quality of the Sucuru River alluvial aquifer in the municipality of Sumé-PB, Brazil. *RBRH* [Online] 23(23), 1–13. Available from: https://doi.org/10.1590/2318-0331.231820160052 [Accessed 21th July 2020].

Sarma, D., & Xu, Y. (2017) The recharge process in alluvial strip aquifers in arid Namibia and implication for artificial recharge. *Hydrogeology Journal* [Online] 25(1), 123–134. Available from: https://doi.org/10.1007/s10040-016-1474-z [Accessed 20th July 2020].

Schimmelpfennig, S., Meon, G., Schoniger, M., Tsuyuguchi, B. B., Braga, A. C. R., Pontes Filho, J. D. A., ... Bertrand, G. F. (2018) Hydrogeological modelling (results from WP2). In A. Abels, M. Freitas, J. Pinnekamp, & B. Rusteberg. (Eds.), *BRAMAR PROJECT: Water Scarcity Mitigation in Northeast Brazil* (pp. 38–56). Aachen: RWTW Aachen University.

Silva, L.F.D. (2006) *Avaliação de unidades produtivas da agricultura familiar no perímetro irrigado de Sumé-PB*. Master thesis. Universidade Federal de Campina Grande. Campina Grande, Brazil.

Silva, S. A. F. (2016) *Análise multicritério espacial no gerenciamento dos recursos hídricos no perímetro irrigado de Sumé-PB*. Master thesis. Universidade Federal de Campina Grande. Campina Grande, Brasil.

Soares, S., Souza, L. De, Silva, W. J., & Silveira, F. G. (2016) Poverty profile: the rural North and Northeast regions of Brazil. *International Policy Centre for Inclusive Growth* [Online] (50), 1–3. Available from: https://ipcig.org/pub/eng/PRB50_Poverty_profile_the_rural_North_Northeast_regions_of_Brazil.pdf [Accessed 20th July 2020].

Sorman, A. U., & Abdulrazzak, M. J. (1993) Infiltration-recharge through wadi beds in arid regions. *Hydrological Sciences Journal* [Online] 38(3), 173–186. Available from: https://doi.org/10.1080/02626669309492661 [Accessed 20th July 2020].

Souza, M. do S. de. (2013) *Irrigação e sustentabilidade: o desenho institucional do Perímetro Irrigado de Sumé*. Undergraduate thesis. Universidade Federal de Campina Grande. Sumé, Brazil.

Taher, T., Bruns, B., Bamaga, O., Al-weshali, A., & Steenbergen, F. Van. (2012) Local groundwater governance in Yemen: building on traditions and enabling communities to craft new rules. *Hydrogeology Journal* [Online] 20, 1177–1188. Available from: https://doi.org/10.1007/s10040-012-0863-1 [Accessed 20th July 2020].

Tsuyuguchi, B. B., Braga, A. C. R., Pontes Filho, J. D. de A., Costa, M. R., Rêgo, J. C., & Galvão, C. de O. (2017) Variability of hydraulic conductivity in alluvial aquifer in the Brazilian semiarid region. In *XXII Water Resources Brazilian Symposium* (pp. 1–8).

Tsuyuguchi, B. B., Morgan, E. A., Rêgo, J. C., & Oliveira Galvão, C. de. (2020) Governance of alluvial aquifers and community participation: a social-ecological systems analysis of the Brazilian semi-arid region. *Hydrogeology Journal*. [Online] 1539–1552. Available from: https://doi.org/10.1007/s10040-020-02160-8 [Accessed 18th July 2020].

Vieira, L. (2002) *Emprego de um modelo matemático de simulação do fluxo subterrâneo para definição de alternativas de explotação de um aquífero aluvial*. Master Thesis. Universidade Federal da Paraíba. Campina Grande, Brazil.

Vieira, R. M. S. P., Tomasella, J., Alvalá, R. C. S., Sestini, M. F., Affonso, A. G., Rodriguez, D. A., ... Santana, M. O. (2015) Identifying areas susceptible to desertification in the Brazilian northeast. *Solid Earth* [Online] 6(1), 347–360. Available from: https://doi.org/10.5194/se-6-347-2015 [Accessed 18th June 2020].

Chapter 7

Stubble burning in northwestern India

Is it related to groundwater overexploitation?

D. Saha
Ministry of Jal Shakti

M. Chakraborty and A. Chowdhury
Partners in Prosperity

CONTENTS

7.1 Introduction 123
7.2 Why stubble burning? 125
    7.2.1 Stubble burning and groundwater overexploitation 125
7.3 Potential strategies and approaches 126
    7.3.1 Reduce area under paddy and pursue crop diversification 128
    7.3.2 Expanding basmati cultivation during *kharif* 128
    7.3.3 A relook into The Haryana Preservation of Subsoil Water Act, 2009 128
    7.3.4 Technological innovation 129
    7.3.5 Finding other uses of crop residues 129
7.4 Stubble burning – a village view 129
7.5 Way forward 132
Acknowledgments 133
References 133

7.1 INTRODUCTION

Stubble burning is rapidly becoming a major contentious issue in agriculture, water and environment in India (Singh et al. 2019). This practice is rampant in the northwestern states of Punjab, Haryana and also spreading in western Uttar Pradesh. The entire region is known for the most intensive rice–wheat cultivation in the country. The state of Haryana, being adjacent to the National Capital, Delhi, always attracts attention whenever this issue is discussed. Haryana, although geographically a small state in India (occupies only 1.5% geographical area of the country), contributes 15% of India's agriculture produce. Eighty-six percent of the state's geographical area is arable, of which 96% area is cultivated and 76% of the cultivated area is under assured irrigation. Paddy rice (monsoon or *kharif* season) and wheat (winter or *rabi* season) are the two main crops, both of which have witnessed a quantum jump in production during the *Green Revolution* initiated in the mid-1960s. During the period from 1966–1967 to 2011–2012, the area under paddy cultivation has increased 6.5 times, to 1.235 million ha, while the area under wheat has increased nearly 4-fold, to 2.531 million ha (Yadav 2017).

DOI: 10.1201/9781003024101-7

The *Green Revolution* has saved India from perpetual hunger and external dependency for food grains, but it has also brought in several unanticipated collateral damages. The most important one is overexploitation of groundwater resources, as irrigation is now predominantly dependent on the wells tapping underlying aquifers. The overexploitation is manifested by fast declining groundwater levels, dwindling well yields and rapidly depleting aquifers, all of them pointing to serious concerns on the water security of the state (Saha et al. 2017). As the groundwater level is declining, energy consumption to lift groundwater is increasing proportionally, putting immense pressure on the carbon footprint over the region. The stressed groundwater resources of the region were initially highlighted by a landmark publication by Rodell et al. in *Nature* (2009), using Gravity Recovery and Climate Experiment Satellite (GRACE) data. Similarly, GRACE-based data have indicated groundwater overexploitation in the Punjab province of Pakistan, representing contiguous Indus Basin aquifers (Iqbal et al. 2016). Groundwater extraction is highly energy-dependent (either electricity or diesel). The staggering 20.5 million wells operational for irrigation in the country (Saha et al. 2017) resulted in an emission of ~45.3 million metric tons (MMT) of carbon into the atmosphere, accounting for 6% of total carbon emissions from India (Rajana et al. 2020). Declining groundwater needs more energy to extract, which is corroborated by the fact that the carbon emission from tube wells has doubled between 2000 and 2013. Haryana has contributed to 2.95 MMT of carbon emission into the atmosphere from its irrigation wells in 2013 (Rajana et al. 2020).

The other concern that stems from intensive agriculture in this region is deteriorating groundwater quality, resulting from fertilizers, pesticides, insecticides, and rodenticides (Kumar et al. 2007; Banerjee et al. 2012). There are also reports of increase in soil and groundwater salinity due to rising water levels in certain canal command areas in the state (Lorenzen et al. 2012).

In the recent past, stubble burning has become the most controversial issue related to agriculture in this region. Although presently restricted to Punjab and Haryana and to some extent in western Uttar Pradesh and adjoining foothill region of Uttarakhand, it is spreading gradually in the other parts of India. This practice in Punjab and Haryana, has been cited as a major contributor to the massive air pollution in Delhi and adjoining urban agglomeration. Although stubble burning is observed for both the main cropping seasons, the intense burning takes place during post-*kharif* harvesting, which corresponds to the period of October–November every year. Farmers from these two states burn more than 23 MMT of crop waste (Thakur 2017; Singh et al. 2019). The generally stated reason is that the burning is a low-cost means of straw disposal, which helps reduce the turnaround time between harvesting of *kharif* paddy and sowing *rabi* wheat. The smoke from the burning of crop residues produces a toxic cloud of particulate matter and gases, which is unable to disperse and envelope the region because of the climatic conditions prevailing during the time in northwest India (NASA 2017). The *kharif* paddy stubble burning contributes 12-22% of poor air quality in Delhi (Government of India 2019), putting an immense health impact. There is an estimated 16,000 premature deaths every year in the National Capital Region (NCR), with an aggregate reduction of life expectancy of 6 years (Singh et al. 2019). The *National Green Tribunal*, the watchdog of environment in India, took the crop burning and related pollution issues with utmost seriousness and even imposed a hefty fine on the administration of NCR for failing to file an action plan to tackle the air pollution, by providing incentives and infrastructural assistance to farmers (Hindustan Times 2018). The *Supreme Court of India* also lashed out at the state governments of Punjab, Haryana, and Uttar Pradesh on November 6,

2019 for not taking enough measures to curb stubble burning (OutlookIndia.com 2019). The Court has also asked the states to spread awareness and extend monetarily reward to farmers to refrain from stubble burning. The Punjab and Haryana governments have adopted a slew of measures including an incentive of INR 2,500/acre for small and marginal farmers for not burning crop residues (Business Standard 2019). In 2018, a *Central Sector Scheme* is launched on "Promotion of agriculture mechanization for in situ management of crop residue in the states of Punjab, Haryana, Uttar Pradesh and NCT Delhi" in an aim to reduce crop burning, where 50%–80% on the cost of machinery is provided as subsidy to the farmers. There are claims that such measures helped in reducing fire counts considerably (Government of India 2019). Recently, Government of Haryana has launched a scheme, where INR 7,000/acre subsidy is offered to the farmers for adopting alternate less water-intensive crops in *kharif* season in preference to paddy (Sarkariyojana.com 2020). The scheme has been initiated in nineteen blocks of the state where the depth of groundwater levels has exceeded 40 m below ground level (bgl).

Besides creating air pollution, the stubble burning impacts soil nutrition and in turn affects the crop yield also. Burning 1 ton of paddy straw accounts for a loss of 5.5 kg of nitrogen, 2.3 kg of phosphorus, 2.5 kg of potassium, and 0.5 kg of sulfur and organic carbon (Sharma and Mishra 2001). The authors also reported that the heat generated kills the microorganism and burrowing creatures such as snails, which as such are also affected because of overdose of fertilizers and pesticides. The burning of crop residues also adversely affects the soil temperature, impacting germination and growth.

7.2 WHY STUBBLE BURNING?

Stubble burning is observed in both post-*kharif* and post-*rabi* harvesting, although the post-*kharif* is practiced widely. Various reasons are being attributed for farmers adopting stubble burning, the most commonly discussed one is that the manual harvesting of the crops is replaced by fast and less labor-intensive mechanized methods like using *Combined Harvester* in 1980. In manual harvesting, crop is cut from its base, while *Combined Harvesters* leave stubble of about 10–14 inches, the removal of which is labor-intensive work. Till a few years ago, migrant laborers were available aplenty and manual harvesting was common particularly for *kharif* paddy. Increased mechanization has dampened the demand for laborers and in turn migrant laborers shifted to other occupations, creating a shortage. For sowing *rabi* wheat, the farmers have different options. The available options are: mulching by cutting the stubble and on-field distribution; bailing and removal from the field; mixing with soil by tilling; or seeding of wheat using zero till seeders which drill wheat seeds without removal of paddy stubble. Despite all these options, the farmers prefer burning the *kharif* paddy stubble to minimize the turnaround time between *kharif* and *rabi* to ensure quick wheat sowing and maintain the yield (Singh et al. 2019). Many farmers are reluctant to use these technologies, citing additional cost which is difficult to bear (Anand 2016).

7.2.1 Stubble burning and groundwater overexploitation

An argument has developed that stubble burning has increased because of promulgation of The *Haryana Preservation of Subsoil Water Act, 2009* (Government of Haryana 2009). This Act aims to reduce huge pre-monsoon pumping for paddy which is considered to be the main reason for extensive groundwater overexploitation in Haryana.

The Central Ground Water Board assesses the usable groundwater resources of the country in collaboration with the state governments, where the extent of extraction with respect to available recharge is expressed as the Stage of Groundwater Development (SGD), expressed as equation 7.1

$$SGD = \text{gross groundwater extraction/Net groundwater availability} \times 100 \quad (7.1)$$

In Haryana state, the increasing overexploitation is reflected by the rising SGD over the years, from 109% in 2004 to 137% in 2017 (Saha et al. 2019). The latest estimate (2017) reveals that 61% of the assessment units of Haryana are overexploited (Figure 7.1) and the major cause is groundwater pumping for irrigation. Eighty-nine percent of the total groundwater extraction of the state is consumed by the agriculture sector - the major consumer of groundwater is *kharif* paddy.

As per the *Haryana Preservation of Subsoil Water Act, 2009*, the farmers are forbidden from sowing *kharif* paddy before May 10 and transplanting before June 10. The earlier practice was to start for nursery by April 20 and transplanting used to initiate by May 28. The monsoon arrives by June 10 and the preceding weeks experience the most severe phase of high atmospheric temperature and evapotranspiration. Thus huge groundwater is pumped during the nursery and the initial part of transplanting period, before the monsoon arrives. The Act envisages delay in the *kharif* paddy sowing and transplanting by about 20 days in order to reduce the pre-monsoon groundwater pumping.

Now, the late transplanting also delays the harvesting for *kharif* paddy (at the end of October) resulting in a very narrow window to prepare their fields for the following *rabi* wheat. There is a strong argument that such a narrow window compels the farmers to use intense mechanized harvesting and burning stubbles as a quick-fix solution.

7.3 POTENTIAL STRATEGIES AND APPROACHES

In the earlier days, paddy straw was considered to be the main fodder for the cattle. Now, with the massive quantity of straw generated in Haryana, the supply is in excess of the demand from the livestock. Now even the farmers are increasingly reluctant to feed paddy straw to their cattle and buffaloes, as they hold the view that this might result in reduced volume and quality of milk produced, owing to the generous use of fertilizers and pesticides in fields. For feeding livestock, farmers are turning to high-quality cattle feed and generous use of green fodder together with small quantity of wheat straw. The farmers are also reluctant to invest further to cut the leftover stubble, using a shredder, as this will involve additional expenditure (INR 2,000–2,500/acre), over and above the amount incurred for mechanized farming. The farmers need a lot of awareness generation and training to adopt new technologies. Various organizations are working alongside the state government in this direction, in an attempt to both increase grain productivity and reduce economic and labor inputs required by the farmers. To get out of this problem, well thought-out strategies need to be adopted, depending upon the cropping practice and calendar, landholding of the farmers, pattern of water use and climatic conditions of the area. The major strategies are outlined in Figure 7.2.

Stubble burning in northwestern India 127

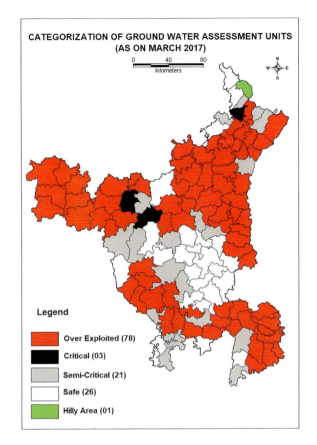

Figure 7.1 Stage of groundwater development as of March 2017 for Haryana state.

Figure 7.2 Sowing of wheat by Happy Seeder, post-Kharif harvesting by combined harvester.

7.3.1 Reduce area under paddy and pursue crop diversification

The farmers in Haryana were historically not traditional rice cultivators. They initiated rice cultivation, largely in response to the national policy to make India self-sufficient in food grains. The Working Group (2013) on *Productivity Enhancement of Crops in Haryana* (Government of Haryana 2013) highlighted the need to replace the water-guzzling paddy by less water-intensive maize, gaur, pulses and oilseeds. However, such attempts did not really take off because of the difference in net farm returns and market risks. Both rice and wheat give assured return through Minimum Support Price and backed by procurement by the Government. Although the Government is taking steps like, incentivizing INR 7,000/acre for shifting from paddy to less water-intensive crops. Such efforts are now restricted to areas with deep groundwater levels (>40 m bgl) spread over nineteen blocks. A rice farmer earns ~INR 57,000/ha, whereas maize in a maize–wheat combination would set them back by about INR 15,000–17,000/ha. An estimate by some experts suggests that a compensation of INR 12,000/ha for such an initiative would be more acceptable, keeping the power savings in mind.

7.3.2 Expanding basmati cultivation during *kharif*

Basmati harvesting is generally done manually, thus hardly any crop residue or stubble is left which completely obviates the need for burning of residues. The data of Government of India (2019) say that only 7%–10% of the area under basmati is burnt. The farmers also feel that the straw from basmati rice serves as better quality fodder. In 2018–2019, of 1.445 million ha under paddy cultivation, 0.795 million ha was under basmati. The state government may further promote organic basmati, which has a growing demand from national and international markets. Basmati also consumes the same water as normal paddy but recent experiments with drip irrigation are showing encouraging results at several test sites like in Karnal district of Haryana.

7.3.3 A relook into The Haryana Preservation of Subsoil Water Act, 2009

Is it possible to relax the Act and allow plantation of paddy before June 10, so that the gap between *kharif* harvesting and sowing of *rabi* widens? A strong argument exists in favor of the Act, as the legislation is conceived to reduce huge pre-monsoon groundwater pumping. The CGWB resource assessment data reveal that the SGD of the state has increased from 109% to 135% between 2005 and 2013, but further between 2013 and 2017 it has increased by only 2%. A more detailed assessment of the groundwater saved on account of implementation of the Act is required to be done. Tripathi et al. (2016) studied the impact of the promulgation of the similar act in the adjoining Punjab state in 2009. Punjab represents a similar hydrogeological condition as the entire region is located in the Indo-Gangetic Plains. Despite significant increase in area under *kharif* paddy in the post-2009 years in Punjab, the average water level decline has reduced from 0.9 m/year between 2000–2001 and 2008–2009 to 0.7 m/year during the 2008–2009 to 2012–2013 period. The elephant in the room; however, is the free/subsidized power for pumping, which is contributing to overexploitation. Some experts suggest that the power subsidy amount can be provided to the farmers as DBT (direct benefit transfer), while they can be asked to bear the actual costs of energy charges. A lot of awareness is to be created among farmers to clear confusion and build confidence in this matter.

7.3.4 Technological innovation

Technological solutions, involving direct seeding (without removing the paddy stubble left by combined harvesters) by machines like *Happy Seeder* are the most talked about. Research by Shyamsundar et al. (2019) reveals that *Happy Seeders* are on an average 10%–20% more profitable than the other practices that involve land preparation and sowing of *rabi* wheat. The higher profits come mainly from enhanced yield and lower input of land preparation. However, it does have limitations - the effectiveness of this technique depends upon the moisture level (not too moist, not too dry) of the soil at the time of seeding. The agronomic practices will need to change particularly with regard to application of fertilizer and irrigation and more innovations are needed in this field. Another issue is the requirement of a large number of machines for use during only 10–15-days a year. The Central Sector Scheme on promotion of agriculture mechanization for *in situ* management of crop residue launched in 2018 tries to address this issue, where provisions for providing subsidy to the individual farmers of cooperative society exists is a welcome move.

7.3.5 Finding other uses of crop residues

One option is to use the paddy straw for power generation. Although this is supported by the Government with some incentives, there remains much to be done. Although not yet practiced significantly in Haryana, in the adjoining Punjab, crop residue is used to produce about 0.5% of the state electricity (Shayansundar et al. 2019). The major hindrance is the large investment required for straw-based power generation. We need to initiate new researches for potential alternate use of paddy straw, such as for manure production or mixing with soil itself or in paper or cardboard and packing industries, etc. If the farmers manage to get a suitable price for the straw they produce, this will serve as an incentive to use the shredder for cutting the stubble.

7.4 STUBBLE BURNING – A VILLAGE VIEW

A survey was conducted in a village called Pindarsi in Thaneashwar administrative block, Kurukshetra district, to understand the issue. The village has 106 agriculture-dependent families, of which 95 families with a landholding of 4–6 acres, whereas six families held more than 30 acres. The village has transformed itself from rainfed cultivation (mainly wheat and gram) to groundwater-based robust agriculture (paddy and wheat). The groundwater-based irrigation was initiated by dug-cum-bore wells during 1970–1975, where dug wells were 5–8 m deep and bore wells at its base penetrated another 7–10 m. The first bore well (of 25 m depth) was drilled in the village in 1975. As the water level started declining with increasing pumping, deeper bore wells were sunk and the capacity of submersible pumps increased simultaneously. Recently constructed bore wells are often 100–120 m deep, and water levels are encountered in the 35–40 m below ground level (bgl). The pumps installed in wells are presently of 15–30 HP capacity and are placed at 40–50 m below ground. To curb extraction of groundwater, during 2008 the Haryana Government has restricted electricity supply for irrigation use for 8 hours a day. Canal-based irrigation was also initiated in 1970–1972 but they are not as effective. Groundwater now caters about 60% of the cultivated area and bore wells are also rampant in the canal command areas. However, the canals benefit mostly the large landed plots, as they are conveniently located in low-lying areas and get the benefit of gravity flow (Table 7.1).

The paddy cultivation started during 1972–1974, once the farmers realized the abundant and round the year availability of groundwater and its easy exploitation with subsidized power. After 1990, all the farmers in the village fully switched over to paddy (which also included basmati variety) during *kharif*. Presently 60% of *kharif* cultivation is under non-basmati rice, while 20% is under basmati and the remaining area is under Jowar/Bajra. The *rabi* crop is predominantly wheat. Burning of hay/husk started in 1998–1999 mostly during post-*kharif*, when manually harvested hay/husk was burnt to reduce the work. Mechanized harvesting by *combined harvesters* arrived in the village in 2002–2004 and by 2008, it became a routine practice. Stubble burning has become a standard practice with the arrival of *combine harvesters,* and is now mostly prevalent for paddy (*kharif*), but also practiced in about 10%–15% for wheat (*rabi*) crop. However, basmati rice grown in *kharif* is harvested manually (80% of total) and hardly any stubble burning is reported from those areas (Figure 7.3).

The villagers conceded that before the enactment of *The Haryana Subsoil Water Preservation Act, 2009*, huge groundwater pumping was undertaken for *kharif* paddy as the nursery used to start around April 25 and transplantation was initiated by May 20. The monsoon arrives around June 15–20, resulting in considerable pumping until the onset of monsoon. The farmers have a tendency to keep the pump switched on throughout the day because of two reasons: electricity is available only for a few hours without any schedule and electricity is highly subsidized. Such a situation led to wasteful pumping of groundwater. Post-enactment of the Act, the paddy nursery starts around May 20–25 and the transplantation by June 15–20, which by and large

Table 7.1 The changing groundwater scenario of the village over the decades between 1970 and 2020

	1970–1980	1980–1990	1990–2000	2000–2010	2010–2020
Number of wells	5–6 (dug well)	30–35 (dug-cum-bore well)	35–40 (dug-cum-bore well)	100 (bore well average depth 70 m)	100–110 (borewell average depth- 120 m)
Irrigation pump capacity	Manual	3 HP electric	5–7.5 HP Electric	Few pumps of 15 HP	15–25 HP
Water level (bgl)	8 m	10–15 m	15–20 m	25–30 m	35–40 m

HP, horsepower.

Figure 7.3 Wasteful pumping observed in Pindarsi Village during field visit.

coincides with the onset of monsoon. An assessment has been made on annual savings of groundwater pumping in recent years as the Act is enforced. The average area under *kharif* paddy is 720 acres, where groundwater pumping contributes 350 mm of irrigation (remaining 400 mm requirement is contributed by monsoon). The data collected on pumping hours of well and their discharge reveal that by deferring the nursery as well as transplanting, about 15% of total pumping hours is saved. Thus annually 52,000 m^3 of groundwater is saved during *kharif* season in the village.

The survey outlined a number of direct and indirect causes for stubble burning and these include financial, behavioral, labor and environment-related issues:

1. The number of cattle at the farm level has reduced, so also there is less demand for fodder. Besides, the farmers felt that the pesticides and fertilizers used in farming made the fodder unhealthy for cattle, resulting in reduced production of milk and sickness. The farmers are thus inclined to burn the crop residues, rather than using the same as fodder.
2. Even after the arrival of the *combined harvester* (during 2001–2002, till 2008), 50% of the *kharif* crop was machine harvested and the remaining 50% manually. In case of *rabi* crops, 80% of wheat was harvested manually until about 2000–2002. Manual harvesting leaves only about 1.5–2 inches of stubble, which is suitable for plowing and incidents of stubble burning were less common.
3. Post-2007–2008, stubble burning was rampant mainly for *kharif* and also for *rabi* crops (about 40% of the cropped area) as *combined harvesters* were started to be used widely.

4. The stubble left by Combined Harvesters can be removed by attaching an additional shredder to the machine, but it cost around INR 4,000–5,000 per acre.
5. Previously all the family members were involved in farm-related work and manual harvesting was more prevalent. With better economic conditions, the farmers shifted to engagement of migrant laborers. Now the migrant laborers are in shortage. Introduction of mechanized farming on a large scale triggered higher number of incidents of crop burning. Sometimes, the farmers resort to burning of stubble after the *rabi* harvesting in order to make their fields "clean". They also believe that burning would kill all the pests and weed seeds and main reason for post-*rabi* harvesting stubble burning.
6. Rarely, the fire in stubble/matured paddy or wheat field can also be caused by short circuits in electric lines. The last of such a major incident happed in this village in 2006, when 30–35 acres of standing wheat crop was burnt.
7. Some progressive farmers are adopting *Happy Seeders* (which can be attached to a tractor), but the general observation is that it would be popular with more subsidy and incentive to farmers.
8. Generally, the farmers are reluctant for crop diversification during *kharif* season, unless there is good procurement price and assured procurement. About 30% of the farmland adopted vegetables and other non-wheat crops during *rabi*.

7.5 WAY FORWARD

The way out from this practice warrants multi-pronged strategies and actions involving all stakeholders at different levels. It is obvious that there are multiple and wide-ranging issues connected with stubble burning in northwestern India, particularly in Haryana. Regulations and incentives are in place to reduce it but stubble burning continues because of uncertainty and lack of political will in implementing the policies and also issues related to access to and return from alternative technologies. It is not proper to put the blame solely on the implementation of *The Haryana Preservation of Subsoil Water Act, 2009* in the state. In fact, the Act has impacted positively on the severely stressed groundwater resources of the state. An assessment made in the studied village reveals that during 2018 *kharif* cultivation, by adhering to the deferred schedule for paddy as mentioned in the Act, the village could save 0.052 million cubic meters groundwater. In the adjoining state of Punjab the research shows slow decline in water level in post-2009 period, after promulgation of similar act. There should be a detailed evaluation on incremental improvement in groundwater resources in the entire Haryana, while at the same time it is important to fine-tune the Act vis-à-vis agronomic practices, if required. There are other social, economic and behavioral issues, which force the farmers towards stubble burning. The technological innovations available to get rid of the stubble left behind by mechanized farming, is required to be adopted by the farmers which will require awareness generation, training and provision of subsidy to farmers. Agro-climate-based crop diversification issues will need to be evaluated and supported by proper Minimum Support Price and assured procurement. Understanding the issues with clarity warrants a two-pronged approach – state-level analyses based on secondary data and in-depth village-level data collection and farmer-based study. Any viable alternative to stubble burning should be at least as profitable and also scalable so that it is adopted widely among 2.5 million farmers practicing rice–wheat farming in northwestern India.

ACKNOWLEDGMENTS

They gratefully acknowledge the support from Water Productivity for Food Security (WAPRO) project in India, funded by the Swiss Agency for Development and Cooperation (SDC) through the Helvetas The authors acknowledge valuable suggestions rendered by Dr. K.R. Viswanathan and the other colleagues for their help during the study. Special thanks are extended to the farmers at Pindersi village and particularly Mr. Joginder Singh for extending all support during the field visit. The authors express their thanks to the reviewer, whose advice has helped to improve the chapter.

REFERENCES

Anand, G. (2016) Farmers' unchecked crop burning fuels India's air pollution, *The New York Times*. [Online]. Available from: https://www.nytimes.com/2016/11/03/world/asia/farmers-unchecked-crop-burning-fuels-indias-air-pollution.html [Accessed 9th November 2017].

Banerjee, D.M., Mukherjee, A., Acharyya, S.K., Chatterjee, D., Mahanta, C., Saha, D., Kumar, S., Singh, M., Sarkar, A., Sengupta, S. and Dubey, C.S. (2012) Contemporary groundwater pollution studies in India. *Proceedings of the Indian National Science Academy* [Online] 78(3), 333–342. Available from: https://www.researchgate.net/publication/262419613_Contemporary_Groundwater_Pollution_Studies_in_India_A_Review [Accessed 28th April 2021].

Business Standard (2019) Punjab, Haryana announce Rs 2,500 an acre incentive to stop stubble burning. [Online] Available from: https://www.business-standard.com/article/economy-policy/punjab-haryana-announce-rs-2-500-an-acre-incentive-to-stop-stubble-burning 119111500057_1.html [Accessed 28th April 2021].

Government of India (2019) *National Clean Air Programme (NCAP)*. Ministry of Environment, Forest and Climate Change, Government of India. Available from: https://moef.gov.in/wp-content/uploads/2019/05/NCAP_Report.pdf [Accessed 28th April 2021].

Government of Haryana (2009) *Haryana Preservation of Subsoil Water Act 2009*. Available from: http://www.ielrc.org/content/e0921.pdf.

Government of Haryana (2013) *Working Group Report on Productivity Enhancement of Crops in Haryana*. Haryana Kishan Aayog. Available from: http://www.haryanakisanayog.org/Reports/WG%20Report%20on%20Productivity%20Enhancement%20of%20Crops%20in%20 Haryana.pdf.

Government of Haryana (2019) Report of the Committee on Review of the scheme "promotion of agricultural mechanisation for in-situ management of crop residue in states of Punjab, Haryana, Uttar Pradesh and NCT of Delhi". Government of India, Ministry of Agriculture and Farmers Welfare, Govt of India. Available from: https://farmech.dac.gov.in/revised/1.1.2019/REPORT%20OF%20THE%20COMMITTEE-FINAL(CORRECTED).pdf.

http://blog.jains.com/JAINIRRIGATIONSYSTEMSLTDsIntegratedAutomationSolution-JAINLOGICwillhelpfarmerstransformthewayAgricultureisdone.htm.

Iqbal, N., Hossain, F., Lee, H. and Akhter, G. (2016) Satellite gravimetric estimation of groundwater storage variations over Indus Basin in Pakistan. *IEEE Journal of Selected Topics in Applied Earth Observations and Remote Sensing* [Online] 9(8). Available from: https://doi.org/10.1109/JSTARS.2016.2574378 [Accessed 28th April 2021].

Kumar, M., Kumari, K. and Ramanathan, A. (2007) A comparative evaluation of groundwater suitability for irrigation and drinking purposes in two intensively cultivated districts of Punjab, India. *EnvironGeol* [Online] 53, 553–574. Available from: https://doi.org/10.1007/s00254-007-0672-3 [Accessed 28th April 2012].

Lorenzen, G., Sprenger, C., Baudron, P., Gupta, D. and Pekdeger, A. (2012) Origin and dynamics of groundwater salinity in the alluvial plains of western Delhi and adjacent territories of Haryana State, India. *Hydrological Processes* [Online] 26(15), 2333–2345. Available from: https://doi.org/10.1002/hyp.8311 [Accessed 28th April 2021].

NASA (2017) Stubble Burning in Northern India, Earth Observatory. https://earthobservatory. nasa.gov/images/82409/stubble-burning-in-northern-india [Accessed 9 November 2017].

OutlookIndia.com (2019) 'Crop stubble burning must stop immediately': SC Tells Punjab, Haryana, UP. *Outlook*. [Online] Available from: https://www.outlookindia.com/website/story/india-news-this-cant-happen-in-a-civilised-country-supreme-court-pulls-up-centre-delhi-govt-over-pollution/341672 [Accessed 28th April 2021].

Rajana, A., Ghosh, K. and Shah, A. (2020) Carbon footprint of India's groundwater irrigation. *Carbon Management* [Online] 11(3). Available from: https://doi.org/10.1080/17583004.2020.1750265 [Accessed 28th April 2021].

Rodell, M., Velicogna, I. and Famiglietti, J.S. (2009) Satellite-based estimates of groundwater depletion in India. *Nature* [Online] 460, 999–1002. Available from: http://dx.doi.org/10.1038/nature08238 [Accessed 28th April 2021].

Saha, D., Marwaha, S. and Dwivedi, S.N. (2019) In Singh, A., Saha, D. and Tyagi, A.C. (Eds.), *National Aquifer Mapping and Management Plan- A Step towards Water Security in India- in Water Governance: Challenges and Prospects*. Available from: http://doi.org/10.1007/978-981-13-2700-1_3.

Saha, D., Marwaha, S. and Mukherjee, A. (2017) Groundwater resources and sustainable management issues in India. In Saha, D., Marawaha, S. and Mukherjee, A. (Eds.) *Clean and Sustainable Groundwater in India*, Springer. [Online] ISBN 978-981-10-4551-6. Available from: https://doi.org/10.1007/978-981-10-4552-3_1 [Accessed 28th April 2021].

Sarkariyojana.com (2020) Haryana Mara Pani Meri Virasat Scheme- Rs. 7000/ Acre Incentive to Farmers to Switch from Paddy. [Online] Available from: http://sarkariyojana.com/haryana-mera-pani-virasat-scheme/ [Accessed 28th April 2021].

Sharma, P.K. and Mishra, B. (2001) Effect of burning rice and wheat crop residues: loss of N, P, K and S from soil and changes in the nutrient availability. *Journal of the Indian Society of Soil Science*, 49(3), 425–429. ISSN: 0019-638X.

Shyamsundar, P., Springer, N.P., Tallis, H., Polasky, S., Jat, M.L., Sidhu, H.S., ... Cummins, J. (2019). Fields on fire: alternatives to crop residue burning in India. *Science* [Online] 365(6453), 536–538. Available from: https://www.researchgate.net/publication/335074478_Fields_on_fire_Alternatives_to_crop_residue_burning_in_India [Accessed 28th April 2021].

Singh, B., McDonald, A.J., Srivastava, A.K. and Gerard, B. (2019). Tradeoffs between groundwater conservation and air pollution from agricultural fires in northwest India. *Nature Sustainability*, 2(7), 580–583.

Thakur, J. (2017) Brace for air pollution in Delhi as crop burning starts in neighbouring states: agricultural stubble running into millions of tonnes is burnt by farmers in northern India every October. An estimated 35 million tonnes are set afire in Punjab and Haryana alone. *Hindustan Times* [Online] Available from: https://www.hindustantimes.com/delhi-news/delhi-s-pollution-nightmare-crop-burning-in-nearby-states-begins/story-djXJY8W0Ugzm-8dgsxmbN0K.html [Accessed 28th April 2021].

Tripathi, A., Mishra, A.K. and Verma, G. (2016) Impact of preservation of subsoil water act on groundwater depletion: the case of Punjab, India. *Environmental Management*, 58(1), 48–59. [Online] Available from: https://pubmed.ncbi.nlm.nih.gov/27015967/ [Accessed 28th April 2021].

Yadav, A. (2017) Rice and wheat cultivation in Haryana: a spatial temporal analysis. *International Journal for Research in Applied Science & Engineering Technology* (IJRASET). ISSN: 2321-9653; 5(12) Available from: https://www.ijraset.com/fileserve.php?FID=12559 [Accessed 28th April 2021].

Chapter 8

Groundwater recharge through landscape restoration and surface water harvesting for climate resilience

The case of upper Tekeze river basin, Northern Ethiopia

K. Woldearegay
Mekelle University

L. Tamene
International Centre for Tropical Agriculture (CIAT)

F. van Steenbergen
MetaMeta Research

K. Mekonnen
International Livestock Research Institute (ILRI)

CONTENTS

8.1	Introduction	136
8.2	Materials and methods	136
	8.2.1 Characteristics of the study area	136
	8.2.2 Research approach	137
8.3	Results and discussion	138
	8.3.1 Properties of soils and rocks in Tigray	138
	8.3.2 Landscape restoration and groundwater recharge in Abreha Weatsbeha Watershed	140
	8.3.3 Landscape restoration and groundwater recharge in the Sero Watershed	143
	8.3.4 Landscape restoration and groundwater recharge in Daero Weyni watershed	146
	8.3.5 Dam construction and groundwater recharge in the May Demu watershed	148
	8.3.6 Emerging dynamics in surface water and groundwater use in Korir Dam	151
8.4	Implications for climate resilience and opportunities for conjunctive use	153
8.5	Conclusions	154
Acknowledgments		155
References		155

DOI: 10.1201/9781003024101-8

8.1 INTRODUCTION

Poor land and water management practices coupled with climate change and low adaptive capacities to weather-related shocks are challenging rainfed agriculture in Sub-Saharan Africa (SSA). IPCC (2008) predicted that by 2020, 75–250 million people may be exposed to increased water stress due to the combined effects of climate change and increased demand. The economic and/or absolute water scarcity in SSA manifests in 315 million people remaining without access to improved drinking water (UNDESA, 2015). The projected combined impacts of climate change and population growth suggest an alarming increase in water scarcity for many African countries, with twenty-two of the twenty-eight countries considered likely to face water scarcity or water stress by 2025 (UNEP, 1999). The strategic importance of groundwater as a resource for global water and food security is likely to intensify under climate change as more frequent and intense climate extremes increase variability in precipitation, soil moisture, and surface water (Taylor et al., 2013). Despite the wealth of regional resources, groundwater use is thought to remain under 5% of sustainable yield for most countries in SSA, meaning that significant renewable resources are currently dormant (Cobbing and Hiller, 2019).

Ethiopia is one of the countries in SSA, which is hugely affected by changing climatic variables. The northern part of the country was for instance the scene of the horrendous famine of 1984. Even recently some parts of Ethiopia were affected by droughts whereby: (a) due to El-Niño about 10 million people were affected in the year 2015 (EMoANR, 2015) and (b) due to low rainfall about 7 million people suffered from droughts in the lowland part of the country (EMoANR, 2016).

In order to enhance soil moisture, reduce land degradation and ensure food security, landscape restoration efforts have been implemented in Ethiopia since the 1970s. Especially in northern Ethiopia, the regional government of Tigray has been implementing not only soil and water conservation measures but also the construction of different water harvesting schemes including dams, ponds, percolation pits, stream/river diversion weirs, and groundwater wells. In recent years, road water management (runoff from culverts, roadside ditches, and bridges) was systematically introduced and upscaled in the region (Woldearegay et al., 2015; van Steenbergen et al., 2018). Although the benefits of the different interventions are multi-dimensional, this chapter highlights the effects of landscape restoration and surface water harvesting (using dams) on groundwater recharge in the Tigray Regional State, Ethiopia, with five representative sites as case studies (Figure 8.1).

8.2 MATERIALS AND METHODS

8.2.1 Characteristics of the study area

The Tigray region hosts about 6 million people and receives highly variable rainfall both spatially and temporally. According to ENMSA (2018), the average annual rainfall in the region dominantly ranges from 500 to 800 mm; with limited parts of the western region reaching up to 1,200 mm. Rainfall variability ranges from 20% in the western part to 40% in the eastern part (mainly close to the western escarpments of the Main Ethiopian Rift System) (ENMSA, 2018). The landform

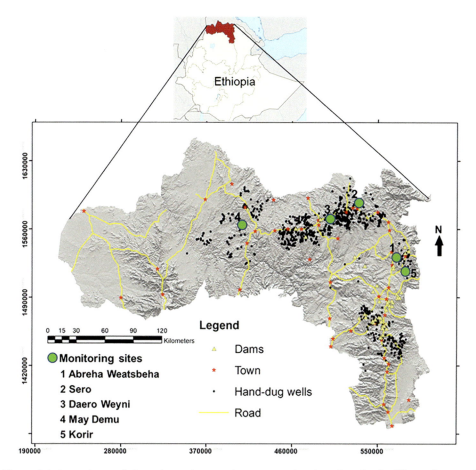

Figure 8.1 Locations of the selected groundwater monitoring sites in the Tigray Regional State, Ethiopia.

is also variable: highlands (in the range of 2,300–3,200 m a.s.l.), moderate relief hills (1,500–2,200 m a.s.l.), lowland plains (with an altitude range of <500–1,500 m a.s.l.), and mountain peaks (as high as 3,935 m a.s.l.). The main rainy season is in the months of June to September. The average annual temperature of the Tigray region varies from 10°C to 25°C (ENMSA, 2018).

8.2.2 Research approach

This research involved both qualitative and semi-quantitative approaches, which include the following:

- *Review of previous studies*: technical reports, published articles, annual government performance reports, and regional/federal governments' strategy documents were collected and reviewed.

138 Groundwater for Sustainable Livelihoods

- *Field survey and participatory evaluation of the various sites*: in order to assess the current status (in comparison with previous conditions), an inventory of different types of landscape restoration technologies implemented along the successful landscape continuum was carried out for the selected watersheds.
- *Groundwater level monitoring*: to evaluate changes in groundwater condition, static groundwater level measurements were carried every month (after the interventions) for the years 2010–2017 and compared with previously recorded groundwater levels in the year 1994/1995 (before the intervention).
- *Groundwater quality evaluations*: in order to assess the changes in groundwater quality as a result of the interventions, groundwater samples were collected and analyzed for their Total Dissolved Solids (TDS) (mg/L) and compared with previous records of groundwater quality before the intervention. In most of the Tigray region, the maximum and minimum groundwater levels are often achieved in October and May, respectively (Woldearegay and van Steenbergen, 2015), and measurement was done in these months. The TDS was determined in the laboratory using the gravimetric method.
- *Evaluating the hydrogeological conditions of the watersheds*: based on the comprehensive assessment of the types and distributions of soils and rocks in the selected sites and evaluations of their hydraulic properties, generalized models which depict the hydrogeological setting of the sites are developed.

Although there are many sites with remarkable positive changes due to landscape restoration effects and dam construction, this research has focused on five sites (Figure 8.1): (a) three sites that are recognized nationally and internationally for their successful landscape restoration (namely, Abreha Weatsbeha, Sero, Daero Weyni watersheds) and (b) two sites on water storage embankment dams. The sites were selected for long-term monitoring for several reasons: (a) there were groundwater data (level and quality) before the implementation of the interventions, (b) the sites (especially Abreha Weatsbeha and Sero) were among the most degraded landscapes with a critical shortage of water and associated food insecurity; a lesson how landscapes could be changed through landscape restorations, (c) different linked/complementary technologies were implemented at scale which resulted in groundwater recharge; an opportunity to systematically integrate watershed management with groundwater recharge beyond the current practice, (d) the sites represent the major hydrogeological settings of northern Ethiopia, and (e) several challenges are emerging in northern Ethiopia, which include an increase in water demand and dynamics in water utilization. The two dam sites are typical examples of such cases, which demand options for enhancing the availability of water and optimum utilization of water resources that integrates surface and groundwater.

8.3   RESULTS AND DISCUSSION

8.3.1   Properties of soils and rocks in Tigray

The geology and hydrogeology of northern Ethiopia have been studied by different authors (e.g. Mohr, 1967; Dow et al., 1971; Kazmin, 1972, 1975; Mohr and Zanettin, 1988; EGS, 2002) and the major rock types (from the oldest to the youngest) include

Enhancing recharge in Northern Ethiopia   139

Metamorphic rocks, Glacial Tillite, Adigrat Sandstone, Antalo Limestone, Agula Shale, Ambaradom Sandstone, Intrusive rocks, and Trap volcanic rock. Different types of soils which include residual, alluvial, and colluvial deposits also occur in the study areas.

Woldearegay and van Steenbergen (2015) assessed the hydrogeology and shallow groundwater irrigation practices in Tigray and reported that the region is dominated by rocks and soils with variable hydraulic properties whereby: (a) the cliffs/steep slopes are mostly rocks which act as recharge areas, (b) the intermediate slopes are dominated by soils/weathered rock of variable hydraulic properties which dominantly range from medium to high, and (c) the flat valley floors are mostly dominated by soils and depending on their hydraulic behaviors these areas act as groundwater storages; a source of shallow groundwater.

A summary of the general hydraulic characteristics of the rocks/soils in the Tigray region is summarized in Table 8.1.

Table 8.1 Summary of the characteristics of the major rocks and soils in Tigray, Northern Ethiopia

No	Rock/soil type	General characteristics
I	Metamorphic rocks	These rocks include metasediments and metavolcanics. The metasediments have low permeability (unweathered) and moderate permeability (fractured/weathered), while the metavolcanics have relatively higher permeability due to fracturing at show depths (mostly not more than 40 m).
2	Palaeozoic sediments	The lower unit (Enticho Sandstone) has moderate to high permeability when fractured and is categorized as a rock with good aquifer characteristics when weathered. The upper unit (siltstone/mudstone) is represented by low permeability, except when weathered to shallow depth (mostly less than 40 m).
3	Adigrat Sandstone	These rocks are represented by high permeability because of fracturing. The geomorphological expression of these rocks (which is often ridge forming) dictates their suitability as recharge zones to the regional groundwater system.
4	Antalo Limestone	These rocks include hard limestone and shale intercalations. The hard limestones are mostly well jointed and are represented by high permeability but the shales have low permeability. As a unit, the Antalo Limestone is classified as a rock with moderate permeability.
5	Agula Shale	These rocks are dominated by shale with some limestone intercalations. The shales are represented by low permeability. Due to the fractures in the limestone component, these rocks contain shallow groundwater providing mostly low yield.
6	Ambaradom Sandstone	It is characterized by a high degree of fracturing, resulting in medium to high permeability. But due to its ridge forming morphological expression, it acts as a recharge to the regional groundwater system.

(Continued)

Table 8.1(Continued) Summary of the characteristics of the major rocks and soils in Tigray, Northern Ethiopia

No	Rock/soil type	General characteristics
7	Intrusive rocks	These rocks have a high degree of fracturing at shallow depths (often not exceeding 40 m). With an increase in depth, these rocks tend to reduce in their degree of fracturing and hence their permeability.
8	Trap volcanics	These volcanic rocks have medium to high permeability when fractured: if ridge forming they act as recharge, but when they occur in flat areas they act as potentials for shallow groundwater development (mostly with depths less than 40 m).
9	Unconsolidated deposits (alluvial and residual)	Alluvial soils have moderate to high permeability: their permeability and productivity vary from place to place depending on grain size, sorting, and thickness; mostly have moderate to high permeability. Residual soils: have variable permeability; clay/silt-dominated ones have poor aquifer characteristics, but coarse-grained ones (sand and gravel) have good aquifer characteristics.

Source: Adopted from Chernet (1993) and EGS (2002).

8.3.2 Landscape restoration and groundwater recharge in Abreha Weatsbeha Watershed

Drilling of boreholes for water supply was carried out in Abreha Weatsbeha in the year 1995 to a depth of 50 m and the well dried during pump-test (Woldearegay, 1995). The well was monitored in the year 1995 in order to evaluate the dynamics of groundwater level and quality. Over 90% of the people in the Abreha Weatsbeha area were under the productive safety net program until about 2005 (TBoARD, 2006). To reverse these challenges, integrated landscape restoration (e.g. Plate 8.1) was carried out over the

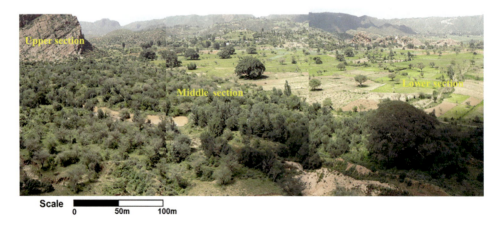

Plate 8.1 Panoramic view of the landscape restoration efforts in Abreha Weatsbeha watershed, Tigray, Ethiopia. (Photo: Kifle Woldearegay.)

years, mainly since 2005, and the area has become one of the most successful sites in terms of landscape restoration in the world (e.g. Tuinhof et al., 2012).

Different landscape restoration techniques have been implemented along the landscape that contributed to groundwater recharge (Figure 8.2; Plate 8.2):

- The upper section of the landscape, which is dominated by fractured and moderately to highly weathered sandstone (with higher permeability) is treated with deep trenches, percolation pits, afforestation, and area closures.
- The middle section, which is dominated by weathered sandstone and debris/colluvial materials with relatively higher permeability is treated with a series of percolation ponds (Plate 8.2a), check-dams (Plate 8.2b), deep trenches with bunds (Plate 8.2c), and afforestation. The interventions at the upper and middle sections of the landscape represent recharge zones to the downstream areas.

The lower section of the landscape, which is dominated by soils (silt and sand) with thickness up to 8 m and weathered sandstone (thickness up to 30 m) acts as a groundwater storage system. A total number of 350 shallow hand-dug wells with depths not exceeding 15 m (e.g. Plate 8.2d) are developed and used for small-scale irrigation, drinking, and livestock watering.

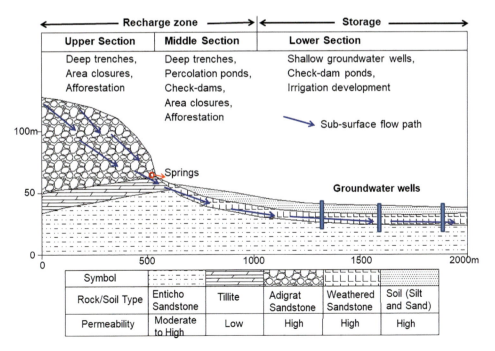

Figure 8.2 Generalized hydrogeological model and different landscape restoration and water harvesting techniques implemented along the landscape continuum in Abreha Weatsbeha area, Tigray, Ethiopia.

Plate 8.2 Different landscape restoration and water harvesting techniques implemented in Abreha Weatsbeha, Tigray, Ethiopia: (a) percolation ponds, (b) check-dams (gabion), (c) deep trenches at farmlands, and (d) shallow groundwater wells (hand-dug) at lower sections of the landscapes.

In the year 1995, before the intervention, the groundwater in Abreha Weatsbeha was having a low yield (0.5 L/s during pumping test but dried later) with a static water level not less than 20 m below the surface. As a result of the landscape restoration, the groundwater level rose to a depth close to the surface (Figure 8.3a). The Abreha Weatsbeha area has received an average annual rainfall of 375 and 395.6 mm in the years 2013 and 2015, respectively (Figure 8.3b), which is below the average rainfall of 520 mm/year (from 1994 to 2017) (ENMSA, 2018). In these 2 years with low rainfall, the region was affected by the worst drought in 50 years (TBoARD, 2015). However, the decline in groundwater level was not significant as compared to the other years after the intervention; this has ensured the availability of water in the area even during a low rainfall period.

The TDS monitoring results (Figure 8.4) show that, as compared to the groundwater quality before the intervention (in 1995), the TDS of groundwater has reduced after the intervention (2010–2017); this is believed to be due to the enhanced recharge of rainwater that replenishes the groundwater.

The factors that contributed to the effective groundwater recharge in Abreha Weatsbeha could be attributed to the following:

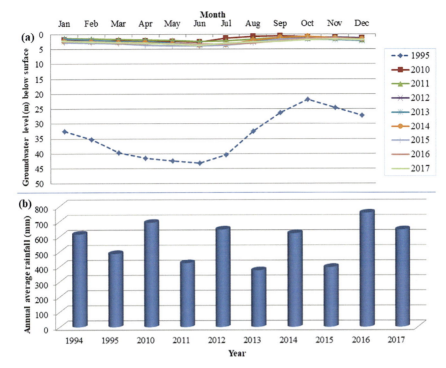

Figure 8.3 Effects of landscape restoration on groundwater level in Abreha Weatsbeha watershed, Tigray, Ethiopia: (a) static groundwater level over the years (1995 and 2010–2017). Note: major landscape restoration was carried in the period of 2005–2009. (b) Variations in annual average rainfall over the years. (Rainfall data: ENMSA (2018); Wukro Station.)

- Implementation of different landscape restoration techniques along the landscape continuum that reduce flooding and enhance groundwater recharge; the upper and middle sections act as recharge zones to the downstream areas.
- Presence of suitable aquifers at a shallow depth for groundwater storage (Figure 8.2): (a) the highly permeable soils and weathered rock (*in situ* weathering of sandstone) at the lower part of the landscape act as groundwater storage, (b) the less permeable rocks that underlie the pervious soil and rock masses act as barriers for deep percolation of groundwater and enhance the lateral flow of water to downslope areas, and (c) long stretch of groundwater storage area (more than 2 km) with fully treated gullies (using check-dams) which act as subsurface barriers to enhance groundwater storage.

8.3.3 Landscape restoration and groundwater recharge in the Sero Watershed

One of the watersheds which was associated with serious land degradation and with a critical shortage of water (even for drinking purposes) was the Sero watershed. Prior

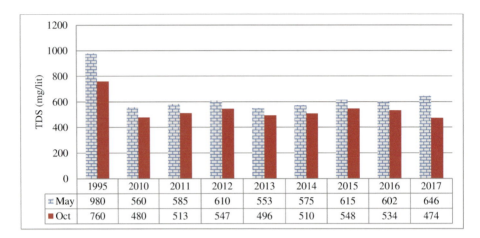

Figure 8.4 Variations in TDS (mg/L) over the years (1995 and 2010–2017) in Abreha Weatsbeha watershed, Tigray, Ethiopia.

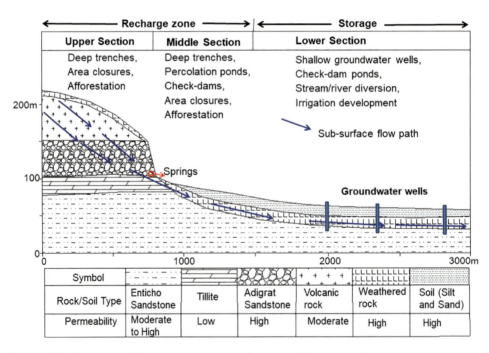

Figure 8.5 Generalized hydrogeological model and different landscape restoration and water harvesting techniques implemented along the landscape continuum in the Sero watershed, Tigray, Ethiopia.

to 2005, over 80% of the local communities were not only water insecure but also food insecure (TBoARD, 2005). To reverse these problems, landscape restorations were carried out along the landscape continuum (Figure 8.5):

- Deep trenches with stone bunds, check-dams, and percolation pits coupled with afforestation at the upper sections of the landscapes.
- Percolation ponds, check-dams, sediment storage dams, deep trenches with soil/stone bunds, and afforestation at the middle sections of the landscapes.
- Water harvesting (using shallow groundwater wells, check-dam ponds, spring development and stream/river diversions) and associated irrigation development in the lower sections of the areas.

Similar to the case of Abreha Weatsbeha, the hydrogeological setting of the Sero watershed has typical characteristics that favor shallow groundwater recharge and storage (Figure 8.5):

- The fractured rocks (basaltic flow and sandstone) that dominate the upper parts of the landscapes have high permeability and enhance infiltration of rainwater into the downstream areas.
- The soils (up to 6 m thickness) and weathered rocks (up to 35 m depth) at the lower sections of the landscapes have high permeability and act as storage for groundwater.
- The less permeable rocks that underlie the permeable soil and rocks act as a barrier for deep percolation of water; enhancing the lateral flow of water to downstream areas.

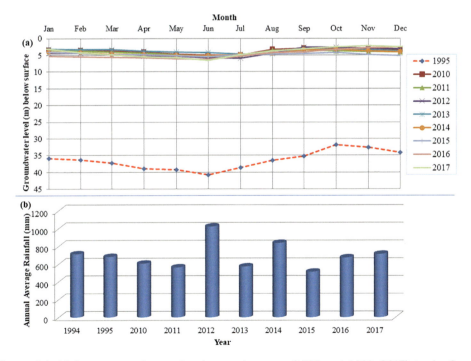

Figure 8.6 (a) Static groundwater level over the years (1995 and 2010–2017) in the Sero watershed, Tigray, Ethiopia. Note: major landscape restoration was carried in the period 2005–2008. (b) Variations in annual average rainfall over the years in the Sero watershed, Tigray, Ethiopia. (Data: ENMSA (2018); Enticho Station.)

146 Groundwater for Sustainable Livelihoods

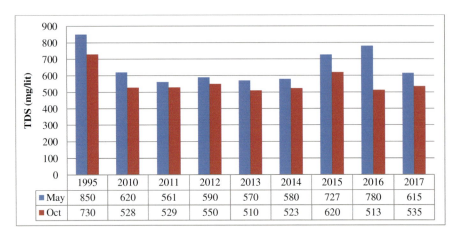

Figure 8.7 Variations in groundwater quality mainly Total Dissolved Solids (TDS) (mg/L) over the years (1995 and 2010–2017) in the Sero watershed, Tigray, Ethiopia.

As can be noted from Figure 8.6, despite rainfall variability over the years, the groundwater level rose to a depth close to the surface (up to a depth of about 2 m below surface) after the intervention (2010–2017) as compared to the year 1995 (before intervention). No major decline in groundwater level was recorded even during the worst drought period of 2015; indicating the effects of integrated landscape restoration in creating climate-resilient watersheds.

In relation to groundwater quality, due to replenishment from rainwater, the TDS was low when compared to the quality before the intervention but remained nearly of similar quality after the intervention (Figure 8.7).

8.3.4 Landscape restoration and groundwater recharge in Daero Weyni watershed

Flooding, gully erosion, and associated land management problems have been challenging the Daero Weyni watershed (Mariam Shweito area). Gullies with depths up to 20 m and width up to 50 m have acted as drainage systems whereby they have drained the whole valley floor resulting in a shortage of water.

Similar to the Abreha Weatsbeha and Sero watersheds, different landscape restoration techniques have been implemented along the landscape continuum (Figure 8.8; Plate 8.3). Gullies were treated with different techniques including physical and biological measures. As a result of the landscape restoration program, which was implemented after 2008, the valley was changed from moisture depleted to the moisture-enriched zone. The weathered basement rocks (depths up to 30 m) at the lower sections of the landscapes act as water storage systems and are hydraulically connected to the recharge zones (fractured and weathered rocks at the upper and middle sections of the landscapes). Shallow groundwater wells are developed for small-scale irrigation, drinking, and livestock watering in the catchment.

Enhancing recharge in Northern Ethiopia 147

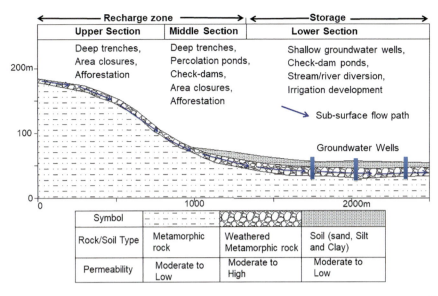

Figure 8.8 Generalized hydrogeological model and different landscape restoration efforts implemented along the landscape continuum in the Daero Weyni (Mariam Shewito) area, Tigray, Ethiopia.

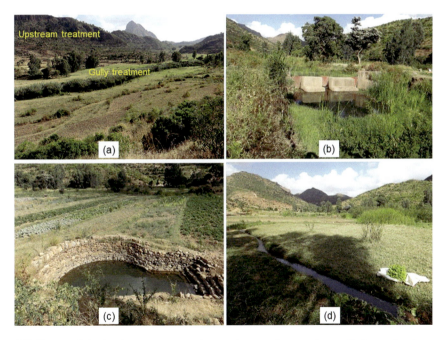

Plate 8.3 Some of the landscape restoration and water harvesting techniques implemented in the Daero Weyni watershed: (a) view of the watershed, (b) check-dams and biological measures along streams, (c) shallow groundwater wells used for irrigation, and (d) stream diversion for irrigation purpose. Note: the photograph was taken in May 2016 (during the worst drought period in the Tigray region).

8.3.5 Dam construction and groundwater recharge in the May Demu watershed

Since 1994, the regional government of Tigray has been implementing various water harvesting techniques including constructions of embankment dams of various sizes for small-scale irrigation, drinking, and livestock watering. Until the year 2017, more than 150 small to medium size dams have been constructed in Tigray (TBoWR, 2018). Various authors (e.g. Tamene et al., 2006, 2011; Berhane et al., 2016) have reported that siltation and leakages are some of the problems associated with the dams. Haregeweyn et al. (2006) studied the performances of fifty-four micro-dam reservoirs (MDRs) constructed in sedimentary terrains and reported that the dams are affected by different problems: 60% by serious leakages, 61% by insufficient inflow, and 35% by sedimentation problems. On the other hand, some authors (e.g. Woldearegay et al., 2006) reported on the benefits of dam construction for groundwater recharge and hence the opportunities for conjunctive use of surface water and groundwater.

One of the MDRs that is completed in 2013 is the May Demu dam, which is designed for both small-scale irrigation and drinking water source. To reduce siltation problems of the dam, different physical structures (deep trenches, gabion check-dams, and percolation pits), as well as biological treatments (afforestation), have been implemented in the whole catchment of the dam (Plate 8.4). To assess the effects of dam construction on groundwater recharge, monitoring was carried out at the downstream of May Demu dam.

Plate 8.4 View of May Demu Dam, the upstream landscape restoration, the command area (irrigated area), and shallow groundwater monitoring well, Tigray, Ethiopia.

A generalized hydrogeological model is presented in Figure 8.9 and different interventions have been implemented along the landscape:

- Deep trenches, afforestation, and area closures at the upper part: this was to reduce sediment transport into the reservoir and to enhance groundwater recharge.
- Construction of water storage dam with the primary purpose of surface water storage for small-scale irrigation and water supply for drinking. The weathered basement rock is characterized by high permeability and the thickness varies from about 30 to 40 m as confirmed during site investigation for the dam. This highly permeable-weathered zone downstream of the dam is hydraulically connected to the reservoir water as a positive cutoff was not provided for the dam. Seepage water from this reservoir is believed to be contributing to groundwater recharge.

To assess the effects of dam construction and upstream landscape restoration on groundwater recharge in the area, the groundwater level before the intervention was compared with the level after the intervention. Upstream landscape restoration coupled with seepage water from the dam has raised the static water level of shallow wells (downstream of the dam) from a depth of 28 m in 2012 to a depth of about 4 m after the reservoir filing of the dam in 2014 and maintained nearly the same level until 2017 (Figure 8.10).

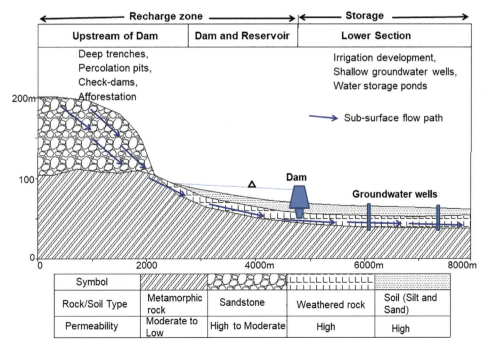

Figure 8.9 Generalized hydrogeological model and different landscape restoration and surface water harvesting techniques (including dams) implemented along the landscape continuum in May Demu watershed, Tigray, Ethiopia.

150 Groundwater for Sustainable Livelihoods

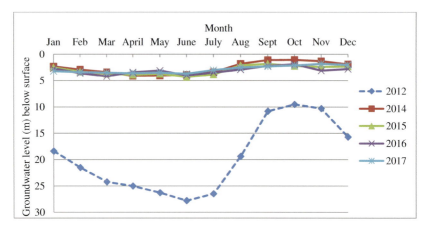

Figure 8.10 Static groundwater level over the years (2012 and 2014–2017) at the downstream position of May Demu dam, Tigray, Ethiopia. The dam was constructed in the period 2012–2013.

Other issues noticed at the downstream position of the May Demu dam are waterlogging (leading to loss of cultivable land) and salinity problems (Plate 8.5). On the other hand, because of the limited reservoir capacity of the dam, the land which is irrigated is less than the potentially irrigable land at the downstream side of the dam. Moreover, the dam is also designed to be a source of water for the nearby towns. Through conjunctive use of surface water and shallow groundwater, it is possible to convert the challenges/problems into opportunities: shallow groundwater development at the downstream of the dam can reduce waterlogging problems and provide additional water for irrigation purposes. This approach could be applied not only to May Demu but also to other dam sites with similar conditions.

Plate 8.5 Emerging issues and opportunities at downstream of May Demu dam, Tigray, Ethiopia: water logging problems due to seepage (a) and shallow hand-dug well at downstream of May Demu dam (b).

8.3.6 Emerging dynamics in surface water and groundwater use in Korir Dam

In recent years, there has been a rapid increase in water demand in northern Ethiopia. A recent report by TBoWR (2019) indicates the emerging competition for water among sectors and users including rural–urban and irrigation drinking. In order to assess the water use dynamics in micro-dams, this research focused on Korir dam, as an example.

Designed to irrigate about 80 ha of land (using gravity method), the Korir dam (Plate 8.6) was constructed in 1998 (CoSAERT, 2002). In 1995, two boreholes were drilled for rural water supply purposes at the downstream position of Korir dam and were found to be dry (Woldearegay, 1996). After the construction of the dam, the following were observed (CoSAERT, 2002): (a) leakage through the dam foundation creating perennial streamflow downstream (Plate 8.6a) and enhancing groundwater recharge, (b) despite the availability of water, many of the farmers were reluctant to practice small-scale irrigation, and (c) there was poor irrigation water management practice leading to waterlogging and low water productivity. Over the years, however, a number of changes have emerged in terms of water use in the Korir dam.

As can be noted from Figure 8.11, after 2009, the total irrigated land has exceeded the 80 ha, which was originally designed to be irrigated because of the following reasons: (a) use of different sources of water including surface water from the dam (using gravity method), (b) pumping seepage water from the streams at downstream, and (c) using shallow wells at downstream of the dam. The more efficient use of water is driven by the fact that most of the people involved in irrigation are those who leased/rented land from the landowners (Figure 8.12); a new trend in irrigation development in northern Ethiopia. Relatively small land was irrigated in 2014 due to the fact that the annual average rainfall was small; in the same year, Korir dam stored water nearly equal to its dead storage (Woldearegay, 2016). Most of the irrigation in 2014 was using shallow groundwater wells and pumping water from seepage in streams at the downstream side of the dam. Although the area received highly variable rainfall over the years (Figure 8.11), farmers managed to implement off-season irrigation through different water management practices.

Plate 8.6 Panoramic view of Korir dam, its command area and the monitoring site, Tigray, Northern Ethiopia.

152   Groundwater for Sustainable Livelihoods

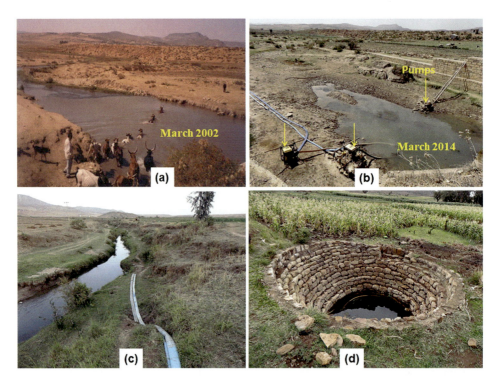

Plate 8.7 Some examples of the dynamics of water management at the downstream of Korir dam, Northern Ethiopia: (a) unutilized seepage water from the dam in 2002, (b) competition for water (three pumps operating at the same time in 2014), (c) pumping water from a stream, and (d) shallow groundwater well (recharged by the dam) for irrigation.

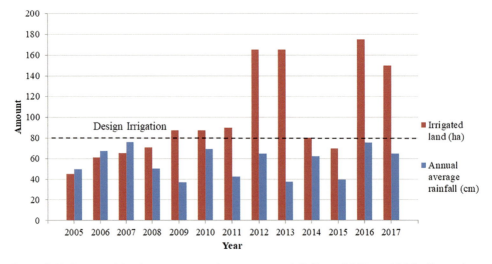

Figure 8.11 Irrigated land versus annual average rainfall (from 2005 to 2017), Korir dam, northern Ethiopia.

8.4 IMPLICATIONS FOR CLIMATE RESILIENCE AND OPPORTUNITIES FOR CONJUNCTIVE USE

Woldearegay and van Steenbergen (2015) reported on the status of shallow groundwater irrigation in Tigray and stressed the benefits of soil and water conservation as well as dam construction on groundwater recharge. Despite the extensive landscape restoration and water harvesting techniques implemented in Tigray, some parts of the region were affected by El-Niño and associated rainfall variability in 2015. According to TBoANR (2016), the major sources of water for irrigation in 2015/2016 were from shallow groundwater wells and from dams. Assessment of the performances of different water harvesting techniques in Tigray during the drought period (Woldearegay, 2016) revealed that due to the low rainfall in the region: (a) many of the embankment dams have recorded the lowest reservoir levels, (b) the majority of stream diversions and springs have dried, and (c) many of the shallow hand-dug wells have reduced their water levels especially those with small recharge systems (storages or catchment size).

The performances of shallow groundwater wells were found to be variable depending on: (a) size of recharge zone and type of hydrogeological settings, (b) type and appropriateness of the implemented landscape restoration along a landscape, and (c) availability of embankment dams at upstream of groundwater wells.

Shallow groundwater wells downstream of the May Demu and Korir embankment dams have not dried during the low rainfall; these wells remained the main sources of water for small-scale irrigation, drinking, and livestock watering during the drought periods. In the catchments where no dams have been constructed but where landscape restoration was implemented, the best performing shallow groundwater wells were found to have fulfilled the following conditions:

- Suitable hydrogeological settings to favor groundwater recharge: (a) presence of large recharge area or surface water storages at the upper sections of the landscape, (b) availability of good storage sites at the downslope areas, and (c) presence of low permeability rock (that act as a barrier) below a pervious zone which retards deep percolation of water and enhances lateral movement that extends for longer distances.

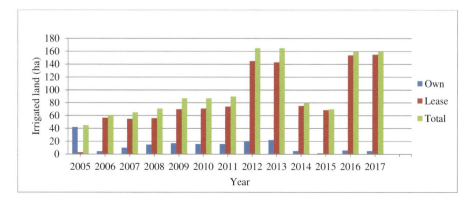

Figure 8.12 Leased versus owned land irrigated (from 2005 to 2017), Korir dam, northern Ethiopia.

154 Groundwater for Sustainable Livelihoods

- Implementation of linked and complementary landscape restoration techniques at the upper and middle sections of the slopes; these include deep trenches, percolation pit/ponds, check-dams, etc.
- No gully erosion and implementation of proper gully treatment/gully plug using physical measures (e.g. check-dams and subsurface dams) combined with biological measures along the valley floors of treated catchments.

8.5 CONCLUSIONS

The following conclusions are drawn from this study:

- Despite rainfall variability and associated droughts, through proper landscape restoration and surface water harvesting, it was possible to facilitate groundwater recharge and enhance existing ones; a lesson for promoting climate resilience.
- Suitable hydrogeological settings (recharge, storage, and hydrological barriers) coupled with the implementation of linked/complementary land restoration and water harvesting technologies along the landscape continuum have contributed to the sustainability and effectiveness of shallow groundwater development in many areas in northern Ethiopia.
- Upstream integrated landscape level of interventions have resulted in an increase in infiltration of rainwater leading not only to raise in shallow groundwater levels, but also to the improvement of groundwater quality (reduction in TDS) due to recharge from rainwater.
- The construction of dams is found to have a positive effect on groundwater recharge. In some cases, however, there is still a challenge of waterlogging which is created at the downstream areas of the dams. On the other hand, there is an emerging increase in water demand. Hence, through conjunctive use of surface water and groundwater, it is possible to promote efficient utilization of water and land resources in the region.
- Overexploitation of groundwater resources coupled with climate change is generally considered to be a major threat to groundwater depletion and quality deterioration. However, the efforts are done and the results achieved related to landscape restoration and water harvesting in the Tigray Regional State, Ethiopia, one of the drought-prone regions of the country, demonstrate that proper land management and water resources development are an integral part of sustainable water resources development.
- Shallow groundwater (depths not more than 30–40 m) is often given less attention in many parts of Ethiopia despite its huge potential for ensuring water security at household levels. These shallow groundwater storages need to be tapped through the implementation of groundwater recharge mechanisms, shallow well development, and the introduction of low-cost water-lifting technologies.
- One of the water management approaches which need to be promoted by linking landscape restoration, water harvesting, and irrigation development is the use of runoff water from roads and from other infrastructures.
- Based on the lessons learned so far (from successful interventions), the major groundwater recharge in northern Ethiopia has been through either landscape

restoration and/or construction of surface water storage systems (embankment dams, check-dams, and ponds). Although these efforts have contributed positively to the groundwater resources, considering the hydrogeology of the region, the availability of excess runoff, and the emerging competition/increase in demand for water, implementation of well-planned groundwater recharge mechanisms could be the next frontier to enhance the availability of water in Tigray region and in other areas with similar hydrogeological settings.

- The major factors that need to be fulfilled for enhanced shallow groundwater recharge are: (a) availability of enough surface runoff (upstream recharge) through either surface water harvesting and/or landscape restoration, (b) presence of downstream storage systems with longer stretch and adequate depth but well connected to the recharging systems, (c) availability of underlying barrier (below the aquifer systems) to reduce deep percolation, and (d) no gully development at downstream areas or provision of proper gully plug/treatment, if there are any, in order to keep the groundwater within the landscape.
- The data presented in this research have focused more on the effects of landscape restoration and/or dam constructions on shallow groundwater recharge. Although, these interventions are expected to enhance the recharge of deep groundwater aquifers, however, this requires further monitoring and evaluations to quantify the impact.

ACKNOWLEDGMENTS

The groundwater monitoring result is part of an on-going long-term shallow groundwater monitoring which is supported by Mekelle University, and Tigray Bureau of Water Resources, Ethiopia. The authors would like to acknowledge for the financial support from (a) The Water, Land and Ecosystems (WLE) program of the CGIAR, (b) Africa RISING which is a program financed by the United States Agency for International Development (USAID) as part of the United States Government's Feed the Future Initiative, and (c) EU-IFAD project under CCAFS. The Ethiopian Geological Surveys (EGS), Relief Society of Tigray (REST), Tigray Bureau of Agriculture and Rural Development are duly acknowledged for their support. The content of this chapter is solely the responsibility of the authors and does not necessarily represent the official views of the financiers and organizations who supported this research. We would like to thank the anonymous two reviewers for their constructive comments and suggestions on the manuscript.

REFERENCES

Berhane, G., Gebreyohannes, T., Martens, K. & Walraevens, K. (2016). Overview of micro-dam reservoirs (MDR) in Tigray (northern Ethiopia): challenges and benefits. *Journal of African Earth Sciences* 123:210–222. doi:10.1016/j.jafrearsci.2016.07.022.

Chernet, T. (1993). *Hydrogeology of Ethiopia and Water Resources Development*. Ethiopian Institute of Geological Surveys, Addis Ababa, p. 222.

Cobbing, J. & Hiller, B. (2019). Waking a sleeping giant: realizing the potential of groundwater in Sub-Saharan Africa. *World Development* 122:597–613.

CoSAERT (Commission for Sustainable Agriculture and Rehabilitation of Tigray) (1996). Geological Investigation for Korir dam, Tigray. A technical Report. p. 30.

CoSAERT (Commission for Sustainable Agriculture and Rehabilitation of Tigray) (2002). Performances of micro-dams in Tigray. A technical Report. p. 65.

Dow, D. B., Beyth, M. & Hailu, T. (1971). Paleozoic glacial rocks recently discovered in Northern Ethiopia. *Geological Magazine* 108(1):53–60.

EGS (Ethiopian Geological Survey) (2002). *Hydrogeological Map of Northern Ethiopia*. Geological Survey, Addis Ababa, Ethiopia.

EMoANR (Ethiopian Ministry of Agriculture and Natural Resources) (2015). Annual report on the performance of the EMoANR for the 2014/2015 fiscal year.

EMoANR (Ethiopian Ministry of Agriculture and Natural Resources) (2016). Annual report on the performance of the EMoANR for the 2015/2016 fiscal year.

ENMSA (Ethiopian National Meteorology Service Agency) (2018). Daily rainfall data for Tigray region, Ethiopia.

Haregeweyn, N., Poesen, J., Nyssen, J., De Wit, J. Mitiku Haile, Govers, G. & Deckers, S. (2006). Reservoirs in Tigray (Northern Ethiopia): characteristics and sediment deposition problems. *Land Degradation & Development* 17:211–230.

IPCC (Intergovernmental Panel on Climate Change) (2007). Climate change and water. Technical Paper IV. p. 214.

Kazmin, V. 1972. *Geological Map of Ethiopia, 1:2,000,000 Scale*. Ethiopian Institute of Geological Survey, Addis Ababa.

Kazmin, V. 1975. Explanatory note to the geology of Ethiopia. Ethiopian Institute of Geological Survey, Bulletin no. 2, Addis Ababa.

Lobell, D. B., Burke M. B, Tebaldi C., Mastrandrea M. D., Falcon, W. P. & Naylor, R. L. (2008). Prioritizing climate change adaptation. *Science* 319:607. doi:10.1126/science.1152339.

Mohr, P. & Zanettin, B. (1988). The Ethiopian flood basalt province. In McDougall, J.D. (Ed.), *Continental Flood Basalts*. Kluwer Acad. Publ., Dordrecht, pp. 63–110.

Mohr, P.A. (1967). Review of the geology of the siemen mountains. *Bulletin of Geophysical Observatory, Addis Ababa University, Ethiopia* 10:79–93.

Tamene, L., Assefa Abegaz, A., Aynekulu, E., Woldearegay, K. & Vlek, P.L.G. (2011). Estimating sediment yield risk of reservoirs in northern Ethiopia using expert knowledge and semi-quantitative approaches. *Lakes & Reservoirs: Research and Management* 16:293–305.

Tamene, L., Park, S.J., Dikau, R. & Vlek, P.L.G. (2006). Reservoir siltation in the semi-arid highlands of northern Ethiopia: sediment yield-catchment area relationship and a semi-quantitative approach for predicting sediment yield. *Earth Surface Processes and Landforms* 31:1364–1383. doi:10.1002/esp.1338.

Taylor, R., Scanlon, B., Döll, P. et al. (2013). Ground water and climate change. *Nature Climate Change* 3:322–329. doi:10.1038/nclimate1744.

TBoARD (Tigray Bureau of Agriculture and Rural Development) (2005). Annual performance report of the TBoARD for the year 2004/2005, Ethiopia.

TBoARD (Tigray Bureau of Agriculture and Rural Development) (2006). Annual performance report of the TBoARD for the year 2005/2006, Ethiopia.

TBoARD (Tigray Bureau of Agriculture and Rural Development) (2015). Annual performance report of the TBoARD for the year 2014/2015, Ethiopia.

TBoWR (Tigray Bureau of Water Resources) (2018). Performance report of the Tigray bureau of water resources for the year 2018.

TBoWR (Tigray Bureau of Water Resources) (2019). A report: water resources development challenges and implementation plans for 2020.

Tuinhof A., Van Steenbergen F., Vos P. & Tolk, L. (2012). Soil and water conservation at scale, Tigray, Ethiopia. Chapter of book Profit from storage: the costs and benefits of water buffering (3R). pp. 86–90.

UNDESA (United Nations Department of Economic and Social Affairs) (2015). 2005–2015 Decade of water for life.

UNEP (United Nations Environment Programme) (1999). *Global Environment Outlook 2000.* Earthscan, London.

Van Steenbergen, F., Woldearegay, K, Agujetas Perez, M., Manjur, K. & Abdullah Al-Abyadh, M, (2018). Roads: instruments for rainwater harvesting, food security and climate resilience in arid and semi-arid areas. In Leal Filho, W. and de Trincheria Gomez, J. (Eds.), *Rainwater-Smart Agriculture in Arid and Semi-Arid Areas.* pp. 121–144. Springer International Publishing AG 2018. doi:10.1007/978-3-319-66239-8_7.

Woldearegay, K. (1995). Groundwater exploration in Mekelle basin, Northern Ethiopia. A technical report. p. 65.

Woldearegay, K. (1996). Technical Report on groundwater resources potential for Genfel watershed. Ethiopian Geological Survey Report. p. 35.

Woldearegay, K. (2011). Effects of dam construction on groundwater recharge in Tigray, Ethiopia: the case of Korir and May Demu dams. *A Paper Presented at a National Conference on Climate Adaptation.* February 16–20, 2011. Addis Ababa, Ethiopia.

Woldearegay, K. (2013). Promoting small-scale irrigation using shallow groundwater wells in Tigray Ethiopia: the case of Abreha Weatsbeha area. *Proceedings of the MU Research Review,* Mekelle, Ethiopia.

Woldearegay, K. (2014). Hydrological benefits of SWC in Abreha Weatsbeha, Sero and Belesa watersheds in Tigray, Ethiopia. *A Paper Presented at the International Conference on Food Security,* September 28–30, 2014. Mekelle, Ethiopia.

Woldearegay, K. (2016). Assessment on the performance of water harvesting technologies in Tigray. A technical report. p. 52.

Woldearegay, K., Behailu, M. & Tamene, L. (2006). Conjunctive use of surface and groundwater: a strategic option for water security in the northern highlands of Ethiopia. *Proceedings of the Highland 2006 Symposium,* Mekelle University, Ethiopia, September 08–25, 2006.

Woldearegay, K., Tamene, L., Mekonnen, K., Kizito, F. & Bossio, D. (2018). Fostering food security and climate resilience through integrated landscape restoration practices and rainwater harvesting/management in arid and semi-arid areas of Ethiopia. In Leal Filho, W. and de Trincheria Gomez, J. (Eds.) *Rainwater-Smart Agriculture for Food Security, Poverty Alleviation and Climate Resilience in Arid and Semi-Arid Regions, Rainwater-Smart Agriculture in Arid and Semi-Arid Areas.* Springer. pp. 37–57. doi:10.1007/978-3-319-66239-8_3.

Woldearegay, K. & van Steenbergen, F. (2015). Shallow groundwater irrigation in Tigray, Northern Ethiopia: practices and issues. In Lollino, G. et al. (Eds.), *Engineering Geology for Society and Territory* – Volume 3, pp. 505–509. doi:10.1007/978-3-319-09054-2_103. © Springer International Publishing Switzerland.

Woldearegay, K., van Steenbergen, F., Agujetas Perez, M., Grum, B. & van Beusekom, M. (2015). Water harvesting from roads: climate resilience in Tigray, Ethiopia. *Proceedings of IRF Europe & Central Asia Regional Congress September,* 15–18, 2015 – Istanbul Turkey.

Chapter 9

The Quaternary aquifer

An affordable resource to address water scarcity in the northern part of the Lake Chad basin

B. Collignon
Urbaconsulting

C. Estienne
Hydroconseil

C. Masse
Urbaconsulting

I.A. Nassour
Hydroconseil

CONTENTS

9.1　Introduction... 160
　　9.1.1　A working area of 300,000 km^2... 160
　　9.1.2　Surface water resources.. 160
　　9.1.3　A booming water demand 160
　　9.1.4　The challenge of exploiting natural resources 160
　　9.1.5　The security issue 161
9.2　The development of groundwater resources 162
　　9.2.1　The Quaternary aquifer... 162
　　9.2.2　Well productivity 163
　　9.2.3　The various options for groundwater intake and abstraction 165
　　9.2.4　The water depth... 166
　　9.2.5　The cost of water 166
9.3　Modelling aquifer water balance and long-term sustainability...................... 168
　　9.3.1　Purpose of the model... 168
　　9.3.2　Building the model... 169
　　9.3.3　Calibrating the model.. 170
　　9.3.4　Groundwater flow patterns.. 170
　　9.3.5　The water balance.. 171
　　9.3.6　Simulation of the most likely developments 171
9.4　The water quality issue 171
　　9.4.1　Salty waters... 171
　　9.4.2　Natron 173
　　9.4.3　Nitrates.. 173

DOI: 10.1201/9781003024101-9

160 Groundwater for Sustainable Livelihoods

9.5 Management of the water service 174
    9.5.1   Accessing the water resource is not the main issue.............................. 174
    9.5.2   The cost of water abstracted from the Quaternary aquifer 174
    9.5.3   Households' ability to pay 175
    9.5.4   Decentralised management of water distribution systems 175
9.6 Summary 175
References.. 177

9.1 INTRODUCTION

9.1.1 A working area of 300,000 km^2

The Lake Chad watershed covers 2.4 million km^2 in central Africa (Figure 9.3). This river basin has been endorheic since at least the Miocene, and a thick sandy-clay sedimentary complex has accumulated. These sediments contain several aquifers of regional importance. The most extensive and most used is the Quaternary aquifer.

Our study area corresponds to the Chadian part of this aquifer, between isohyets of 50 and 300 mm/year, which corresponds to the most arid half of the Sahel and the southern fringe of the Sahara (Figure 9.3).

The area covers 300,000 km^2 that, despite its aridity, is home to 2.5 million people and has a population growth rate of over 2%/year. In such an arid region, rainfed agriculture is highly unpredictable and thus is limited to flood-recession crops grown in clay-covered depressions located between ancient dunes. However, it is a magnificent pastoral breeding region, and there are estimated to be over 30 million heads of ruminant livestock (cattle, goats, sheep and camels). This is the main source of income for most households.

9.1.2 Surface water resources

Surface water resources are limited to Lake Chad and the large rivers that feed it (Chari and Bahr Erguig) at the southwestern edges of the study area. This is the only area where it has been possible to develop irrigated agriculture. Everywhere else, the only water resource available to rural populations and livestock is the Quaternary aquifer.

9.1.3 A booming water demand

Demand for water is booming as a consequence of both rapid population growth and the rise in sedentary lifestyles that are increasing the per capita demand for drinking water (Figure 9.1).

9.1.4 The challenge of exploiting natural resources

The Chadian Sahel is occupied by many communities belonging to different ethnic groups (Arabs, Goranes, Fulani, etc.). They share natural resources (water, land, and pastures) in accordance with long-established rules of use that reflect both the

Figure 9.1 Typical urban growth in Batha Province.

demographic weight of each group and their economic activities (transhumant or non-transhumant livestock farming, flood-recession agriculture, handicrafts, etc.).

The use of pastures is highly dependent on access to water points. This is why the process of constructing new wells and determining their management model must always involve in-depth negotiations with the communities likely to be affected in order to limit the risk of subsequent conflicts. The hydrogeologist cannot work alone. It is imperative that he seeks the expertise of sociologists and agro-economists who are familiar with the local social and economic context.

9.1.5 The security issue

The insurgency led by rebel groups commonly referred to as "Boko Haram" is having a major impact on the Lake Chad region. This is causing great insecurity in areas within a 100 km radius of Lake Chad and along the border with Niger. It also has consequences for the implementation of water supply programmes:

- In order to flee the insecurity in the region, some of the people living in the island areas of the lake have moved further north where new wells are needed;
- Most livestock is taken to market "on foot," i.e. after having walked 500–1,000 km to remote cattle markets in Cameroon and Nigeria; the herders have been forced to change their trade routes, which also means additional water points are required;
- In order to properly maintain the pumps, fuel and spare parts supply chains are required; these supply chains have now been weakened.

The current situation in the Lake Chad region thus creates a vicious circle: the degradation of essential public services (such as water supply) encourages young people to join the ranks of Boko Haram, and the growing influence of Boko Haram weakens the organisation of public services.

Although the insecurity in the region makes it more difficult (and more costly) to implement projects, it is nonetheless essential to continue to invest in additional water facilities (boreholes, water towers, and distribution networks).

9.2 THE DEVELOPMENT OF GROUNDWATER RESOURCES

9.2.1 The Quaternary aquifer

The upper part of the Lake Chad Basin (LCB) sedimentary series consists of 50–100 m of Quaternary-dated permeable sands (Figure 9.2). They contain a very large aquifer that is the main water resource for 5 million people in Chad, Niger and Nigeria and throughout the entire study area.

The Quaternary aquifer is an abundant and easily accessible water resource, and for this reason, it has been exploited for thousands of years and is indispensable for all pastoral life.
It is a transboundary aquifer, extending over an area of 1.5 million km² across Chad, Niger and Nigeria. It is superimposed on an aquiclude (Pliocene clays), which outcrops on the margins of Quaternary outcrops (Figure 9.3).

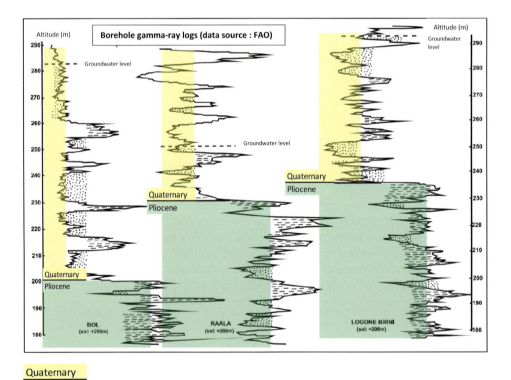

Figure 9.2 Gamma-ray borehole log highlighting the sandy Quaternary aquifer. (From Schneider, Relations entre le Lac Tchad et la nappe phréatique (1969) and (Hamit, 2012) (modified).)

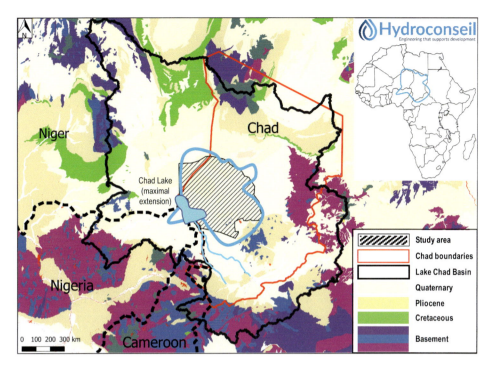

Figure 9.3 Quaternary aquifer extension and the study area.

9.2.2 Well productivity

The Quaternary aquifer is highly productive (Figure 9.4). For a production target of 100 m^3/day (enough to supply a settlement of 5,000 inhabitants), the success rate of 60-m-deep drilling exceeds 70% and for a production target of 5 m^3/day (enough to supply a hand pump), the success rate of 60-m-deep drilling exceeds 85% (it is harder to miss a borehole than it is to succeed). Drawdowns are modest, which reduces operating costs.

Due to its ease of access, this aquifer has long been exploited by all population groups in the region (Fulanis, Arabs, Kanembous, Goranes, Baguirmis, etc.). They have dug thousands of traditional wells, which have been used to supply water to both villages and herds.

The two main limitations of this traditional method of exploitation are: (a) the gradual filling of the wells with sand, which makes it necessary to carry out exhausting dredging operations; and (b) the limited depth to which the well-diggers can dig (due to the lack of suitable pumping equipment). As a result, these wells often dry up at the end of the dry season (March or April) or even earlier if annual rainfall is lower than normal.

As, for a long period, the nomads managed to build water points on their own, the Chadian government invested relatively little in developing water resources and this aquifer has not been systematically studied by hydrogeologists.

However, over the past 20 years, the national Millennium Development Goal (MDG) strategy has required hundreds of new water points to be installed to meet the

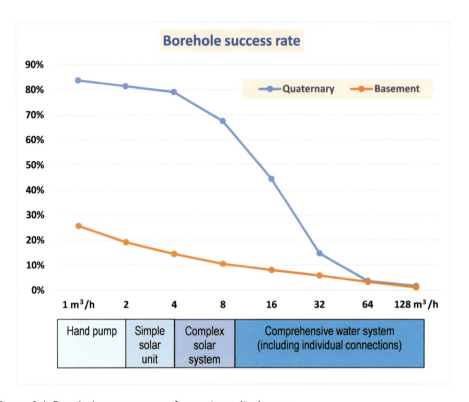

Figure 9.4 Borehole success rate for various discharges.

> The figure illustrates the high borehole success rate in the Quaternary aquifer of the LCB compared to the nearby metamorphic zone (Guera). In the LCB, 85% of the boreholes have a sufficient yield to install a hand pump (compared to only 25% in the Guera) and 60% produce enough water to supply a comprehensive motorised system (compared to only 10% in the Guera).

demand of a rapidly growing population and urbanisation. As a result, more than 400 new boreholes have been built in the last 9 years, completely transforming our understanding of how this aquifer works (Table 9.1).

Table 9.1 Facilities that were completed over the past 9 years to mobilise the Quaternary aquifer in the Chadian Sahel

Programme	Region	Hand pump	Solar pump house	Diesel water network	Solar water network	Serviced population
FED 10 (2012–2018)	Kanem, Bahr el Ghazal, Batha	300	100	20	20	250,000
RESTE (2019–2021)	Lac, Hadjer Lamis	120	40	10	10	100,000
Total		420	140	30	30	350,000

9.2.3 The various options for groundwater intake and abstraction

There is a range of options available for abstracting water from the Quaternary aquifer.

The most suitable options for meeting the water demand of the population and livestock depend on the main water use and the size of the settlement (Figure 9.5 and Table 9.2).

For centuries, the only option available was shallow and deep wells dug by the communities themselves. These works, consolidated with tree branches, have a limited life span if they are not regularly cleaned and deepened. Most of them dry up before the end of the dry season.

- Since 1950, the use of reinforced concrete casings and powerful pumps has made it possible to build more durable and sustainable wells. These works are very appreciated by nomadic pastoralists, but they are very expensive and it is more and more difficult to find competent craftsmen ready to complete these difficult and dangerous works.

Figure 9.5 Various options for water intake and abstraction from the Quaternary aquifer in Chad: (a) dug well, (b) borehole and hand pump, (c) solar pumping station and small iron tank, (d) large size water network with concrete water tower (e). (Photos Hydroconseil)

166 Groundwater for Sustainable Livelihoods

Table 9.2 Various options for water intake and abstraction

	Main purpose	Population serviced	Advantages	Constraints limitations	CAPEX($)
Dug well	Population and livestock	200	Community self-construction	Limited water reserve, limited lifetime	2,000
Concrete well	Population and livestock	500	Long-lasting	Expensive	50,000
Hand Pump	Population	200	Cost-effective	Poor reliability	20,000
Solar pump house	Population	800	Reliable	Difficult to repair	100,000
Solar water network	Population and livestock	2,000	Reliable	Difficult to repair	200,000
Diesel-powered water network	Population and livestock	>2,000	Cost-effective can be managed by local mechanic	Requires a sustainable fuel supply chain	500,000

- Since 1990, mechanical boreholes have made it possible to considerably extend the service coverage in villages. Small-diameter boreholes are not so expensive and can be equipped with hand pumps (suitable for simple repairs at the village level). However, the long-term maintenance of these hand pumps is difficult and these pumps are not suitable for livestock feeding.
- Since 2000, more and more boreholes have been equipped with submersible electric pumps, which can then be used for feeding large settlements or herds. Solar-powered pumps are well suited for medium-sized localities (1,000–3,000 inhabitants) and diesel generators for larger localities and major pastoral water points.

9.2.4 The water depth

In this region, the piezometric surface is often deep (more than 40-m deep in 40% of the wells) and it is even deeper in some specific zones (central Batha, eastern Barh el Ghazal and central Chari Baguirmi). This constitutes a hazard when drilling boreholes with mud/rotary method because in a region where the piezometry is poorly known, there is a risk of stopping the drilling with insufficient water height.

To limit this risk, we have constructed piezometric maps of the working area by remote sensing (Figure 9.6), using a new method, particularly suitable for pastoral regions (Collignon, 2020).

9.2.5 The cost of water

The fact that many settlements are extremely remote increases the cost of drilling boreholes in the Chadian Sahel. It takes 2 days to reach villages in the north of the

Water scarcity in the Lake Chad basin   167

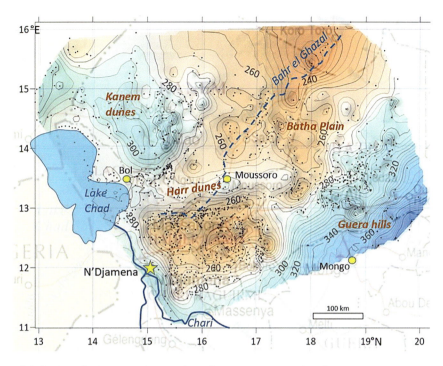

Figure 9.6 Regional piezometric map based on remote sensing data.

study area and vehicles frequently break down due to the sandy conditions. This explains why the average cost of drilling is higher in Chad than in Cameroon or Nigeria. We averaged actual drilling cost as $15,000 euros for a 60-m borehole equipped with 5″ PVC casing. After adding the cost of hand pumps (including installation) and the cost of hydrogeological surveys, a successful well with a hand pump in this region costs between $20,000 and $25,000.

This price is still much lower than in the nearby granite zones, where the drilling success rate is less than 50% for a flow rate of 1 m³/hour (Figure 9.7 and Table 9.4).

Table 9.4 Cost of water abstracted from Quaternary aquifer compared with saprolite aquifer

		Quaternary aquifer	Saprolite aquifer
Depreciation (CAPEX)			
Borehole average cost	$	14,000	19,000
Success rate 10 m³/hour–80 m³/day		55%	10%
Cost for a successful borehole	$	25,455	190,000
Yearly depreciation (CAPEX)	$/y	1,697	12,667
CAPEX	$/m³	0.06	0.43
Running cost (OPEX)			
Pumping head	M	40	50
OPEX	$/m³	0.21	0.25

168   Groundwater for Sustainable Livelihoods

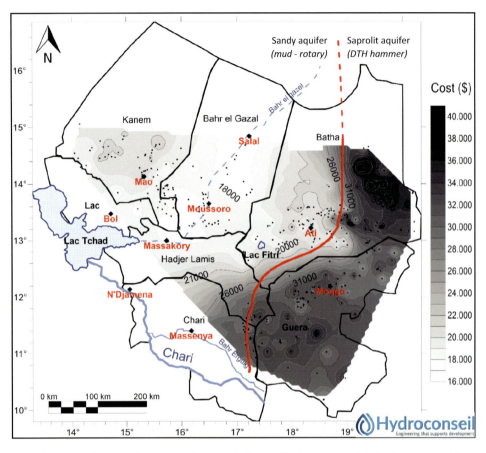

Figure 9.7 Actual cost of a successful borehole (well + hand pump). The Lake Chad Basin Quaternary aquifer is located on the west of the red line.

9.3   MODELLING AQUIFER WATER BALANCE AND LONG-TERM SUSTAINABILITY

9.3.1   Purpose of the model

The population in the study area is estimated to be 2.1 million inhabitants (and is growing at 2%/year). Taking the cattle water demand into account, annual water consumption is estimated as being 20–30 million m^3/year.

The Quaternary aquifer provides a tremendous opportunity for meeting this growing water demand as it is a shallow, extensive and highly productive water resource. The relatively easy access to this resource makes it vulnerable to overexploitation in adverse conditions, such as a diminution in precipitation and, therefore, in groundwater recharge, as predicted by some climate change models (Ben Booth, 2010; Biasutti, 2019).

To assess the resilience of the aquifer, it is necessary to determine its recharge.

The water balances that have been proposed to date are based on estimations of recharge made using methods that involve high uncertainties (Thornthwaite-type hydrologic balance, complex interpretation of ^{2}H and ^{18}O contents, etc.) (Gaëlle Gauthier, 2003; Hamit, 2012).

All this does not constitute a sufficiently robust hydrogeological knowledge base on which to make important decisions such as whether to support irrigated farming extension. It is essential to improve the accuracy of the aquifer water balance in order to be able to develop a sustainable management strategy and anticipate the measures to be taken to compensate for the effects of climate change. To this end, we have built a mathematical model of the aquifer to test its sensitivity to three types of disturbance:

- Increases in demand and water abstraction linked to population growth and urbanisation;
- The consequences of climate change on recharge;
- The potential development of irrigated farming.

9.3.2 Building the model

The model was built based on data from the many boreholes drilled over the past 15 years. The main features of the model are as follows:

- It is a steady-state model, processed under Modflow (finite differences).
- The model boundaries were defined using hydrogeological criteria (Table 9.3 and Figure 9.8).
- The transmissivity in each cell was deduced from pumping tests carried out on village hydraulic boreholes. The limitation of this method is that these are usually short-duration tests, which produce a high level of uncertainty. Thus, because the spatial distribution of transmissivity is complex, as is generally the case for continental sediments, we sought to improve the calibration of the model by initiating a campaign to conduct additional pumping tests of longer duration.
- The quantity of water abstracted from each cell was calculated based on population (with a ratio of 30 L per day per inhabitant, to cover both domestic and livestock needs).
- Infiltration recharge is considered an unknown (the purpose of the model is precisely to calculate this recharge).

Table 9.3 Aquifer model boundaries

Side	Location	Boundary type
South	Chari river	Constant head boundary
	Lake Chad	Constant head boundary
West	100 km west of Niger/Chad border	No flow boundary
North	Bodele depression	No flow boundary
East	Eastern Batha and Guera provinces	No flow boundary

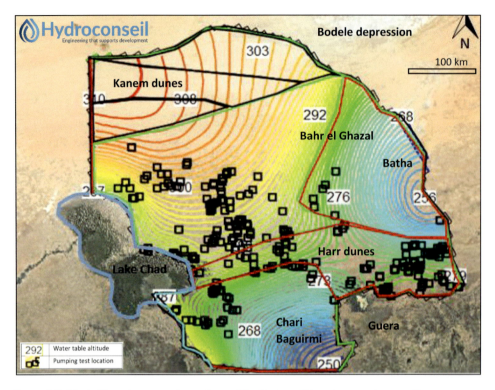

Figure 9.8 Modelling piezometric surface altitude (m asl.) and groundwater flow patterns.

9.3.3 Calibrating the model

The study area was divided into large geomorphological units (Kanem and Harr dunes, Bodele depression, Barh el Ghazal, Chari Baguirmi, Lake Chad and Chari shores). Calibration was carried out by modifying step by step the infiltration value in the different units until the model reproduces as closely as possible the piezometric surface.

9.3.4 Groundwater flow patterns

The aquifer communicates directly with Lake Chad and its piezometric surface connects with that of the lake (Djoret and Favreau, 2014). The piezometric surface is marked by a series of domes (Kanem and Harr sand dunes) and depressions (Chari Baguirmi, Barh el Ghazal) whose origin, based on chemical tracers, is discussed in this chapter.

Piezometric surface depressions were already described in preceding regional hydrological studies (Schneider, 1966, 1969; CBLT, 1972). An updated piezometric map was constructed from level measurements in 400 boreholes and using satellite imagery of pastoral wells, according to a new method developed as part of this study (Collignon, 2020, 2021). Since these piezometric domes and depressions have been very stable for more than 50 years, it is logical to assume that they reflect aquifer equilibrium conditions on a secular or even millennial scale. The domes (such as the Kanem dune zone) correspond to active recharge zones, where rainfall inputs are more important than water abstraction and evaporation. Depressions (such as the Bahr el Ghazal, Bodele or Chari Baguirmi) correspond to zones where the water balance is negative (rainfall does not compensate for abstraction and evaporation).

9.3.5 The water balance

We used the model as a tool to determine the water balance of the aquifer in the different zones and the efficiency of its recharge by precipitation. In Kanem, recharge is positive (1–3 mm/year), which is lower than the values that had been estimated by other methods (Gaëlle Gauthier, 2003). In Chari Baguirmi and Barh el Ghazal, recharge is negative (between −1 and −2 mm/year) (Collignon, 2019).

An interesting result of this modelling is to show that the rainiest part of the aquifer (400 mm/year in the Chari Baguirmi) is the one where the water balance is the most negative. This is probably linked to the clayey nature of the outcrops, which favours water stagnation after rainfall, plant growth and intense evaporation (Hamit, 2012).

9.3.6 Simulation of the most likely developments

Once it has been calibrated through measurements made during drilling campaigns, the model can be used to test the sensitivity of the aquifer to three types of disturbance:

1. Impact of increasing demand: a doubling of water abstraction is expected within 20 years; the Quaternary aquifer will support it well in Kanem, but the piezometric depressions of the Chari Baguirmi and Barh el Ghazal could deepen by more than 10 m, which will inevitably lead to the loss of many wells where the water level will be too low by the end of the dry season.
2. Climate change (and in particular rising temperatures) will have a negative impact on grazing and will increase livestock water demand; however, it is still difficult to predict the effect it will have on rainfall and aquifer recharge in northern Chad (Ben Booth, 2010).
3. Impact of the development of groundwater irrigated agriculture: this economic model has little future in the region: the only areas where it would be possible to sustainably exploit the aquifer for irrigation are those where surface water recharge occurs (Chari and Lake Chad), and in these areas, direct irrigation from surface water is preferable (as it allows better control of salinity).

9.4 THE WATER QUALITY ISSUE

9.4.1 Salty waters

Generally speaking, the salinity of the water in this aquifer is low (EC < 1,000 μS/cm over 90% of the working area) and the salinity level is compatible with national standards for drinking water distribution.

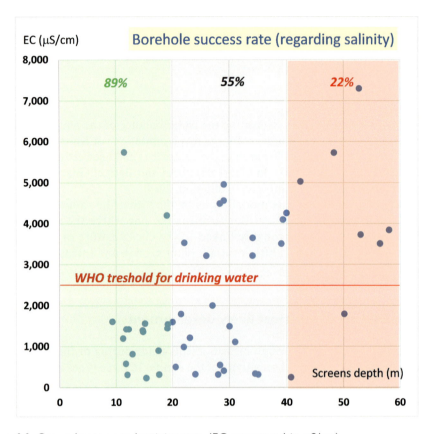

Figure 9.9 Groundwater conductivity map (EC expressed in µS/cm).

We were able to delineate the areas where the water is saltier: northern shore of Lake Chad (Serele, 2017), central Bahr el Ghazal, Kouloudia prefecture (Figure 9.9) and the eastern part of Hadjer Lamis Province (Hamit, 2012). We were also able to correlate this high salinity with intense evapotranspiration phenomena (current or recent) (Collignon et al., 2021).

In areas where groundwater salinity is the highest, pairs of boreholes with different depths demonstrated that salinity increases significantly with depth; it is, therefore, possible to reduce the "water quality" hazard by limiting the depth of the boreholes: for boreholes whose screens are located less than 20 m below the piezometric surface, the success rate (89%) is four times higher than the success rate of boreholes whose screens are located between 40 and 60 m below this surface (Figure 9.10).

However, by limiting the depth of the boreholes, the "climatic" hazard is increased, i.e. the risk that the pump will be out of water in case of persistent rainfall deficit. The art of the hydrogeologist is to achieve the best compromise between these two hazards. In areas where salty water has been reported before, the optimal strategy consists in capturing the aquifer between 20 and 25 m below the piezometric surface.

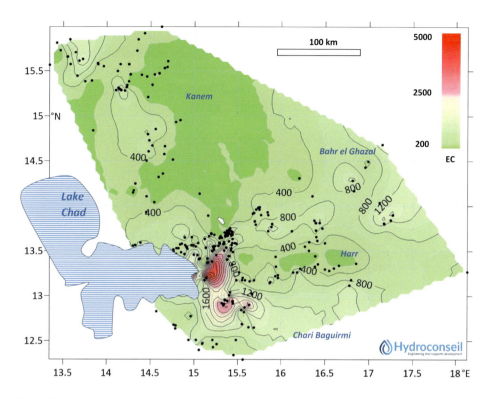

Figure 9.10 Borehole success rate versus screen depth below the water table.

9.4.2 Natron

The chemical composition of the region's groundwater is somewhat peculiar. The water contains high quantities of bicarbonates and sodium, which distinguishes it from the chloride- and sodium-rich water often found in most brackish aquifers. When such water evaporates, it deposits small white crystals of natron (Na_2CO_3–10 H_2O) on the ground. This composition reflects that of the Chari River that has fed the LCB for millions of years (Tardy, 1980). An interesting consequence of this composition is that this water has an economic benefit as, even when it is too salty to be used as drinking water, it helps maintain the health of livestock and so is much appreciated by nomadic herders. This type of water is used in the region as "salt cures" for livestock during the rainy season (when the livestock drinks directly from the pools that form between the dunes) and also during the dry season (when groundwater is exploited through temporary shallow wells).

9.4.3 Nitrates

Nitrate levels are very low almost everywhere and are perfectly compatible with the use of water for village crops and livestock feed. Average nitrate concentration is 6.6 mg/L (NO^3) and most of the drilled wells contain less than 16 mg/L.

174 Groundwater for Sustainable Livelihoods

This can be explained by the low population density (<10 inhabitants/km^2) but it is still good news, as excessive nitrate concentrations have been observed in groundwater in other scarcely populated arid regions of Africa, notably in Namibia and Botswana (Tredouw, 2006), where they are considered to be anthropogenic (population and livestock) in origin.

9.5 MANAGEMENT OF THE WATER SERVICE

9.5.1 Accessing the water resource is not the main issue

Managing the public water service in the rural areas of the Chadian Sahel is a real challenge due to the dispersion of the population and the lack of roads. The national water company (STE – Société Tchadienne des Eaux) is already struggling to provide water to the country's capital and some of the major cities and it is not in a position to serve rural areas (which are not included in the scope of its concession contract). The challenge is further complicated by the extreme poverty of the people living in the Sahel region, which limits households' capacity to pay.

In this difficult context, the Quaternary aquifer is undoubtedly the main and most technically and economically exploitable water resource for supplying decentralised water distribution systems. This resource offers many comparative advantages:

- It is accessible at a depth of less than 100 m in 100% of settlements; it is a local water resource, which means there is no need to transfer water from one area to another (this is an important asset in this insecure region, where long-distance water supply lines are vulnerable to vandalism and attacks);
- Groundwater quality is compatible with direct water supply (without any treatment) in 90% of settlements, which considerably simplifies exploitation (Figure 9.8);
- The productivity of 60% of boreholes is sufficient to supply a settlement of 5,000 inhabitants with a single well (Figure 9.4); for the largest settlements (5,000–50,000 inhabitants), well fields with 2–6 boreholes have been built, such as those recently implemented in Ati and Massakory (two regional capitals with a population of more than 40,000 inhabitants each).

9.5.2 The cost of water abstracted from the Quaternary aquifer

In the calculations below, the production cost includes both the depreciation of equipment (CAPEX) and operating expenses (OPEX). In the Quaternary aquifer, this production cost is about \$0.27/m^3, which is significantly lower than in the neighbouring granitic areas, where drilling costs are higher, as are pumping heights.

This difference in cost is of central importance in such a poor region, where households' ability to pay is very low and pastoralists are always looking for low-cost water points for their livestock.

9.5.3 Households' ability to pay

Such a cost may seem low compared to the price of water in European countries; however, it must be considered in conjunction with households' ability to pay, bearing in mind that the country is at the very bottom of the HDI – Human Development Index world ranking (UNDP, 2019). In addition, the Sahel is one of the poorest regions in Chad, where GDP/capita does not exceed $1 per day.

Due to water scarcity, willingness to pay remains high among the majority of households, who are willing to buy water regularly at standpipes, or even to invest more to obtain a household connection. In the main cities of the region (Ati and Massakory), the customer base for this service has grown rapidly over the past 10 years and there are currently eight connections per 100 inhabitants, a rate equivalent to that of N'Djamena, the capital of Chad.

However, connection to the network does not mean unbridled water consumption. The extreme poverty of these same households translates into minimal water consumption (8–15 L per capita per day) and there is a significant drop in consumption during the rainy season when inhabitants have access to low-cost alternative water sources (rainwater, ponds and traditional wells).

9.5.4 Decentralised management of water distribution systems

Neither the Chadian State nor the national public water company is currently able to manage the 250 or so water distribution systems that have been installed over the past 15 years throughout the Sahel region. The only workable option for their operation is decentralised management within the communities that use these systems.

This is the option that has been chosen by the government in the Water and Sanitation Master Plan – SDEA (MEE, 2003). The management of water production and distribution systems is driven by the communities themselves through Water Users' Associations (WUAs), which are formal institutions registered by the state that can either operate the service themselves or delegate operation to a small local private company.

For larger settlements (such as Ati or Massakory), the operator must be able to manage several thousand individual connections, each equipped with a water metre. This falls under the umbrella of urban water service management, for which Chadian regulations provide for the organisation of a real Public–Private Partnership (PPP) arrangement with a local private company as operator (Figure 9.11).

9.6 SUMMARY

Water supply in the poor rural areas of the Sahel is made difficult by the dispersion of the population, poverty and the poor state of the roads.

In this difficult context, the Quaternary aquifer of the LCB provides a tremendous opportunity for establishing a sustainable, reliable and affordable water service.

Figure 9.11 Management options depend on water system complexity: while the simplest water points (top) can be managed by a simple association of users, more complex systems (bottom) require the implementation of a PPP.

It offers much more potential than the granitic basement areas further east in Chad (Guera, Ouaddaï, Sila, Wadi Fira).

Well construction does not pose many technical difficulties, despite the isolation of many villages. Drilling boreholes is relatively simple and inexpensive, and nearly a thousand have been installed over the past 15 years.

In contrast, the sustainable exploitation of these water points poses two more serious challenges: (a) ensuring sustainability of the resource in such an arid environment where groundwater recharge is very low and (b) ensuring the sustainable exploitation of water distribution systems in a region plagued by poverty and insecurity.

Hydrogeologists can help address the first challenge by improving knowledge of the aquifer, its water balance and its resilience to foreseeable events (population growth and climate change).

Socio-economists can help address the second challenge by providing communities and the government with economic analyses and management tools adapted to the ethnic diversity of this vast region and to pastoral livestock. The pastoral livestock activity poses additional constraints, as access to the water point is a precondition to access pastureland (the most valuable natural resource).

The insecurity that has been growing in the area around Lake Chad since 2013 (Boko Haram) poses an additional challenge: that of making the supply chains for diesel and spare parts more reliable in an unstable context. It is vital that this challenge is met, as ensuring a reliable and sustainable public water service is one of the best ways to combat the roots of insecurity.

REFERENCES

Ben Booth, D. R.-O. (2010). *Climat Sahélien: rétrospective et projections.* Exter: Secrétariat du Club du Sahel et de l'Afrique de l'Ouest (CSAO/OCDE).

Biasutti. (2019). Rainfall trends in the African Sahel. *Wiley Interdisciplinary Reviews: Climate Change* 10:e591.

CBLT. (1972). *Synthèse hydrologique du bassin du Lac Tchad 1966–1970.* Paris: PNUD/UNESCO.

Collignon, B. (2020). A new tool for the remote sensing of groundwater table: satellite images of pastoral wells. *Open Geospatial Data, Software and Standards* 5:4. doi:10.1186/s40965-020-00077-3.

Collignon, B. (2021). Apports de la télédétection optique multisources des puits pastoraux à la cartographie des eaux souterraines du Sahel. *Revie Française de Télédétection et de Photogrammétrie.* n°223, pp. 189–199.

Collignon, B. and Masse, C. (2019). *The Upper Aquifer in Lake Chad Basin: A Tremendous Opportunity to Supply Good Quality Water to People and Cattle in an Extremely Arid Region.* Malaga: IAH Congress. Abstract 675.

Collignon, B., Gröschke, M., Nassour, I., Vassolo, S.I. and Witt, L. (2021). Evapotranspiration in Lake Chad dry environment and its impact on groundwater quality and drinkability. *AGIC 2021 Conference.* Hammamet (Tunisia).

Djoret, D. and Favreau, G. (2014). *Ressources en eau souterraine et relations avec le Lac (Tchad)* (éd. Le développement du Lac Tchad: situation actuelle et futurs possibles). Marseille: IRD.

Gaëlle Gauthier, C. M. (2003). *Hydrogéologie isotopique de la dépression piézométrique de Kadzell* (Niger). Montpellier: IAHS Publ. n°287.

Hamit, A. (2012). *Etude du fonctionnement hydrogéochimique du système aquifère du Chari Baguirmi* (éd. Thèse). Poitiers.

MEE. (2003). *Schema Directeur de l'eau et de l'Assainissement.* N'Djamena: Ministère de l'Eau et de l'Environnement.

Schneider, J. L. (1966). *Carte hydrogéologique de la République du Tchad au 1/500 000- feuille Mao* (éd. LAM 66). Orleans: BRGM.

Schneider, J. L. (1969). *Relations entre le Lac Tchad et la nappe phréatique.* Orléans: BRGM.

Serele, C. (2017). *Cartographie de l'accès et de la qualité des points d'eaudans la zone humanitaire du Lac Tchad.* N'Djamena: UNITAR/UNOSA/Projet ResEau.

Tardy, J. Y. (1980). Géochimie d'un paysage tropical: le bassin du Lac Tchad. In D.Y. Tardy (Ed.), *Géochimie des interactions entre les eaux, les minéraux et les roches* (pp. 199–239). Paris: ORSTOM.

Tredouw, G. and Talma, A. (2006). Nitrate pollution of groundwater in Southern Africa. In Y.X. Usher (Ed.), *Groundwater Pollution in Africa* (pp. 15–36). London: Taylor & Francis.

UNDP. (2019). *Human Developement Index (Per Country)* (Human Development Reports - http://hdr.undp.org/en/data# ed.). New York: UNDP.

Chapter 10

An overview of Karst groundwater springs in Al Jabal Al Akhdar region (North East Libya)

S.M. Hamad and A. El Hasia
University of Omar Al-Mukhtar

CONTENTS

10.1 Introduction 179
10.2 Hydrogeology 180
10.3 Groundwater springs 182
10.4 Springs discharge 183
10.5 Springs water quality 184
10.6 Springs water use 186
    10.6.1 Shahat area (Cyrene) springs 186
    10.6.2 Al Qubah area springs 189
    10.6.3 Derna area springs 189
10.7 Conclusion 193
References 194

10.1 INTRODUCTION

Al Jabal Al Akhdar (Green Mountain) is an upland area along the northeast coast of northern Libya, as shown in Figure 10.1. It is bounded by Sirte Gulf on the west, Sirte Basin on the southwest, Bomba Gulf on the east, and Marmarica on the southeast. In its central part, Al Jabal Al Akhdar is a crescent-shaped ridge, which reaches more than 870 m above mean sea level. The northern flank is made up of step-like plateaus bordered by two main escarpments, further apart in the west but gradually drawing eastward closer together, both roughly parallel to the coast. A large portion of the two benches, particularly the second, is dissected by wadis, giving Al Jabal Al Akhdar a predominantly hilly to mountainous appearance. In some parts of the benches, too, the low local relief has helped to develop substantial karstification. The southern flanks dip gently toward a depression stretching from AJdabyah to Jaghbub, marked by several large Sebkhas. In addition, the coastal plains are well developed between the foot of the first escarpment and the sea in the east, and mostly in the west (Arghin, 1980; Hamad, 2012).

    Al Jabal Al Akhdar has been the focus of the attention of various civilizations since ancient times. The great civilizations of the Greeks and Romans established and built cities such as Cyrene and Apollonia, characterized by high urbanization planning in terms of leading social, legal, and administrative institutions, which competed with the oldest major cities in ancient history (e.g., Rome and Athens). The population

DOI: 10.1201/9781003024101-10

180 Groundwater for Sustainable Livelihoods

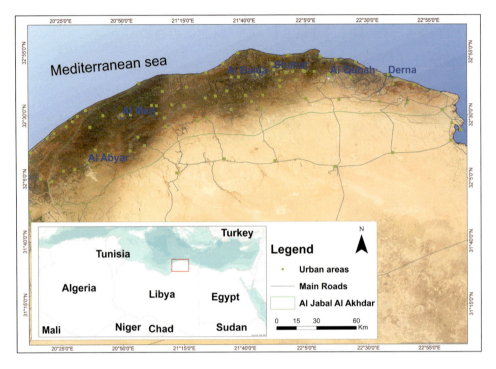

Figure 10.1 Location map.

activity was mainly in agriculture and grazing, as is evident from the old infrastructure such as wells, tanks, dams, and irrigation canals scattered throughout the area, as well as water and soil conservation, olive presses, and marine ports. Therefore, in those times, Al Jabal Al Akhdar represented a diverse food basket for the cities of Greece and Rome. Concerning present water use in the Al Jabal Al Akhdar area, groundwater represents approximately 70% of total other water resources, including those from desalination plants and seasonal streams; groundwater is used primarily for domestic use, agriculture, industry, and livestock watering (GWA, 2006; NWRE, 2019). Groundwater is obtained from springs and drilled water wells, ranging from 20 to 600m in depth, depending on location in the region. This chapter presents an overview of karst groundwater springs in the Al Jabal Al Akhdar region, North East Libya, by highlighting their hydrogeological characteristics and their significance for local communities' livelihoods. Moreover, since these springs represent a hydrogeological system that is prevalent in the Mediterranean, the dissemination of more information will, therefore, be of great benefit to specialists in the field.

10.2 HYDROGEOLOGY

Stratigraphy and surface geology of the exposed rocks in the Al Jabal Al Akhdar area consists mainly of sedimentary marine carbonate units ranging from late Cretaceous to late Miocene in age, divided into cycles of sedimentation (El Hawat and

Abdulsamad, 2004). The tectonic and structural geology of Al Jabal Al Akhdar regions represents the only folded and faulted mountain chain in northern Libya, unlike other areas of Libya belonging to the Sahara platform, where this mountain chain is an isolated large highland area occupying much of northern Cyrenaica. It stretches about 300 km from east to west, with crest-line rising to more than 870 m above sea level (Hamad, 2008). Tertiary and Upper Cretaceous carbonate rocks are the major constituents of the principal aquifers in the Al Jabal Al Akhdar area, while other aquifers generally (perched or local) are constituted by fluvial deposits of the Quaternary. The lithological nature of the aquifers is predominately made of chalk, calcarenite, and algal limestone. Also, karst processes play a prominent role in the presence of water bodies throughout the area, in which they expressed the development of macrokarst features. Moreover, groundwater flow mainly relates to a system of micro-fissures and micropores. Aquifer recharge is due to direct rain infiltration from the runoff along the wadi beds during rainy seasons, and the aquifer natural discharge is either through inland springs or, in the northern flank, into the Mediterranean Sea (Arghin and Hamad, 2007). North East Libya, as shown in Figure 10.2, is divided into hydrogeological units; because of the scarcity of detailed studies, limited exploratory drilling, and the lack of data integration between the various institutions working in the region, the boundaries of these units are not precisely defined. In addition, the hydrogeological units share common aquifers, as shown in Table 10.1. The occurrence, spatial distribution, and depth of the aquifers vary due to the influence of the tectonic geology. Furthermore, the Eocene and Upper Cretaceous aquifers, among other aquifers, have the highest water potential in terms of quality and quantity, and their spatial extension is primarily around the central parts of Al Jabal Al Akhdar (GWA, 2006; Hamad, 2012).

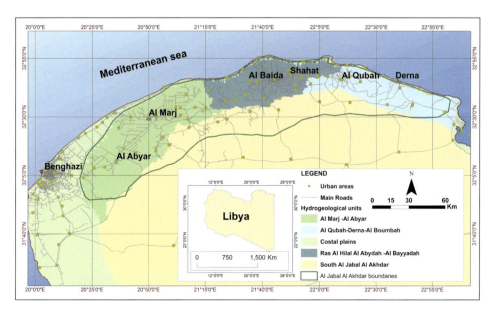

Figure 10.2 The spatial extension of Al Jabal Al Akhdar hydrogeological units was modified after (GWA, 2006; Hamad, 2012).

Table 10.1 Main aquifer characteristics of Al Jabal Al Akhdar (GWA, 2006 and Hamad, 2012)

Aquifer	Depth (m below ground)	Main constitutes
Quaternary	From 10 to 50	Fluvial deposits
Miocene	From 100 to 150	Marly limestone
Oligocene	From 200 to 250	Calcarenitic limestone
Eocene	From 250 to 350	Nummlitic limestone
Upper cretaceous	From 350 to 550	Dolomitic limestone

10.3 GROUNDWATER SPRINGS

More than 220 karst water springs occur in the Al Jabal Al Akhdar region, mostly spatially distributed in the northeastern parts, as shown in Figure 10.3. However, their occurrence and distribution are related and controlled by the following factors:

1. The Al Jabal Al Akhdar region is characterized by the highest rainfall rate compared to the rest of the Libyan territory. Therefore, groundwater recharge occurs and thus the discharge is sustainable.
2. The geological structures and the configuration of the geological formations, where permeable formations are based on impermeable layers.
3. The occurrence of karstification processes, as the geological formations mainly consist of marine sedimentary carbonate.
4. The relationship between the regional faults and the wadis: the connection of springs with fault lines and wadis is an apparent phenomenon in the Al Jabal Al Akhdar where there is a dense network of fractures and wadis in some zones cut aquifers deeply inside.

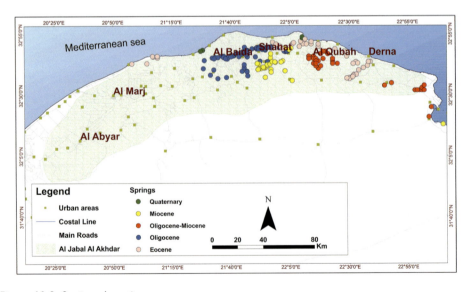

Figure 10.3 Springs location map.

Table 10.2 Summary of Al Jabal Al Akhdar springs

Age of the carbonate formations	Number of springs	Average discharge (m³/d)	Electrical conductivity (μS/cm)
Quaternary	10	2,700	3,000–10,000
Miocene	30	6,480	280–600
Oligocene-Miocene	49	13,824	370–1200
Oligocene	90	15,552	370–1,000
Eocene	49	38,794	400–800

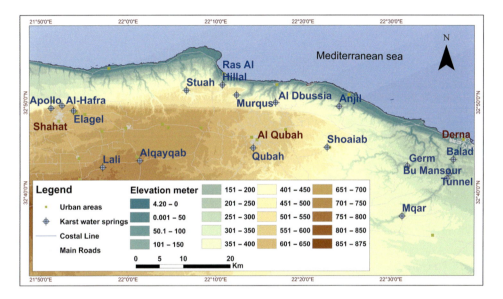

Figure 10.4 Selected springs location map.

Springs occur at various onshore, offshore, and inland locations and vary in discharge from 1 to 200 L per second. Table 10.2 shows the summary of Al Jabal Al Akhdar springs, where the annual total discharge is about 8,232,750 m³. Some of the springs also flow temporarily during the rainy seasons, while the discharge increases at the permanent springs. The following sections will discuss in detail the characteristics of seventeen karst water springs shown in Figure 10.4 and their importance and impact on local communities.

10.4 SPRINGS DISCHARGE

As shown in Table 10.3, springs are located at different altitudes; they flow as a result of the groundwater level intersection with the ground surface or wadi side. Also, most of the springs flow from the contact surface between geological formations, often representing non-conformity surfaces made up of weathered unconsolidated materials. More than 50% of the springs flow from the Eocene formations and represent the highest discharge among the other springs.

184 Groundwater for Sustainable Livelihoods

Table 10.3 Spring characteristics

No	Springs	Longitude	Latitude	Elevation AMSL	Geology	Discharge m^3/day
1	Apollo	21.85237	32.823	571.7	Oligocene	1,296
2	Al-Hafra	21.87366	32.828	525.5	Oligocene	1,728
3	Elagel	21.89585	32.818	591.8	Oligocene	1,036.8
4	Alqayqab	22.02207	32.726	706.3	Miocene	518.4
5	Lali	21.95145	32.713	709.5	Miocene	432
6	Qubah	22.2381	32.749	582.1	Oligocene-Miocene	777.6
7	Al Dbussia	22.2806	32.834	283.1	Eocene	15,552
8	Stuah	22.1103	32.856	309.3	Eocene	1,296
9	Shoaib	22.3786	32.7500	425.1	Oligocene-Miocene	6,912
10	Murqus	22.206	32.848	386.4	Eocene	432
11	Ras Al Hillal	22.17897	32.866	291.4	Eocene	1,728
12	Anjil	22.40126	32.826	241.1	Eocene	2,592
13	Germ	22.5306	32.716	167.7	Eocene	345.6
14	Wadi Dernah Floor	22.6072	32.693	135.2	Eocene	5,184
15	Tunnel & Bu Mansour	22.6111	32.702	126.3	Eocene	13,824
16	Balad	22.6193	32.728	68.6	Eocene	4,320
17	Mqar	22.5205	32.623	266.2	Eocene	864

10.5 SPRINGS WATER QUALITY

The hydrochemical properties of groundwater springs are shown in Table 10.4. Data represent the results of chemical analysis of groundwaters sampled during the preparation of this chapter by the Department of Soil and Water Laboratory at Omar

Table 10.4 Hydrochemical characteristics

No	Springs	EC (μ/cm)	TDS (ppm)	pH	Total hardness	SAR	Alkalinity	Water type
1	Apollo	683	374	6.9	253	1.1	165	Ca–Na–HCO_3–Cl
2	Al-Hafra	695	465	7.2	251	0.906	180.5	Ca–Na–HCO_3–Cl
3	Elagel	946	485	7.25	360	1.186	248	Ca–Na–HCO_3–Cl
4	Alqayqab	963	495	7.93	355	1.458	240	Ca–Na–HCO_3–Cl
5	Lali	530	283	7.93	205	0.838	145	Ca–Na–HCO_3–Cl
6	Qubah	703	357	7.25	268	1.034	180	Ca–Na–HCO_3–Cl
7	Al Dbussia	794	433	7.24	335	0.929	291	Ca–Mg–Na–HCO_3–Cl
8	Stuah	778	480	7.29	315	1.239	273	Ca–Na–HCO_3-Cl
9	Shoaib	705	358	8.06	278	1.076	180	Ca–Na–Mg–HCO_3–Cl
10	Murqus	754	446	7.43	290	1.76	310	Ca–Na–HCO_3–Cl
11	Ras Al Hillal	958	616	7.15	383.7	0.91	285.5	Ca–Na–HCO_3–Cl
12	Anjil	886	527	7.4	383.7	1.136	220.7	Ca–Na–HCO_3–Cl
13	Germ	926	496	8.54	340	1.57	205	Mg–Ca–Na–HCO_3–Cl
14	Wadi Dernah Floor	960	536	7.17	330	1.65	230	Ca–Na–Mg–HCO_3–Cl
15	Wadi Dernah Tunnel	942	520	7.36	310	2.04	216	Na–Mg–Cl–HCO_3
16	Balad	999	532	7.83	320	2.07	213	Na–Ca–Mg–Cl–HCO_3
17	Mqar	1,859	1,014	7.76	330	6.33	120	Na–Ca–Cl

Al-Mukhtar University. The water of all springs is fresh, except for the Maqar spring that shows a high value of Total Dissolved Solids (TDS) compared to other springs. Total hardness and alkalinity values indicate that water of some springs such as Ras Al Hillal and Anjil is hard. No trace elements above the permissible levels have been recorded from previous studies by Hydrogeo (1992) and Benischke and Hamad (2008). Evidence of microbiological contamination was reported only for Apollo and Al-Hafra springs (Reem, 2017). According to the Piper diagram, as shown in Figure 10.5, the common type of water is Ca–Na–HCO$_3$–Cl, while the Wilcox diagram as shown in Figure 10.6 indicates that the water from these springs is suitable for agriculture.

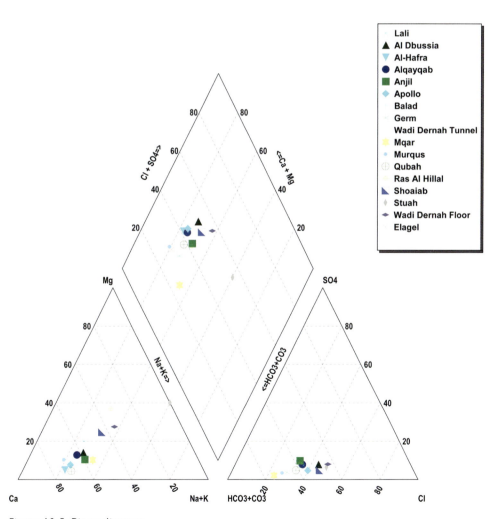

Figure 10.5 Piper diagram.

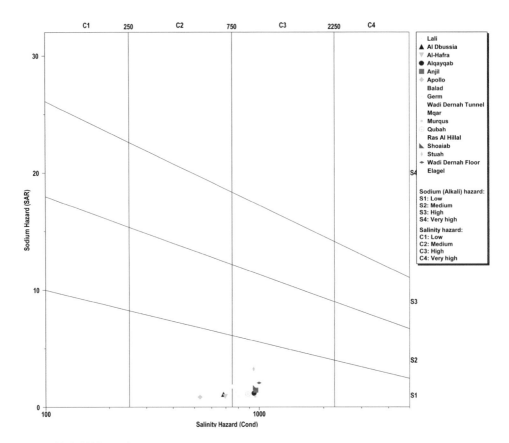

Figure 10.6 Wilcox diagram.

10.6 SPRINGS WATER USE

10.6.1 Shahat area (Cyrene) springs

Cyrene, a designated UNESCO World Heritage Site (WHS) in modern eastern Libya, was the leading city of the Libyan Pentapolis. Settled by Greek colonists toward the end of the 7th century B.C., it remained an active Graeco-Roman town of distinctly Hellenic character until the time of Islamic conquest (Cuttler et al., 2009). During the Greek period, Cyrene city was supplied by water mainly from springs and rain, especially over the winter (Gregory, 1916), where the spring water was transferred through aqueducts built or cut into the rock to feed various areas and buildings.

Cyrene city is currently known as Shahat, a city with more than 35,000 inhabitants. Springs from an Oligocene formation known as Shahat marl member of Al Baydah formation surround the city and the neighbouring villages. Most of these springs are seasonal, yielding between 0.5 and 15 L per second. Also, the springs are exploited for domestic, irrigation, and watering of animals by the local communities from ancient times to the present. Perhaps the most important event of history related to these springs dates back when the population of the Cyrene town, which was dependent on the water from Apollo spring (Figure 10.7) for water supply, increased during the Roman era and Emperor Marcus Aurelius issued a decree to supply the town with water from outside. Engineers built a canal from Al Qayqab village 15 km southeast to the town of Cyrene, and created a lift station in the Safsaf area, which consists of two vast reservoirs (Figures 10.8 and 10.9) to collect water that then opened to supply Cyrene city (Lloyd and Lewis, 1977).

Shahat City is presently relying on water supply from groundwater wells and a desalination plant on the Susa coast, about 15 km northeast of Shahat City. Currently, due to microbiological contamination, Apollo spring is not used by the local community, and water continues to flow through dedicated channels within the historic city of Cyrene, creating a tourist attraction. The rest of the springs are used by local communities outside the city, such as Ain Al-Hafra spring (Figure 10.10), which is exploited and managed by local communities to meet their irrigation and animal watering demands according to a well-managed water distribution system.

Figure 10.7 Apollo spring.

Figure 10.8 Safsaf historical reservoirs.

188 Groundwater for Sustainable Livelihoods

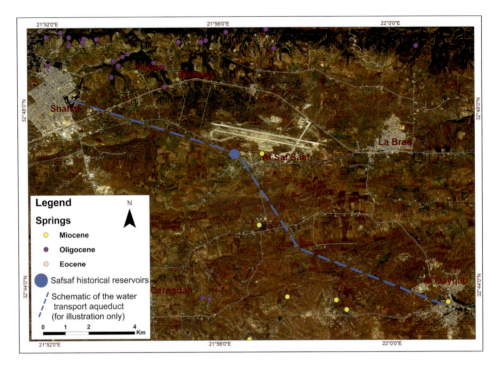

Figure 10.9 Historical water transport to Cyrene city.

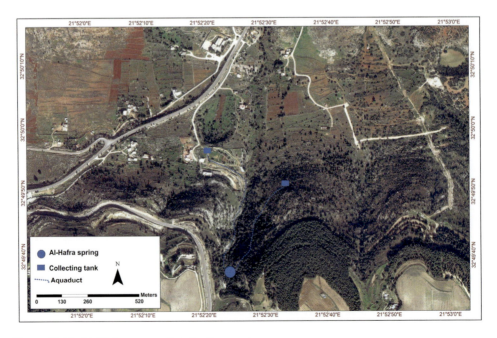

Figure 10.10 Al-Hafra water distribution system.

10.6.2 Al Qubah area springs

Springs are scattered around the town of Al Qubah city in all directions; Al Dbussia is the most famous spring in Al Jabal Al Akhdar because of the highest discharge rate among all the springs (Figure 10.11). During the 1960s through the 1980s, Al Dbussia spring was a significant source of water, where the Libyan government set up a vast network of spring water transportation pipelines to cover the domestic water demand of Sha-hat, Al Abydah, and Al Marj and other surrounding villages. Currently, the spring only covers the water demands for the area of the municipality of Al Qubah and its suburbs, which is home to around 150,000 people. Moreover, Shoaib spring (Figure 10.12), which is characterized by a high discharge, is of critical importance for domestic and agricultural use of Ayn Mara town. Furthermore, there are other springs of great significance have contributed to the livelihood and stability of the population in the Al Qubah area, such as Murqus spring. It is named after St. John Mark (the Evangelist) born in the city of Cyrene according to some historical records, Anjil spring, which lies in a coastal town known as Karsah, and Qubah spring in the vicinity of Al Qubah city (Figure 10.13).

10.6.3 Derna area springs

The springs in the Derna area have been an essential source of water for farms on the banks of wadi Derna throughout history, where fruits and seasonal crops are cultivated. Derna city was formed during the Hellenistic period, in which the ancient town of Darnis was part of the Pentapolis of Libya. The main water springs in wadi Derna

Figure 10.11 Al Dbussia spring.

190 Groundwater for Sustainable Livelihoods

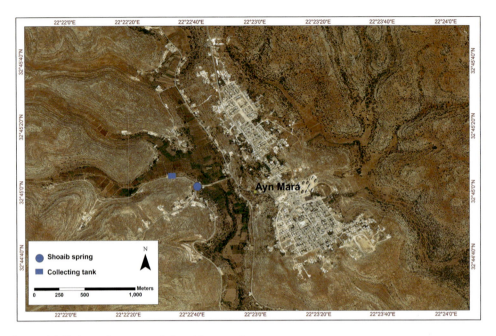

Figure 10.12 Location map of Ayn Mara spring.

Figure 10.13 Qubah spring.

(Figures 10.14 and 10.15) are Wadi Derna Tunnel, Bu Mansour, and Balad, characterized by high discharge and good water quality. In addition, some springs flow from the floor of the wadi, collected in a concrete canal. During the 1950s through the 1970s, in addition to agriculture, the spring water became an essential source of water for domestic use for Derna city, as well as to Tobruk city, located 250 km to the east. Currently, as an inevitable result of population growth and hence the increasing demand for water, these springs only partially cover the water demand of the city of Derna and nearby towns. During the 1990s, the Libyan state installed pump stations and built aqueducts, pipelines, and concrete channels to take advantage of all the flows from these springs and transported it through the wadi to the city of Derna, which currently has a population of around 100,000. Despite the establishment of a desalination plant in Derna city, dependence on these springs still exists to cover about 40% of the water demand of the town. For nearby villages, Mqar and Germ springs (Figures 10.16 and 10.17) are of great importance, currently used for animal watering.

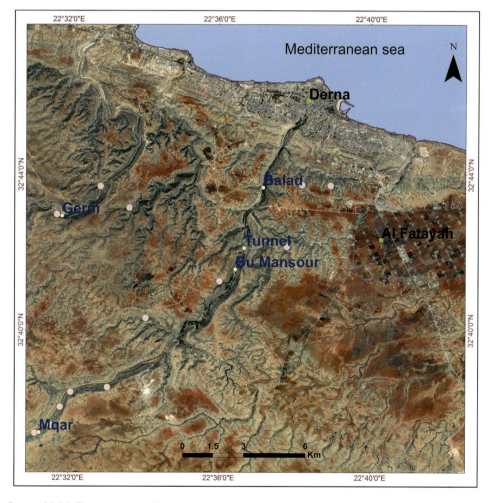

Figure 10.14 Derna springs location map.

Figure 10.15 Wadi Dernah tunnel (a), Balad spring (b).

Figure 10.16 Germ springs.

Figure 10.17 Mqar spring.

10.7 CONCLUSION

At the beginning of their development, the ancient villages and towns of Al Jabal Al Akhdar stood around permanent springs, where obtaining freshwater did not require much efforts or means that were not available at the time. Nowadays, with population growth, the water volumes from the springs are unable to meet the increasing demand. Hence, the shortage is compensated by drilling water wells and building structures for water harvesting. Nevertheless, Al Jabal Al Akhdar's springs continue to flow, covering local communities' water requirements. Because of their distance from pollution sources, most of the springs are still not affected by pollution, except for a few polluted springs due to their proximity to urban areas.

Furthermore, to ensure springs sustainability, a new inventory programme should be implemented, as well as periodic quantity and quality monitoring are crucial. In addition, more pressing than ever is the importance and significance of providing a different plan of action to meet the ever-increasing demand for water, particularly in urban areas near the springs. Water scarcity has a different definition in these locations, where it only means that running water supply to households is not sustainable and continuous. Natural springs are a viable and proven sustainable water supply source, but no real efforts have been made to correctly exploit and use them. Throughout the past, the presence of natural water springs has been a vital source of water in the region and remains so to smaller-scale local communities; however, considering the scenarios in tourism and other related industries, more could be done in using its production. Water, indeed, could be a catalyst for a healthier local economy in adjacent cities and villages.

REFERENCES

Arghin, S.S. (1980) *Hydrogeology of El Marj basin.* MSc Thesis, University of Nevada Reno USA.

Arghin, S.S. and Hamad, S.M. (2007) *An overview of water resources of North east Libya.* [Lecture] Workshop on Integrated Water Resources Management (IWRM) General Water Authority - Tripoli, Libya, 11–12 April 2007.

Benischke, R. and Hamad, S.M. (2008) *Al Jabal Al Akhdar Karst Water Resources Management,* Joint venture project between Joanneum Research institute and General water authority, Karst springs study of Al Jabal Al Khdar area – Libya. Working Report:1.

Cuttler, R., Gaffney, C., Gaffney, V., Goodchild, H., Howard, A., and Sears, G. (2009) Changing perspectives on the city of Cyrene, Libya: remote sensing and the management of the buried archaeological resource. *ArchéoSciences* [Online] 33 (65–67). Available from: https://doi.org/10.4000/archeosciences.1295 [Accessed 7th June 2020].

El Hawat, A.S. and Abdulsamad, E.O. (2004) The geology and archaeology of cyrenaica, short note and guide book on the geology of Al Jabal Al Akhdar. In L. Guerrieri, I. Rischia and L. Serva (eds.), *ITALIA 2004, Proceedings of the International Symposium of 32nd International Geological Congress,* August 20–28, 2004, Florence – Italy. Volume n° 1- from PR01 to B15. Published by: APAT – Italian Agency for the Environmental Protection and Technical Services, Rome, Italy.

Gregory, J. (1916) Cyrenaica. *The Geographical Journal* [Online] 47 (321–342). Available from: www.jstor.org/stable/1779632. doi:10.2307/1779632 [Accessed 7th June 2020].

GWA, General Water Authority Libya. (2006) *Water status of Libya.* Annual Report:2.

Hamad, S.M. (2008(*Spatial analysis of groundwater level and hydrochemistry in the south Al Jabal Al Akhdar area using GIS.* MSc Thesis, Salzburg University Austria.

Hamad, S.M. (2012) Status of groundwater resource of Al Jabal Al Akhdar region, North East Libya. *International Journal of Environment and Water* [Online] 2 (68–78). Available from: http://ijew.ewdr.org/ [Accessed 7th June 2020].

Hydrogeo. (1992) *Groundwater resource evaluation of Al Baydah - Al Bayadah Area.* General Water Authority Benghazi-Libya. Final report.

Lloyd, J. A. and Lewis, P. R. (1977). Water supply and urban population in Roman Cyrenaica. *Annual Report - Society for Libyan Studies* [Online] 8 (35–40). Available from: https://doi.org/10.1017/S0263718900000777 [Accessed 7th June 2020].

NWRE, National Water Resources Establishment. (2017) *Water supply in Al Jabal Al Akhdar.* Department of planning and data center. Report:19.

Reem, M. (2017) *Microbiological investigation for five groundwater springs in Shahat area.* Faculty of Natural Resources & Environmental Sciences, University of Omar Al-Mukhtar, Al Baydah, Libya. Lab Report:10.

Chapter 11

The governance and water security of groundwater obtained from private domestic wells in periurban areas in Brazil

A case study on the Guandu river basin in the metropolitan region of Rio de Janeiro, Brazil

D. Tubbs Filho and A.S. Schueler
Universidade Federal Rural do Rio de Janeiro – UFRRJ

S.Y. Pereira
Universidade Estadual de Campinas – UNICAMP

CONTENTS

11.1 Introduction.. 195
11.2 Governance of groundwater 197
    11.2.1 The "four layers" of groundwater management at the municipal
        level – the origin of the problem 198
11.3 The intermittency of the water supply and the water crisis 199
11.4 Environmental characteristics of the Guandu river basin........................... 200
    11.4.1 Geology and groundwater occurrence in the Guandu river basin 200
    11.4.2 Groundwater hydrochemistry and contamination in the
        Guandu river basin... 202
        11.4.2.1 Groundwater hydrochemistry.. 202
        11.4.2.2 Groundwater contamination 203
11.5 Individual alternative systems as a source of supply for the periurban
    population during water crisis.. 204
11.6 The resilience of periurban communities 206
11.7 What is being done to improve governance and water security.................... 208
11.8 Conclusion.. 209
Notes 209
References... 209

11.1 INTRODUCTION

Groundwater has been a vital source since the first urban settlements when water was obtained from springs and shallow wells. Groundwater wells, therefore, were a fundamental factor in the establishment of human sedentary lifestyle and the formation of the first cities. There is now considerable evidence to suggest an increasing dependence

DOI: 10.1201/9781003024101-11

on groundwater for water supply in developing cities and metropolitan regions. This occurs in response to population growth, accelerated urbanization, increased water use and higher ambient temperatures, also contributing to insecurity regarding possible climate changes (Foster et al., 2010). The study area is located in the metropolitan region of Rio de Janeiro and is part of the Guandu river basin. It has intense economic development with numerous large industries, power generation, ports, shipbuilding (including the construction of nuclear submarines) as well as mining activity and tourism. Urban and periurban expansion in these peripheral municipalities is now an unquestionable reality. However, the area stands out for being historically characterized by marked socio-spatial inequality, where environmental risk factors and lack of basic sanitation coexist with a population of more than 1,800,000 inhabitants. In this region flows the Guandu River, its main drainage, where the largest water treatment plant in the world is situated. This region also has the Committee of the Guandu, Guarda, and Guandu Mirim River Basins – Guandu Committee. This committee, characterized as the water parliament, is co-responsible for managing raw waters, which after being treated supply more than 10 million inhabitants from a large part of the Rio de Janeiro city and its metropolitan region. It is one of the Brazilian river basin committees that have advanced the most in the management of water resources and in the development of management instruments provided by Federal Law No. 9433/1997 (BRASIL, 1997). This committee is cited in several academic works and by international organizations as an example of a successful experience in the management of water resources in Brazil (OECD, 2015). Figure 11.1 shows the map of the Hydrographic Region II of the Rio de Janeiro state, area of competence of the Guandu Committee, and focus of this text.

Despite the existence of significant financial resources derived from the charging for the use of raw water from rivers and groundwater of this basin and economic development, the region has a low rate of sewage treatment and does not guarantee water supply to all periurban communities (BRASIL, 2017). There is also no water supply

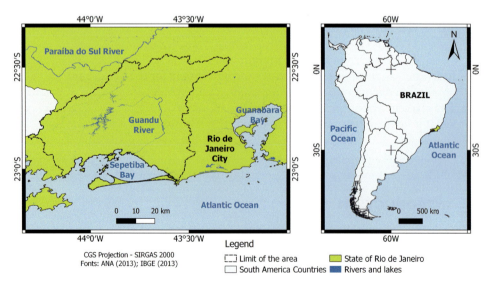

Figure 11.1 Location of the study region.

in the region for all periurban communities, with high rates of intermittency (absence of water supply for more than 6 hours without notice).[1] In response to this reality, communities living in these urban fringes and peripheral territories seek their own self-built solutions, by drilling their own domestic wells without technical guidance and institutional control. Therefore, in response to adverse infrastructure conditions, they build solutions that are not always legal or orthodox, clandestine, thus exercising "hydro social resilience." Because of this lack of governance, water insecurity, and the exposure of this population, in the order of hundreds of thousands (only in the hydrographic basin considered), to waterborne diseases overload the SUS (Unified Health System), ANA (2005) and Profil (2017). In Brazil, studies and services in groundwater are generally focused on large aquifer systems, contaminated sites, and drilling of deep tubular wells. Although there are numerous studies dedicated to assessing water quality in shallow wells in periurban and perirural areas, none has considered the issue of governance and institutional control. Another important consideration is due to the lack of knowledge and training of municipal and state public managers on the subject, and mainly, the absence of an adequate legal framework to generate effective actions aimed at governance and water security for these communities. In addition to the previous arguments, this chapter also discusses the importance of groundwater to contribute to the resilience of the population, the actions taken, and the proposals to achieve the requirements of good governance and water security.

11.2 GOVERNANCE OF GROUNDWATER

Throughout the 20th century, the Brazilian water legislation gradually evolved and the "Water Code," of 1934, is a milestone of this process. Specifically in relation to groundwater, this Decree established important guidelines, among which the following stand out: the landowner may appropriate the groundwater existing on his/her land, provided that he/she does not change the conditions of the other users; the drilling of wells should be sufficiently far away so that there is no damage to the neighborhood; it is prohibited to pollute water from the well or spring of someone else; and the drilling of wells in public domain land depends on concession. Since the Federal Constitution of 1988, all water resources in Brazil have been nationalized and groundwater ownership has been attributed to the states and Federal District, although some aquifers exceed state or international political divisions. Thus, according to Art. 26, the assets of the States include surface water or groundwater, flowing, emerging and in deposit, except in this case, as set forth by law, for those arising from works of the Country (BRASIL, 1988). Therefore, the Federal Constitution of 1988 holds the states and the Federal District accountable, attributing groundwater management to federal entities, and allowing the collection of raw water. In Art. 23, the constitution highlights the need for articulation between the Union, States, Federal District and Municipalities for protecting the environment in all its forms, as well as exploiting water and mineral resources. Thus, this article implies the need for complementary legislation related to water resources, environment, sanitation, and mining. Besides, in Art. 200, the constitution establishes the competence of the health system in controlling water quality for human consumption. The National System of Water Resources Management is a responsibility of the Ministry of the Environment, composed of the National Council of Water Resources; State Councils of Water

Resources; federal and state agencies responsible for managing the system of water use licenses; river basin committees; and river basin agencies, which will be the executive offices of these river basin committees. The National Policy of Water Resources in Brazil was established through Law No. 9433 of January 8, 1997 (BRASIL, 1977). This Law confers on water the importance of a public domain asset, limited, of economic value, whose priority use is the consumption by humans and animals, stating that, whenever possible, it should have multiple use, besides defining the river basin as a territorial unit of water resource management. It also determines that, besides the government, there should be participation of users, communities, and civil entities, so that the management is decentralized. Art. 5 deals with the instruments of the National Plan of Water Resources: Plans of Water Resources; classification of water bodies into classes, according to the predominant uses of water; granting of use rights of water resources; charging for the use of water resources; and Water Resources Information System.

11.2.1 The "four layers" of groundwater management at the municipal level – the origin of the problem

Groundwater in Brazil is a responsibility of state governments, therefore, it is up to the state bodies and entities that manage water resources to authorize the drilling of wells and the use of aquifer. The municipal role is vital for protecting groundwater and the communities that use them, especially in more vulnerable areas, whether socially and/or environmentally (Villar 2011). According to the National Water Resources Policy, the hydrographic basin is the spatial unit used for the management and control of water extraction depending on authorization from the state or federal government. However, territorial management, that is, the soil surface is the responsibility of the municipal government. Therefore, it is not possible to provide water security to the population without taking measures to regulate the municipal territory where groundwater (or surface) water is obtained. Unfortunately, the National Water Resources Policy has limitations regarding the integration of the municipalities, as they do not directly participate in the management of water resources or the financial resources generated, so they do not prioritize the protection of groundwater for these reasons and due to the lack of knowledge on the topic.. Although the municipality is responsible for the territorial management and autonomous in various activities, it is not self-sufficient to fulfill all its duties alone. This is another side of municipal autonomy: the existence of an interdependence relationship between local governments, and between them and other governmental spheres. On the normative plane, there are responsibilities that must be shared with other governmental spheres, such as in the areas of health, education, and environment. On the other hand, according to Federal Law No.11445/2007 (BRASIL, 2007a), establishes as of exclusive competence of the municipality the organization and management of public services of which it is the holder, with emphasis the basic sanitation services (supply of treated water, collection and treatment of sewage, collection and disposal of solid waste and urban drainage). Differently, according to Federal Law No. 10257/2001 (BRASIL, 2001), entitled Statute of the Cities, municipalities should include the environmental aspect in the urban planning and the planning of the city should be based on urban laws and environmental legislation. Thus, it is important to ensure that the environmental legislation provides municipalities with the foundations so that they can include groundwater protection in territorial management. Regarding the Surveillance of Water Quality for Human Consumption, the Consolidation Ordinance No. 5 of September 28, 2017, of the Ministry of Health, (BRASIL,2017), establishes the following: (a) all water intended for human consumption, collectively distributed through a conventional system or a collective alternative solution of water supply, must have its quality controlled and monitored by the municipality and (b) all water intended for human consumption coming from an individual alternative solution (Solução Alternativa Individual – SAI) of water supply, regardless of the way it is accessed by the population, is subject to water quality surveillance. Therefore, in the first case, there is an obligation for the municipality to carry out control and surveillance actions in articulation with the state, and sanctions may be applied if this does not occur. For individual alternative solutions, there are no such obligations, and therefore, the government does not cover the populations that use these systems. Although subtle, this small textual difference, but with significant impact, makes these populations that use shallow domestic wells, called individual alternative solutions, become vulnerable in terms of water security. Thus, "four layers" of laws: water resources, basic sanitation, environment, and health constitute these complex and inefficient rules aimed at groundwater governance and management at municipal level. However, in practice these laws do not integrate and hinder groundwater governance in periurban areas.

11.3 THE INTERMITTENCY OF THE WATER SUPPLY AND THE WATER CRISIS

Brazil has the largest land water reserves, but such abundance is uneven and does not guarantee supply to all regions of the country. The southeast region of Brazil, despite having good water potential, with rivers, reservoirs, and rains between 1,200 and 1,500 mm/year, faced a water crisis between 2014 and 2016. The two largest metropolitan regions of the country, those of the cities of São Paulo and Rio de Janeiro, consume 60 and $50\,m^3/s$, respectively, and depend on the Paraíba do Sul River (Britto et al., 2016). Due to this dependence, the two most populated Brazilian states, São Paulo and Rio de Janeiro entered into a federative conflict for the waters of this river, which required the mediation of a minister from the Supreme Federal Court (BRASIL, 2019). The water crisis aggravated this situation, leading the population of these regions to use groundwater intensively, especially in periurban areas (Hirata et al., 2019). Although the universalization in the access to water has substantially evolved, guaranteeing this access in peripheral areas with security, in quantity and quality, is an obstacle to be overcome. According to recent estimates, more than 70 million Brazilians may suffer from lack of water until 2035. The National Plan of Basic Sanitation (*Plano Nacional de Saneamento Básico* – PLANSAB), approved in 2013, pointed out how far the country is from meeting this right, with significant deficits in all components of basic sanitation. Based on analysis of data from the Brazilian Institute of Geography and Statistics of 2010, 33.9% of the country's population still had poor service and 6.8% did not have any service (Britto et al., 2016). Concerning sanitary sewage, 50.7% of the population had poor service – that is, sewage collection, not followed by treatment, or use of rudimentary septic tank, which represents millions of people living in unhealthy environments and exposed to various risks, which may compromise their health (SNIS, 2015). In the studied region, the rates of treatment, in turn, decrease significantly, resulting in average values of 0.8% of the generated sewage and 2.3% of the collected sewage (Profil, 2017). The causes of the water crisis cannot be reduced, however, only to the lowest rainfall rates observed in recent years, as other factors related to the guarantee of water supply and the management of water demand are important to aggravate or mitigate its occurrence. According to Quintslr (2017), the very existence of something that can be termed as a crisis is not unanimous among managers and researchers of the theme, not to mention its causes. Several researchers argue that water crisis (2014–2016), even if considered a warning for climate change, is a "management crisis," since extreme events, such as the prolonged drought in southeastern Brazil, are not unprecedented. In this case, for instance, water supply service providers should have been prepared. However, although the most prosperous regions of the Rio de Janeiro city were not affected by the drought, much of the metropolitan, peripheral and poorer regions was intensely hampered, especially in periurban areas. Water supply rotation system is a rule for several areas of this region, where water is supplied through a complex system of networks, since the volume of water to be supplied to the region is insufficient (Quintslr, 2017). Therefore, the frequent lack of water for these populations cannot be attributed only to drought and the "water crisis." Supply turnover is the rule in several areas of this region, where water supply is carried out through a complex system of networks, since the volume of water to be supplied to the region is insufficient.

In addition, the lack of water in parts of the Metropolitan Region may become even more frequent in the summer, when there are periods of weeks with no water distribution. This intermittence (absence of water supply for more than 6 hours without previous notice) in water distribution is the reality and makes the water crisis familiar to this population, besides increasing health risks. Due to the lack of rainfall and depletion of groundwater levels, many of these communities had to deepen their wells or drill new ones. Therefore, in the regions where water supply is deficient or nonexistent, the communities usually opt for individual alternative solutions, through self-constructed withdrawing systems, by drilling shallow domestic wells and/or using water springs to obtain water from the water table (Tubbs et al., 2019). Also due to the lack of rain during the water crisis and the depletion of water tables, many of these communities had to dig or dig new wells. However, even if the population seeks to overcome this problem with their own solutions, the lack of governance that integrates public actions aimed at ensuring water security, according to Bakker (2012), exposes the population to waterborne diseases and contamination of the water table.

11.4 ENVIRONMENTAL CHARACTERISTICS OF THE GUANDU RIVER BASIN

11.4.1 Geology and groundwater occurrence in the Guandu river basin

The region of this study is in the Mantiqueira Geotectonic Province, which represents a neoproterozoic orogenic system that extends in a Northeast–Southwest direction strip with more than 3,000 km in length and an average width of 200 km, parallel to the

Atlantic coast of southeastern and southern Brazil (Heilbron et al., 2004). There are igneous and metamorphic rocks of Precambrian age, metasedimentary and metavolcanic neoproterozoic rocks, and small intrusions of basic and alkaline rocks of Mesozoic age. Colluvial, marine, and fluvial Cenozoic sediments also appear covering the Precambrian rocks. There is no hydrogeological map in the region, and Figure 11.2 presents a simplified map of the region's aquifers. Precambrian metamorphic igneous crystalline terrains are in green and pleistocene sediments are in yellow.

The Sepetiba sedimentary basin is the most important geological feature in terms of groundwater. It is located near the metropolitan region of Rio de Janeiro, occupying an area of about 4% of the Rio de Janeiro state, and its main tributary is the Guandu River, which originates from the Serra do Mar mountains. The Guandu drainage basin occupies an area of 2,000 km^2, 90% of which has features of alluvial plain deposits (SEMA, 1996). The Guandu River receives water from the Paraíba do Sul River diversion and flows to the Guandu Water Treatment Station. According to Marques et al. (2012), quaternary sediments from alluvial environment (fluvial, fluvial-lacustrine and fluvial-marine) compose the geology of the studied region, deposited on Precambrian basement. These sediments form the Piranema Formation (Góes, 1994) represented by two units. The lower unit has pleistocene sandy facies, with medium to coarse texture and generally basal gravel, and essentially quartz-feldspathic mineralogy. The upper unit, also called alluvial cover, is composed of Holocene silt-clay facies. Sediment cores collected in the region showed average thickness ranging about 35 and 40m, reaching depths greater than 70m in some cases. The Sepetiba sedimentary basin also has some features, like high porosity and good permeability, which provide the conditions for water accumulation and transmission, characterizing the Piranema Formation as an aquifer called Piranema Sedimentary Aquifer (Tubbs, 1999). This sedimentary aquifer system has an area of about 350 km^2 and is located approximately 60 km west from Rio de Janeiro city. The free aquifer system's recharge is distributed upon its occurrence

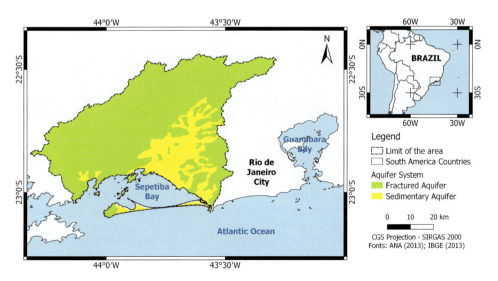

Figure 11.2 Simplified map of the region's aquifers.

202 Groundwater for Sustainable Livelihoods

area, trending to highest potentiometric level as high as the regional topography. So, the flux direction is controlled by topographic irregularity. The water table level ranges from 3 to 7.5 m, depending on the weather season. The soil covers, originated from crystalline rocks (igneous and metamorphic), could generate an aquifer system with similar characteristics to those of a sedimentary porous aquifer (colluvial deposits), and as depth increases and gradually change into fractured systems. Together, the fractured aquifers and soil covers are responsible for 30% of the area. Intercommunication among sedimentary, fractured, and colluvial aquifers could increase the regional groundwater potential and determines the aquifer recharge and flow patterns (Eletrobolt, 2003). The Guandu river basin (and Rio de Janeiro Metropolitan Region) has strong dependence on the water diversion from Paraíba do Sul River. While the average transfer flow is about $166 \, m^3/s$, the contribution of its own basin is about $3.18 \, m^3/s$. With the absence of another significant water body in that region, the Paraíba do Sul River is practically the only water supply source to industrial and domestic use in Rio de Janeiro Metropolitan Region. The Piranema Aquifer, which has an area of about $500 \, km^2$ and an available outflow of about $1.6 \, m^3/s$, is the main water supplier to the Seropédica and Itaguaí district. In the other municipalities, groundwater is obtained from fractured aquifers and cover soil. Despite having limited water availability, compared to the transfer volume of the Guandu River, this sedimentary and fractured aquifer is of strategic importance to the periurban communities of the metropolitan region of Rio de Janeiro. Its water reservoir is the only alternative that can be used during periods of potable water scarcity or environmental accidents when the conventional systems of water supply are threatened. However, the natural vulnerability of aquifers and the lack of sanitation in the region increase the risk of groundwater contamination. However, the natural vulnerability of aquifers, the quality of abstractions, and the lack of sanitation in the region increase the risk of contamination of groundwater.

11.4.2 Groundwater hydrochemistry and contamination in the Guandu river basin

11.4.2.1 Groundwater hydrochemistry

In general, groundwaters have good quality in both aquifers (sedimentary and fractured) and are classified as sodium-bicarbonated and calcium-bicarbonated, followed by sodium- and/or calcium-chloridic waters, besides calcium-sulfated waters. These classifications reproduce the different geological environments in the region. Calcium-bicarbonated waters are typical of crystalline terrains on the mountains, whereas chloridic waters, especially sodic ones, originate in a sedimentary environment near the coast (INEA, 2014). The pH ranges from neutral to slightly basic in waters of the fractured aquifer, while the sedimentary aquifer is characterized by acidic pH. Electrical conductivity values vary around 600 µs/cm for the fractured aquifer and 250 µs/cm for the sedimentary aquifer. However, in waters near the coast the values can exceed 1,500 µs/cm, evolving into brackish waters. Waters associated with the fractured aquifer mostly have reduced values of total dissolved solids, which gives a more pleasant taste for human consumption. Natural occurrences of Ba, F, Mn, and Fe are also observed.

11.4.2.2 Groundwater contamination

Regionally, sand mining is recognized as the main cause of change in the physico-chemical quality of underground springs mainly in the generation of very low pH values (2.5) and increase of sulfate concentrations (acid mine drainage). It is worth pointing out that the decrease in pH is the main consequence of the increase in the concentrations of dissolved aluminum (Marques et al., 2012). Although groundwater usually has good quality, the interaction between soil use and the characteristics of the withdrawals in these communities locally affect water quality. The following factors are relevant: wells drilled in inadequate locations (near bathrooms, garbage dumps, and cemeteries), shallow wells that have been abandoned and used as dumps of garbage and chemical products, and non-compliance with technical standards for drilling wells, such as the absence of protection slab and lid. Thus, in domestic wells of several communities, the presence of bacteria and the occurrence of high concentrations of nitrate and ammonia are frequent; even caffeine has already been found in some wells (Tubbs et al., 2004). Contaminations originated from a chemical products treatment center in the region and gas stations have also been found. Therefore, under these conditions of use of groundwater, the occurrence of diseases due to groundwater contamination can be facilitated, and although these health issues are commonly underreported, some cases of relevance are known, as presented in Table 11.1.

In view of the above, it is clear that the periurban communities of the Rio de Janeiro city metropolitan region that uses individual alternative systems for groundwater withdrawal have no water security.

Table 11.1 Identified causes of groundwater contamination in periurban communities in the metropolitan region of Rio de Janeiro located in the Guandu river basin

Type	Site	Cause	Consequences
Waterborne Hepatitis[a]	Jardim Nova Era (July/2000) Paracambi	Hepatitis A	Eighteen inhabitants got sick and ninty-seven were contaminated
Japeri Syndrom[b]	Tricampeão neighborhood, border between Japeri and Queimados (2000)	Contamination by cemetery leachate, or rotavirus	Nine deaths, wells were closed and inhabitants were removed
Presence of bacteria and nitrate[c]	Piranema (Seropédica and. Itaguaí) and. Paracambi	High concentrations of nitrate and fecal bacteria	Health issues recorded at SUS
Contamination/ hydrocarbons (BETEX)[a]	Itaguaí (1998)[d] Seropédica[a] (2002/2004) Paracambi[a] (2003)	Leaks in gas stations (BTEX)	Intoxication of inhabitants – Itaguaí monitoring by FIOCRUZ
Cancer (1998)[d,e]	Santo Expedito neighborhood municipality of Queimados	Toxic waste leaks at CENTRES	Fourteen diagnoses of cancer

Sources: Adapted by the authors: [a]R 7-8 PERH - GUANDU (2005); [b,c]SILVA (1998); [d]SANTOS (2011); [e]CETEM (2013).

11.5 INDIVIDUAL ALTERNATIVE SYSTEMS AS A SOURCE OF SUPPLY FOR THE PERIURBAN POPULATION DURING WATER CRISIS

In the Guandu river basin, groundwater is mainly obtained with two types of wells: deep and shallow, in addition to springs. In periurban (and perirural) regions, the most used are the shallow wells, which can be a viable and sustainable solution for the supply of these communities, provided that there is technical guidance during their location and construction. The waters obtained in both cases are differentiated from the conventional public supply system and are individualized in Collective Supply Solution (*Solução de Abastecimento Coletivo* – SAC) and Individual Supply Solution (*Solução de Abastecimento Individual* – SAI) (BRASIL, 2021). The conventional public supply system and the collective alternative system have safe specific laws that make it possible to ensure the verification of the quality of water offered to the population. Differently, for individual alternative systems (SAIs) at the municipal level, as presented, the legislation is confusing, and due to their building characteristics and local environmental conditions, usually precarious, these are the systems that most influence the number of health issues in the SUS. An alternative water supply solution, either collective or individual, is any modality that differs from the conventional public supply system, which includes sources, springs, wells, community/residential or not, distribution by transporting vehicle, and horizontal and vertical condominium installations (BRASIL, 2017). The same norm defines an alternative individual solution (SAI) of water supply for human consumption more appropriately as the modality of water supply for human consumption that provides for residences with a single family, including their relatives. Among the SAI, the most used in periurban areas are the shallow wells of domestic use, which are usually dug at depths not exceeding 15 m. These wells enable the withdrawal of volumes around 1,000 L per day and are dug manually or using small drilling machines, manual tools, employing community labor in task force regime, and not considering the cost of drilling, but the benefit it will bring (Tubbs et al., 2019). It should be noted that, according to Brazilian legislation, these well drillers are irregular and are outside the law! It is also common for the population to rely on the "help" of dowsers to locate the well. However, this method of supplying water to the population when carried out through individual alternative systems is generally not considered by conventional regulation and governance systems. Characterized by IBGE (2010) as "shallow wells and/or spring water used for home use," in the state of Rio de Janeiro alone there would be more than 2 million inhabitants consuming water through these alternative systems through domestic wells, INEA (2014). Figures 11.3 and 11.4 show several withdrawals in a community beside the Presidente Dutra highway (which connects the two largest cities in Brazil: Rio de Janeiro and São Paulo) in the Baixada Fluminense section. We can see close proximity to a gas pipeline marker can be seen.

According to the Strategic Plan for Water Resources of the Guandu, Guarda, and Guandu – Mirim river basins, Guandu (2005), approximately 25% of the urban population of the basin does not have supply through public networks. We believe that this portion is supplied with water from alternative sources, which do not go through quality control, so that these people are exposed to waterborne diseases. Therefore, we can emphasize the social importance of alternative systems. In certain municipalities, many communities or even entire neighborhoods are partially or fully supplied by these systems, due to either the deficiency or absence of the conventional system.

Figure 11.3 Wells beside the Presidente Dutra highway in the Baixada Fluminense section. (Authors' Collection.)

Figure 11.4 Wells beside the Presidente Dutra highway in the Baixada Fluminense section. (Authors' Collection.)

Therefore, we can emphasize the social importance of alternative systems. Individual utilization in the Guandu river basin exceeds the order of thousands, because periurban (and perirural) areas also make use of these withdrawals, even if occasionally, due to the intermittence of the conventional supply by the concessionaire. In the area of this river basin, wells commonly have depths ranging from 10 to 15m. They can rarely exceed this depth, depending on local geological conditions, if drilled in soils originated from the alteration of crystalline rocks or constructed in sandy sediments of river origin. They are drilled with diameters ranging from 2 to 3 inches, and less usually with 4 inches, lined with common rigid PVC pipes or those used in sanitary plumbing. Very common are also the manually dug wells of large diameter, called "cacimbas" (*kixima*, from the Angolan dialect Kimbundu), which are rustic and more subject to contamination. There is often no concern about protecting the well, so it is left exposed or simply closed with PET bottle tops. In these drilling procedures, there is neither technical study nor water analysis reports, and frequently, it is possible to observe the construction of alternative systems in very inadequate sites, close to bathrooms, septic tanks, garbage dumps, and even near cemeteries. The proximity of wells to septic tanks and sewage networks and the presence of domestic animals in the residence are factors to be considered, as they facilitate the contact of groundwater with human and animal feces. The water collection system employing manual methods through crank, ropes, and buckets facilitates fecal contamination, as well as the lack of hand hygiene and inadequate water transport. Due to these characteristics pointed out, it is extremely important that this population be accompanied, guided, and protected by public authorities. Waters from springs and collective wells are occasionally examined, and sodium hypochlorite is rarely distributed to residents for treating well water. However, cases of diarrhea, directly linked to the consumption of contaminated water, are not of compulsory recording, thus making it difficult to identify the consequences of consuming this water on the population's health. Other health issues such as hepatitis A are also known but not recorded (see Table 11.1). The lack of specific legislation that integrates public actions aimed at ensuring water security exposes the population to waterborne diseases and to water table contamination. In order to prevent this, it is necessary to create appropriate conditions of governance.

11.6 THE RESILIENCE OF PERIURBAN COMMUNITIES

Thus, it is possible to observe that there is a dichotomy in the Brazilian legislation, between the laws related to water management and the laws related to basic sanitation management, which encompasses the conventional water supply. Therefore, these systems are individualized in legal, political, and institutional terms. The first is the responsibility of the state and refers to the activities of use, conservation, protection, and recovery of raw water, in quantity and quality, which includes groundwater. The second is the responsibility of the municipality and refers to services of drinking water supply, collection and treatment of sewage, and pluvial drainage (Whately, 2017). Consequently, when withdrawing groundwater through domestic wells, the periurban communities would theoretically be subject to specific legislation, the one related to water management. In the state of Rio de Janeiro (and in other states), even the insignificant volumes (less than 5,000 L/day) must be informed and registered with the management body. However, that same body is not obliged to verify whether the environmental conditions of the drilling site are safe. There is no appropriate legislation or institutional program to determine this control. The municipality responsible for environmental, sanitation, and basic health management usually does not have this competence, technical and financial capacity, or the political will to engage in these issues. Therefore, when there are no specific public policies to provide water security for these populations, when the State and the Municipality are absent, this process results in water insecurity and social injustice. Thus, the population's water insecurity results from the distancing (physical, geographical, and legal) of the organ responsible for the management of groundwater (state), organ responsible for sanitation (municipality), or the concessionaire, if it exists. It is also important to highlight the absence of a multi-administrative agenda at the municipal level that integrates the different aspects of water management, for alternative solutions in periurban areas. Thus, the water insecurity of the population results from the distancing (physical, geographical, and legal) of the agency responsible for the management of groundwater (state), the agency responsible for sanitation (municipality), or concessionaire, if any. It is also important to highlight the absence of a multi-administrative agenda at the municipal level that integrates the different aspects of water management, for alternative solutions in periurban areas.

Therefore, it is evident that there is a gap in the water management of municipalities, especially in areas not covered by and/or with a deficiency in the conventional water supply systems. As a result, tired of waiting for assistance from the government and due to the ineffectiveness of the water supply by conventional systems, the populations of many periurban communities seek their own alternative solutions, which are not always orthodox. Figure 11.5 presents a series of self-constructed solutions where pipes that bring groundwater from a set of shallow wells are below the street.

Figure 11.5 Self-constructed solutions in a neighborhood of the Basin.

Thus, these populations build an everyday resistance, and according to Scott (2002), they fight silent battles seeking to find their own solutions and sometimes even resort to stealing water when close to a water main. Thus, wells are indiscriminately dug in any situation, by empirical well diggers without any technical training, who provide a service for the community in the absence of the government, despite not knowing the current norms and being illegal. Finally, it is necessary to find a technical/legal arrangement to improve environmental conditions and provide better living conditions for these communities.

11.7 WHAT IS BEING DONE TO IMPROVE GOVERNANCE AND WATER SECURITY

The actions necessary to achieve the governance and water security of the groundwater of the periurban population, now presented, are developed by Universities and by the Hydrographic Basin Committee. They are divided into four groups: (1) actions taken, (2) actions in progress, (3) actions under development, and (4) scheduled actions.

1. Actions taken
    - Map the supply network, identifying sites with higher urban and periurban density with nonexistent or deficient water supply by conventional systems;
    - Identify and register areas with higher occurrence of SAIs, establishing priorities;
    - Register the points of greatest relevance in terms of sanitary conditions;
2. Actions in progress
    - Assess the environmental risk in each family unit or group of residences in the same place as the previous item;
    - Water quality assessment of the most vulnerable points regarding sanitary conditions;
    - Identify the most vulnerable single-family units to apply emergency solutions;
    - Pilot project to install emergency systems to improve water quality by solar disinfection (SODIS) and Diffusion Chlorinators in the most vulnerable homes.
3. Actions in development
    - Development of a Simplified Risk Assessment Index for the use of Groundwater in Individual Alternative Systems (SAIs) for use by public agents, community agents, and stakeholders through a smartphone application;
    - Qualification and training of public agents and local leaders to continue the project.[2]
4. Actions scheduled
    - Identify and train well builders in the region ("poceiros"), especially those from the community itself;
    - Promote discussion on the criteria for inserting alternative groundwater withdrawal systems in the water resources management system of the Rio de Janeiro state;
    - Promote the consolidation of the legal framework through resolution of the Rio de Janeiro State Council of Water Resources, recommending the integration of the four layers of water management at the municipal level.

11.8 CONCLUSION

Although Brazil has the largest reserves of water on Earth and the universalization of access to it has significantly evolved, in peripheral areas the guarantee of supply with security, in quantity and quality, is an obstacle to be overcome. Brazilian legislation is effective for the protection of collective water supply systems but flawed and confusing for individual alternative systems. Countless people live in periurban areas and depend on the supply of groundwater obtained through residential wells and other alternative systems. They are exposed to environmental weaknesses, climatic variations, and waterborne diseases, but are not reached by the official statistics. Despite such negative situation, periurban populations develop self-constructed strategies, thus guaranteeing the minimum amount of water needed to survive. However, the lack of technical guidance in the use of alternative systems imposes water and food insecurity on these populations. Generally, in Brazil, studies on the use of groundwater in periurban communities are limited to identifying contamination problems and proposing isolated and palliative solutions, despite the long-lasting solutions generated in specific legislation. Despite these difficulties inherent to the Brazilian legal scenario, some advances can be made through the integration of public policies, extension projects carried out by universities, and participation of the river basin committees (including with financial resources) and water concessionaires.

NOTES

1 During the process of revising this text, the region was seriously affected by covid-19 and a lack of water for hygiene was identified as one of the factors responsible for the progress of the disease.
2 Temporarily stopped due to COVID-19.

REFERENCES

ANA (Agência Nacional de Água). (2005). *Plano estratégico de recursos hídricos das bacias hidrográficas dos Rios Guandu, da Guarda e Guandu Mirim: Relatório Gerencial*. Agência Nacional de Águas, Superintendência de Planejamento de Recursos Hídricos. Rio de Janeiro. 2015. Available from: www.comiteguandu.org.br.
Bakker, K. (2012). Water security: research challenges and opportunities. *Science*, 337(6097), 914–915.
BRASIL. (1988).*Constituição da República Federativa do Brasil (1988)*.
BRASIL. (1997). *Lei Federal nº 9.433, de 8 de janeiro de 1997*. Institui a Política Nacional de Recursos Hídricos, cria o Sistema Nacional de Gerenciamento de Recursos Hídricos. Brasília.
BRASIL. (2001). *Lei Federal nº 10.257, de 10 de julho de 2001*. Estabelece diretrizes gerais da política urbana e dá outras providências.
BRASIL. (2007a). *Lei Federal nº 11.445, de 5 de janeiro de 2007*. Estabelece diretrizes nacionais para o saneamento básico. Brasília.
BRASIL. (2017). *Ministério da Saúde. Sistema de Informação de Vigilância da Qualidade da Água para Consumo Humano (Sisagua)*. Brasília: Ministério da Saúde.
BRASIL.(2019).STF.SupremoTribunalFederal.PrimeiraTurma.InteiroTeordoAcórdão-Página1de15. http://redir.stf.jus.br/paginadorpub/paginador.jsp?docTP=TP&docID=14829282 10/06/2019.

BRASIL. (2021). *Ministério da Saúde. Portaria GM/MS N° 888, de 4 de Maio de 2021*

Britto, A.L.; Formiga-Johnsson, R.M.; Carneiro, P.F. (2016). *Abastecimento Público e Escassez Hidrossocial na Metrópole do Rio de Janeiro*. Ambiente & Sociedade São Paulo v. xix, 41p. 185–208, jan.-mar.

Britto, A.L.; Maiello, A.; Quinstsr, S. (2015). Working Paper Vol. 2, No 8 Democratization of Water and Sanitation Governance by Means of Socio-Technical Innovation. Assessment of Appropriate Water and Sanitation Technologies in Vulnerable Communities in the Baixada Fluminense, Rio de Janeiro, Brazil Waterlat Network Working Papers Research Projects Series SPIDES DESAFIO Project ISSN 2056-4864 (Online), December.

CETEM. (2013). *Centres Deixa Passivo Ambiental em Queimados (RJ)*. Dísponível em Available from: http://verbetes.cetem.gov.br/verbetes/ExibeVerbete.aspx?verid=115. 28/10/2019.

Eletrobolt. (2003). Estudos Hidrogeológicos dos Aqüíferos Intergranulares a Oeste do Rio Guandu. Município de Seropédica/RJ. Relatório de Consultoria Técnica. 234 p.

Exame. (2019). Mais de 70 milhões de brasileiros podem sofrer com falta de água até 2035. Disponível em. Available from: https://exame.abril.com.br/brasil/mais-de-70-milhoes-de-brasileiros-podem-sofrer-com-falta-de-agua-ate-2035. 01/08/2019.

Foster, S.; et al. (2010). *Sustainable Groundwater Management Urban Groundwater Use Policy The World Bank*. Strategic Overview Series Number 3. 36p. Washington, DC.

Góes, M.H.B. (1994). *Diagnóstico Ambiental por Geoprocessamento do Município de Itaguaí, RJ*. (DSc Thesis) Geociências, Universidade Estadual Paulista Júlio de Mesquita Filho, Rio Claro.

Heilbron, M.; Machado, N. (2003). Timing of terrane accretion in the Neoproterozoic Eopaleozoic Ribeira belt SE Brazil. *Precambrian Research*, 125, 87–112.

Hirata, R.; Suhogusoff, A.V.; Marcellini, S.S; Villar, P.C.; Marcellini, L. (2019). A revolução silenciosa das águas subterrâneas no Brasil: uma análise da importância do recurso e os riscos pela falta de saneamento. Realizado por: Instituto Trata Brasil, 19 p.

IBGE. (2010). Fundação Instituto Brasileiro de Geografia e Estatística. *Censo Demográfico*.

INEA. (2014). Plano Estadual de Recursos Hídricos do Estado do Rio de Janeiro. Instituto Estadual do Ambiente/SEA. Rio de Janeiro, RJ.

Johnsson, R.M.F.; Farias Junior, J.E.; Costa, L.F. (2015). Segurança hídrica do Estado do Rio de Janeiro Face à Transposição Paulista de Águas da Bacia Paraíba do Sul: Revista Ineana. *Revista Ineana (Revista técnica do Instituto Estadual do Ambiente, RJ)*, 3(1), 48–69.

Marques, E.D; Tubbs, D; Gomes, O.V.; Silva Filho, E. V. (2012). Influence of acid sand pit lakes in surrounding groundwater chemistry, Sepetiba Sedimentary Basin, Rio de Janeiro, Brazil. *Journal of Geochemical Exploration*, 112, 306–321.

OECD (2015). *Governança dos Recursos Hídricos no Brasil*. OECD Publishing, Paris. doi:10.1787/9789264238169-pt.

Profil. (2017). Plano Estratégico de Recursos Hídricos das Bacias Hidrográficas dos Rios Guandu, da Guarda e Guandu-Mirim: Tomo I do Diagnóstico (RP-02).

Quintslr, S. (2017). *Crise hídrica e debate público sobre saneamento*. XVII ENANPUR. São Paulo.

Santos, M.C.B. (2011). Avaliação da Contaminação por Metais em Solos Impactados pela Disposição de Rejeitos Industriais: estudo de caso – CENTRES (Queimados, RJ). Maria Carla Barreto Santos. –. Niterói: [s.n], 2011. 66 f.: il., 30cm. Dissertação (Mestrado em Geociências – Geoquímica Ambiental). Universidade Federal Fluminense, 2011. Available from: https://app.uff.br/riuff/handle/1/4595. Access in 28/10/2019.

SEMA. (1996). *Diagnóstico ambiental da Bacia Hidrográfica da Baia de Sepetiba*: Programa de Zoneamento Econômico-Ecológico do Estado do Rio de Janeiro. CARTOGEO/NCE/UFRJ, Rio de Janeiro, Brasil. 63 p.

Silva, L.C. (2001). *Geologia do Estado do Rio de Janeiro: texto explicativo do mapa geológico do Estado do Rio de Janeiro*. Programa Levantamentos Geológicos Básicos do Brasil. Execução: CPRM - Serviço Geológico do Brasil/Departamento de Recursos Minerais – DRM – RJ. CPRM. 2ª edição revista.

Scott, J.C. (2002). Formas cotidianas da resistência camponesa. *Revista Raízes*, 21(1), 10–31.

Silva, R.L.B. (1998). Contamination of shallow wells in the neighborhood Brisamar, Itaguaí, RJ, for spill of gasoline: concentration of BTEX and evaluation of the quality of the water consumed by the population. Available from: https://www.arca.fiocruz.br/handle/icict/4356.

SNIS. (2015). *Sistema Nacional de Informações sobre Saneamento – Série Histórica 2015*. Disponível em <www.snis.gov.br/> Acesso em: dezembro de 2019.

Tubbs, D. (1999). Ocorrência das Águas Subterrâneas – "Aquífero Piranema" – Município de Seropédica, área da Universidade Rural e Arredores, Estado do Rio de Janeiro, FAPERJ (Fundação de Amparo a Pesquisa do E. Rio Janeiro), Relatório Final de Pesquisa, 123 p.

Tubbs, D.; de Schueler, A.S.; Yoshinaga Pereira, S. (2019). A Água Obtida Através de Sistemas Alternativos Autoconstruídos em Comunidades Periurbanas: O Caso da Área de Atuação do Comitê Guandu, Rio de Janeiro. XVIII Encontro Nacional da ANPUR (Encontro Nacional de Pós – Graduação e Pesquisa em Planejamento Urbano e Regional) Natal 2019.

Tubbs, D.; Freire, R.B.; Yoshinaga, S. (2004). Utilização da Cafeína como Indicador de Contaminação das Águas Subterrâneas por Esgotos Domésticos no Bairro da Piranema – Municípios de Seropédica e Itaguaí, Rio de Janeiro. Revista Águas Subterrâneas. Suplemento - Anais do XIII Congresso Brasileiro de Águas Subterrâneas.

Villar, P.C., (2009). Brazilian Regulatory Process: including groundwater in urban water management. 9th World Wide Workshop for Young Environmental Scientists WWW-YES-Brazil-2009: Urban waters: resource or risks? Oct 2009, Belo Horizonte, MG, Brazil. Ffhal–00593302ff.

Whately, M. (2017). O município e a governança da água: Subsídios para a agenda municipal de cuidado com a água organização de Marussia Whately.

Chapter 12

Groundwater policy, legal and institutional framework situation analysis

Gaps and action plan: the case of Malawi

J. Sauramba
SADC-GMI

T. Mkandawire
Malawi University of Business and Applied Sciences

B. Munyai and M. Majiwa
SADC-GMI

CONTENTS

12.1 Introduction..214
    12.1.1 Background 214
    12.1.2 Groundwater occurrence in Malawi...................................215
12.2 Methodology 218
    12.2.1 PLI framework situation analysis.......................................218
12.3 Policy legal and institutional framework for groundwater management
    in Malawi..220
    12.3.1 Policies to support groundwater management 220
    12.3.2 Legislation to support groundwater management................220
    12.3.3 Strategy and guidelines necessary to support groundwater ...221
    12.3.4 Institutional framework...222
        12.3.4.1 Institutional arrangements to support groundwater
                management...222
        12.3.4.2 Institutional gaps and challenges identified...........224
12.4 Discussion..225
    12.4.1 Must haves...226
    12.4.2 Should haves..227
    12.4.3 Could haves 227
12.5 Conclusions 228
References..228

DOI: 10.1201/9781003024101-12

214 Groundwater for Sustainable Livelihoods

12.1 INTRODUCTION

12.1.1 Background

The Republic of Malawi (Figure 12.1) is categorised as water-stressed, with estimated water availability in 2017 as 1,050 m^3/capita/year. About 80%–90% of Malawi's 17 million population is rural (MoIWD, 2008; GOM/MWP, 2008). Agriculture is the mainstay for the country's economy with subsistence and smallholder farming being

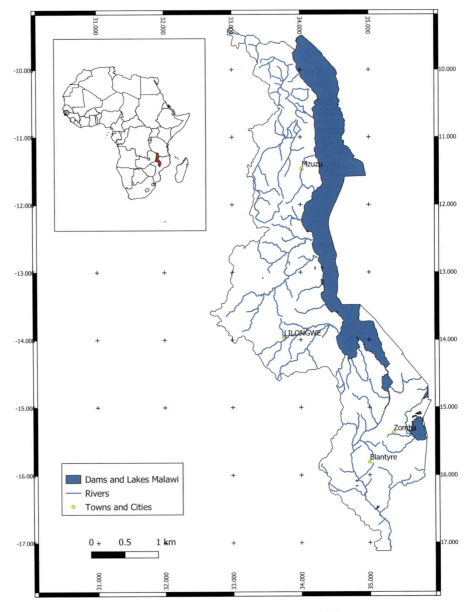

Figure 12.1 Location of the Republic of Malawi (the Case study).

prevalent among the rural population with 50.7% of the population living below the poverty datum line and 25% living in extreme poverty relatively (IMF, 2017).

There is a growing national demand for water resources and concern about their availability, particularly during the dry season. There are increasing calls for better management of the water resources of the country to ensure that it is available and does not limit the social and economic development and poverty alleviation programmes (SADC-GMI, 2019). The government of Malawi places a high priority on water resources management and development to ensure food and water security at the household level by, among other things, enhancing water-harvesting technologies, promoting catchment protection and management, and including disaster risk reduction measures (SADC-GMI, 2019). Groundwater provides an opportunity to enhance water security for the majority of Malawi's population.

In response to the challenges in groundwater governance and the lack of sustainable governance mechanisms, the Southern African Development Community established the Groundwater Management Institute (GMI) to spearhead the region's efforts to elevate the importance of groundwater and improve their groundwater management capabilities. The primary mandate of SADC-GMI, which is drawn from the fourth phase of the SADC Regional Strategic Action Plan (RSAP IV: 2016–2020), revolves around the creation of an enabling Policy, Legal and Institutional (PLI) environment, institutional strengthening and capacity building, knowledge generation and management as well as infrastructure development for sustainable groundwater development and management (SADC, 2016). In pursuance of the focus area of creating an enabling environment, SADC-GMI implemented a project titled "Policy, Legal and Institutional Development for Groundwater Management in the SADC Member States, (GMI-PLI)" whose objective was to identify gaps in the PLI arrangements in all the 16 SADC Member States and at the regional level for groundwater management and to determine ways of closing the identified gaps.

The study presented in this chapter is grounded in the conceptual framework for groundwater governance, which recognises three parts to the groundwater governance system, i.e. the policy level, the strategic level and the local government level and further points at the need for governance tools for enhancing the water security for marginalised rural communities (Wijnen et al., 2012; SADC-GMI, 2019).

The study departs from the Desired Future State (DFS) for policy, legislation and the general framework for effective groundwater management in the SADC Member States (SADC-GMI, 2019). The DFS is anchored in the four aspects of groundwater governance identified in the Global Diagnostic on Groundwater Governance (FAO, 2016) and recognises that there is no "one-size-fits-all" governance arrangement, but that general principles and ethical practices must suit the specific socio-economic and hydrogeological conditions of each Member State (SADC-GMI, 2019). The synergies and commonalities between the DFS of SADC-GMI (2019) and the analytical framework for groundwater governance described in Wijnen et al. (2012) are recognised in this chapter.

12.1.2   Groundwater occurrence in Malawi

Development of groundwater resources has been primarily for drinking water supply for both rural and periurban areas. Malawi has three major aquifer systems, namely, the extensive but low yielding weathered Precambrian Basement Complex aquifer of the plateau area (1–2 L/s), the high yielding alluvial aquifer of the lakeshore plains, the Lower Shire Valley and the Lake Chilwa – Mphalombe Plain (>15 L/s), and the medium yielding aquifer of the fracture zone in the rift valley escarpment (5–7 L/s) (Chilton and Smith-Carrington, 1984; BGS/WaterAid, 2008; Chavula, 2018). The prolonged *in situ* weathering of the crystalline basement rocks has produced a layer of unconsolidated saprolite material, and it is this, which forms an essential source of groundwater for rural domestic requirements (SDNP, 1998). The average yield in the weathered zone of the basement complex lies in the range of 1–2 L/s (Laisi, 2009). Groundwater provides a significant contribution to a rural water supply through hand-pumped boreholes and protected shallow wells. They also contribute towards urban and market centres water supply through motorised-pump equipped boreholes reticulating to piped water supply systems. The contribution of groundwater to the domestic water supply is estimated at 29% (Pietersen and Beekman, 2016).

Groundwater quality in Malawi is highly dependent on aquifer lithology (Figure 12.2); it varies spatially (Chavula, 2012) and is generally suitable for drinking. Nevertheless, some water points yield water of high electrical conductivity, which indicates a level of mineralisation (Chilton and Smith-Carrington, 1984). High sulphate levels occur in some areas of the weathered basement (BGS/WaterAid, 2008). Several boreholes in alluvial aquifers are not in use due to high salinity (Chavula, 2012). Groundwater from boreholes is generally of better microbial quality than that from shallow wells, which is more vulnerable to contamination due to poor siting of sanitation facilities, and in some cases, runoff wastes easily find their way into shallow wells.

Most of the groundwater infrastructure (dispersed wells without defined protection zone) is in a state of disrepair and requires significant rehabilitation due to poor management arrangements (lack of resources for maintenance and lack of skilled or trained human resources to do the required technical work) (Pritchard et al., 2008). About 30% of shallow wells studied in six districts in Malawi in 2007 were not functional (Pritchard et al., 2008). A survey on the functionality of boreholes equipped with hand pumps undertaken in Malawi in 2016 indicated that 74% of hand pump boreholes (HPBs) are functional at any one point. About 66% of HPBs passed the design yield of 0.25 L/s, while 55% of those passing the design yield also experienced less than 1-month downtime within a year. About 43% of HPBs, which passed the design yield and reliability, also passed the World Health Organization (WHO, 2011) standards of water quality. The Malawi Bureau of Standards (MBS, 2005) and WHO prescribe more stringent limits as compared to the temporary guidelines for untreated water developed by the ministry responsible for water (MoWD, 2003) standards and as such the percentage of compliant boreholes are expected to be higher when compared to the MBS (2005). Vandalism, lack of maintenance and inadequate capacity are some of the challenges affecting groundwater infrastructure (MoAIWD, 2017).

More than 57,000 groundwater access points exist across the country, which has helped to increase the access to safe water in rural areas to 85.7% (GoM, 2017). However, some problems remain such as low functional ratios of 77% (MoAIWD, 2017), seasonal variations of the water levels, lack of maintenance funds, and water quality issues amongst others. Drought and poor water quality have meant more people are turning from hand-dug wells to drilled boreholes. Estimates show that

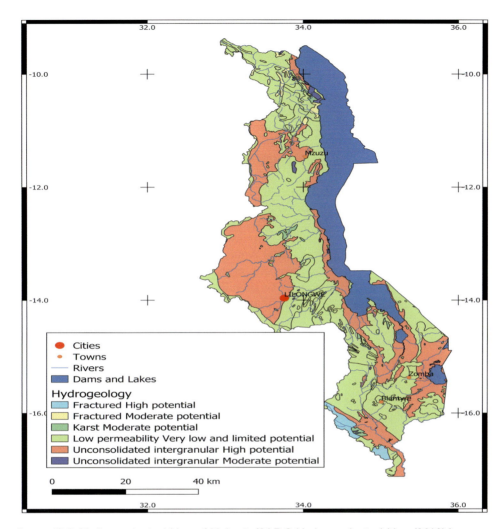

Figure 12.2 Hydrogeological Map of Malawi. (SADC Hydrogeological Map (2010).)

65% of the population depends on groundwater for domestic supply. In rural areas, this rises to 82%, while in urban areas, it is closer to 20% (Chavula, 2018). Some urban centres in Malawi get their water supply from groundwater, e.g., the Madisi (Dowa district), Salima (Salima district), Karonga (Karonga district), Nkhotakota (Nkhotakota district) and Ngabu (Chikwawa district) (Chavula, 2018). It is envisaged that more rural areas and towns in Malawi will get their water supplies from groundwater resources in the future because of the unreliable rainfall and irregularity of river flows, currently abstracted to sustain gravity-fed rural piped water schemes and urban water.

Groundwater use for irrigated agriculture is, at present, mostly confined to growing vegetables and maize in dambo areas during the dry season (Chavula, 2018).

12.2 METHODOLOGY

12.2.1 PLI framework situation analysis

The methodology for conducting the groundwater management PLI gap analysis in Malawi followed the one applied by SADC-GMI across all the 16 SADC Member States as schematically presented in Figure 12.3.

The research was started with a desktop review of available literature on the PLI frameworks in the country. The literature collected for this purpose consisted of relevant policies, legislation, tools and guidelines available on (ground) water resources in Malawi.

This was followed by applying the DFS model developed at SADC level to provide a benchmark for groundwater governance. The DFS sets out the *minimum* governance frameworks that support the delivery of national, regional and international developmental goals, including the Sustainable Development Goals (SDGs), meeting basic human needs to water, energy and food (the Water-Energy-Food (WEF) nexus), and the protection of ecosystems that are dependent on groundwater (SADC-GMI, 2019). This framework provides the vision of what an enabling PLI framework would be required for sustainable groundwater development and management in the country.

The DFS was contextualised for the SADC region, taking into account the following:

- The high levels of groundwater dependency in many SADC countries, in rural areas in particular;
- The variety of hydrogeological contexts;
- High levels of poverty, gender disparities, social exclusion and pollution and
- Relatively low levels of state capacity as assessed through skills, infrastructure and finance.

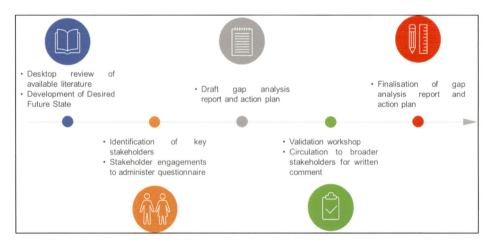

Figure 12.3 Methodology Outline.

Using the literature review and the DFS for the country, a questionnaire was developed based on the data that would be required to identify the gaps. The questionnaire was used in a nationwide survey targeting stakeholders in the Republic of Malawi. Respondents to the questionnaire were identified by the In-Country Expert from the Stakeholder Register. The questionnaire was administered to the identified Stakeholders in the Republic of Malawi by a Consultant based in the Republic of South Africa with the support of the In-Country Expert. The Consultant subsequently used the data from the responses in the questionnaire to evaluate the current state of the PLI groundwater management framework for the country. The resultant draft report was validated by the stakeholders from the Republic of Malawi in a validation workshop.

The MoSCoW method of prioritisation was used to develop the action plan. This method identifies the *Must have*, *Should have*, *Could have* and *Won't have* elements for the Groundwater Management Regulatory Framework. The core actions are set out using the MoSCoW approach towards prioritisation (see Box 12.1 for the description of *MoSCoW* approach).

The results of stakeholder engagements culminated in the elaboration of a draft gap analysis report and action plan for the country. The findings of the PLI situation analysis were validated at Validation Workshops organised by SADC-GMI for each Member State. These workshops involved key groundwater actors from the 16 SADC Member States' experts who provided an opportunity to obtain buy-in and support for the gap analysis report as well as obtaining further inputs. The draft report was also circulated to the key stakeholders in order to gather written comments. The draft gap analysis report was then finalised based on the comments received.

Box 12.1 The MoSCoW method of prioritisation

Upon establishment of a clear set of actions to address the gaps identified for each Member State, it was necessary to rank them in order of importance. The ranking helped to develop an understanding around the level of urgency for each action listed, effectively sequencing actions from the most important to least important. This level of prioritisation for the action plans was achieved using the MoSCoW method.

The acronym MoSCoW stands for Must, Should, Could and Would. In full, the acronym stands for:

M = Must have this action(s) to significantly address existing gaps,
S = Should have this action(s) if possible, but success does not rely on it,
C = Could have this action,
W = Would like to have this action item, but it is not terribly urgent.

The method is a useful tool to help assess and decide on which actions to come first, and which ones may come later. The MoSCoW method was developed by Dai Clegg of Oracle UK in 1994 and has been made popular by exponents of the Dynamic Systems Development Method (DSDM).

Source: https://www.projectsmart.co.uk/moscow-method.php.

220 Groundwater for Sustainable Livelihoods

12.3 POLICY LEGAL AND INSTITUTIONAL FRAMEWORK FOR GROUNDWATER MANAGEMENT IN MALAWI

12.3.1 Policies to support groundwater management

The National Water Policy (NWP) (2005) for Malawi makes provision for the protection of groundwater by preventing pollution and overuse. The government policy is to provide clean potable water to all people to reduce the incidence of water-borne diseases and the time devoted by individuals, particularly women, for water collection. It has general principles, objectives and strategies on water resources development, pollution and water supply including the roles of stakeholders in water management, for example, Integrated Water Resources Management (IWRM) and consideration of cross-cutting issues. The NWP (2005) is consistent with the National Sanitation Policy (2008) as both policies aim at the provision of sanitation services that are equitably accessible to and used by individuals and entrepreneurs, on the promotion of public health and hygiene. Both policies also promote private sector participation in the delivery of water supply and sanitation services and setting standards of treated and untreated water supply services.

Literature acknowledges the role of groundwater as a highly valuable source of domestic and agricultural water supply and a vital resource for poverty alleviation, food security, and the sustainable economic development of rural areas. However, the NWP (2005) considers water resources mostly in general terms and does not

- include exploring and development of deeper high yielding boreholes in rural areas that can be reticulated;
- recognise that other players collect water resources data, hence does not mention the need for the use of harmonised tools in data collection among institutions that collect groundwater data;
- mention the need for comprehensive desktop studies for every water development work, including drilling work;
- mention the need for the identification of potential risks to groundwater as part of field practice;
- recognise gender balance resulting in all objectives, policy statements and actions being gender blind;
- address the biophysical and ecological linkages between ground and surface water for their use, protection and management.

The policy review also speaks to issues around gender and the absence of gender issues at the policy level. The inclusion of these aspects at the policy level has been noted as a necessity to provide a framework for the improved contribution of groundwater in enhancing the access to water by the communities in Malawi. Lastly, groundwater management has not been duly taken into account in other sectoral policies, for example, the National Environmental Policy of 1996 and the National Forest Policy of 2016.

12.3.2 Legislation to support groundwater management

The Water Resources Act of 2013 (WRA, 2013), Water Resources Regulations (2018) and Environmental Management Act (2017) are key pieces of legislation in place that explicitly address the use, management and protection of groundwater and provide the necessary tools for the state to regulate, manage, control, protect and develop groundwater resources in conjunction with surface water resources in Malawi.

The human right to water is mentioned in Sections 37 and 38 of the Water Resources Act of 2013 in terms of groundwater legislation, facilitating prioritisation of drinking water and basic human needs, as well as small-scale users. For example, the development of hand pump wells does not require an Environmental Empact Assessment (EIA). This can be viewed as a consideration for small-scale users not being prohibited from using groundwater for their domestic water uses.

Although this section does not present an exhaustive discussion of the legislative framework for groundwater management in Malawi, the gaps identified in the analysis of the legal framework are presented.

12.3.3 Strategy and guidelines necessary to support groundwater

The first set of technical manuals for Malawi were developed in 2001 by the Ministry of Water Development, which was revised in 2016. The current 2016 Technical Manuals (http://www.rural-water-supply.net/en/resources/details/807) describe borehole aspects associated with groundwater development mainly for rural domestic supply and groundwater monitoring boreholes and the associated groundwater monitoring or management aspects thereof. The revision added a chapter on groundwater monitoring, database management and water permit. The accompanying documents are Standard Operating Procedures (SOPs) for groundwater sampling, aquifer pumping test, groundwater level monitoring, groundwater use permit, drilling and construction of national monitoring boreholes, operation and management of the national groundwater database.

The Ministry of Agriculture, Irrigation and Water Development – through the Water Quality and Groundwater Divisions and District Offices – provides policy guidance on the development and use of groundwater. However, there is no Water Quality and Groundwater Divisions at the district level; hence, the divisions' functions are performed by Water Monitoring Assistants from the Water Supply Department.

When benchmarked against the minimum requirements for the DFS, no significant shortcomings were found relating to the guidelines and standards for groundwater management. Malawi has done reasonably well to develop guidelines (SOPs) for groundwater development, i.e. Technical Manual – Water Wells and Groundwater Monitoring Systems, SOPs for Drilling and Construction of National Monitoring Boreholes, Aquifer Pumping Tests, Groundwater Level Monitoring, Groundwater Sampling, Operation and Management of the National Groundwater Database, Groundwater Use Permitting and SOP for Drilling and Construction of Production Boreholes. There is a need for advocacy and enforcement on the guidance documents amongst stakeholders, through, for example, training of contractors or a certification programme, which can be part of the legislation. It was also noted that there is a lack of technical capacity in the government department at district levels. At the same time, there is capacity at the national level, and the national staff cannot adequately provide support to enable implementation of these guidelines in all the districts, hence the need for building capacity at the district level.

12.3.4 Institutional framework

12.3.4.1 Institutional arrangements to support groundwater management

The Central Government of Malawi is responsible for strategic planning, coordination, quality assurance and technical assistance systems, including collaborative efforts with donors/NGOs and the private sector. The Ministry of Water Development and Irrigation is the lead ministry to provide overall policy direction for water services in Malawi. The Ministries of Health, Environmental Affairs, Finance, Gender, Industry, as well as Local Government and Rural Development are also involved in the water sector (MoAWDI, 2015).

The ministry responsible for water development and irrigation comprises of five departments, of which the Department of Water Supply and the Department of Sanitation are responsible for the rural water supply and sanitation services. The groundwater division is responsible for groundwater research, storage of borehole data, issuance of groundwater abstraction rights through the Water Resources Board, whereas the responsibility to operate and maintain boreholes, rests with the Water Supply Department (Chavula, 2012). More specifically, both the Department of Water Resources (Groundwater Division) and Department of Water Supply are responsible for policy formulation on groundwater development in Malawi. The Water Supply Department has two divisions, with the first division being responsible for the operation, maintenance and monitoring and the second is responsible for planning, design and construction. However, the department is understaffed. The 2012/2013 Sector Performance Report indicates that out of the total 493 established staff posts for Water and Sanitation Department, only 143 were filled (MoAWDI, 2015). The Groundwater Division in the Ministry of Agriculture, Irrigation and Water Developments (Figure 12.4) is responsible for leading capacity-building initiatives and for guiding the curriculum offered at tertiary institutions. Management of groundwater is mostly according to district boundaries, although there are seventeen Water Resources Areas (WRAs), which are the recommended units of management.

The Local Governments are responsible for the provision and management of rural water supply and sanitation services, in liaison with the ministry responsible for water. Local Governments carry out planning, budgeting and resource allocation, community mobilisation and participation, follow-up implementation by the private sector and support the operation and maintenance (O&M) of water services, monitoring and reporting among other duties. However, the capacities of the Local Governments are still inadequate to undertake these tasks more so concerning the technical aspects of water supply from groundwater resources. District Councils do not have a revenue base to fund investment, and they also do not have the funds to contribute to the operations and maintenance of existing facilities. Financing of rural water supply is restricted by the deficient level of fiscal devolution (MoAWDI, 2015), thereby hindering the capacity of the District Councils to sustainably supply water to the communities.

Local leaders such as Village Head (VH), Group Village Head (GVH), and Traditional Authorities (TA) also play roles in the management of rural water supply. Where there is strong local leadership, communities are actively engaged in the implementation, O&M of water supply facilities (MoAWDI, 2015).

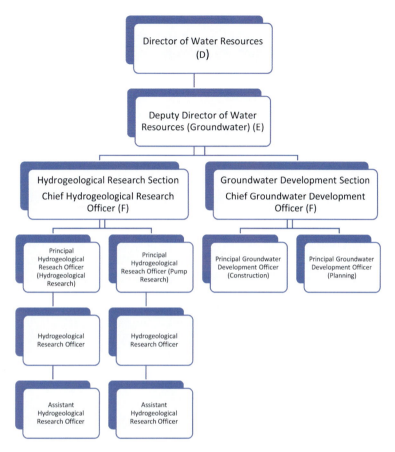

Figure 12.4 Organogram for Water Resources Department: Groundwater Division (Headquarters) under the Director of Water Resources.

Community users are organised in several forms to enable their full participation in the planning and implementation stages of O&M of the water supply facilities. The Water Statute provides the legal platform for the formation of Water and Sanitation Committees, Water User Groups and Water User Associations at the community level that will ensure sustainability and proper management of the facilities (MoAWDI, 2015).

Malawi has received considerable support from development partners for funding the development budget, including rural water supply and sanitation. The major donors include the World Bank, African Development Bank, European Union, Department for International Development (DFID), United Nations Children's Fund (UNICEF) and ASEAN Infrastructure Fund (ACGF). Some donors transfer funds directly to the National Water Development Programme (NWDP), while others manage the funds on their own (MoAWDI, 2015).

About forty-six NGOs are operating in twenty-six districts in the rural water supply sector. They are essential and useful partners of government in the development and are well placed to raise public awareness and build capacity at the local level.

224 Groundwater for Sustainable Livelihoods

In the meantime, most of the works of these NGOs are focused on hardware with very little, if any, on raising awareness and capacity building at the local level.

The Water Resources Act of 2013 (2013) provides for effective implementation, including the mandate, competence and power of the relevant authorities per governance principles. All groundwater development activities are required to be supervised or carried out by competent professional hydrogeologists, drilling companies and other bodies and professionals.

There are Water User Associations (WUAs) in Malawi. These are legal entities, which operate as 'small water boards' at the community level. They are responsible for overseeing the O&M of rural water supply systems. However, most of these WUAs have technical and financial challenges, especially those for rural water supply (mostly gravity-fed). As for the ones, which are under water boards, they are better off but still they face political interference (e.g. they can impose the leadership, they can inform the users that you should not be paying for the water). At the water point level, there are Water Point Committees who are trained under Community-Based Management (CBM) to maintain the facility. At the district level, there are structures such as the District Coordination Team (DCT) (composed of water-related sector Heads), which is the technical arm of the district council that looks into water and sanitation issues. The long-drawn and ongoing decentralisation process in Malawi is not devolving both functions and resources down to the districts. As such even though the structures could be there on paper, they are too weak thus culminating in the glaring technical gaps. Most of the technical capacity comes directly from state level where insufficient resources are allocated from the national budget.

12.3.4.2 Institutional gaps and challenges identified

The main gaps to be addressed on institutional arrangements for groundwater management in Malawi include the following:

i. Inadequate human, financial and technological capacity. There are less than fifteen qualified hydrogeologists in the country, which negatively affects the supervision and monitoring of groundwater development programmes. Capacity building in groundwater development needs immediate redress, as most of the people involved in groundwater development in Malawi are not fully qualified hydrogeologists/groundwater experts. The lack of capacity also applies to drillers, people involved in groundwater exploration, and managers of groundwater projects; recurrent financial resources are usually not adequate to conduct groundwater management unless there are Development Projects; no equipment for deeper drilling and lack of formal drilling vocational training programmes;

ii. Weak enforcement of regulation mainly due to non-operationalisation of the National Water Resources Authority (NWRA), which has a mandate for enforcement of legislation. Its operationalisation would therefore improve enforcement of regulations; Meanwhile, it is only the Board and Secretariat that is in place;

iii. Inadequate data exchange among institutions involved in groundwater development;

iv. Inadequate comprehensive groundwater research;

v. Inadequate comprehensive knowledge in the groundwater availability and quality;
vi. Lack of provision for the creation of a professional body of National hydrogeologists to register and control the participation of hydrogeologists in Malawi who may affect the quality of outputs from national groundwater development activities;
vii. Wellfields Consulting Services (2011) reported that as of 2002, there was no documented routine groundwater monitoring to collect groundwater data. However, over the last decade or so, there has been a great effort to develop a national groundwater monitoring network and a database. Over the years, the monitoring network comprises seventy-five boreholes equipped with automatic data loggers. Data are cross-checked with manual measurements. After borehole completion, groundwater quality checks should be made every 6 months, but many gaps are reported. pH, EC and TDS are measured directly in the field. Major ions are measured in the lab, with a bacteriological analysis when requested. Monitoring data are first processed in and checked in spreadsheets, then entered into WISH and Hydstra databases. However, groundwater abstraction is not monitored;
viii. Lack of a dedicated groundwater level, quality and yield monitoring programme.

12.4 DISCUSSION

Malawi is categorised as water-stressed country, with estimated water availability during 2017 as $1,050 \text{ m}^3$/capita/year (Global Water Partnership, 2009). With the stress expected to increase with increasing demand for water in the light of climate change and population growth, the population of the Republic of Malawi is becoming more reliant on groundwater, which provides a buffer against the impact of climate variability and climate changeHowever, often groundwater is only considered for water supply during times of drought and on an emergency basis, thus compromising the sustainable utilisation of the resource and limiting the benefits accrued from groundwater by the communities. Reasons for the limited focus are myriad, including the absence of an enabling PLI environment. This study has identified the groundwater PLI shortcomings in Malawi as benchmarked against the DFS (FAO, 2016).

Although it is acknowledged that groundwater plays a central role in the livelihoods for most Malawians, there remain some, PLI shortcomings, which hinder the nation from fully benefiting from groundwater, notably: Groundwater is not explicitly included in the water resources statutes, a situation which is not an exception to Malawi. Even the SADC Protocol on Shared Watercourses (2000) does not make explicit reference to groundwater.

Identified gaps at the policy level relate to, lack of policy on the exploration of deep aquifers to supply communities through centralised schemes, limited gender mainstreaming initiatives, inadequate policy direction on biophysical and ecological linkages between ground and surface water for their use, protection and management.

The Republic of Malawi has a significant complement of strategies and guidelines that support groundwater management. Challenges, however, relate to the dissemination and enforcement of the guidelines. There is inadequate sectorial coordination between state institutions.

Pietersen and Beekman (2016), noted that there is room for further development of groundwater resources in Malawi. For a country that relies on groundwater, the less than fifteen qualified hydrogeologists reported in this research are inadequate. A minimum of one Hydrogeologist per district would go a long way to alleviate the problem. As identified by Pietersen and Beekman (2016) and FAO (2016), capacity for groundwater monitoring and data storage is also lacking. Capacity Challenges in Malawi relate to both public and private sectors, with capacity shortcomings in the former being more apparent at the district level.

Before the gap analysis was conducted, the DFS had to be developed. The DFS provides a baseline of what the groundwater management framework should provide as the minimum requirement to support the delivery of national, regional and international developmental goals, including the SDGs, meeting basic human needs to WEF nexus, and the protection of ecosystems that are dependent on groundwater. The DFS was contextualised for Malawi and based on the MoSCoW method of prioritisation a number of actions were identified as follows:

12.4.1   Must haves

Policy

i. Develop a National groundwater management policy with clear objectives.
ii. Include transboundary issues and the establishment of cooperation mechanisms.
iii. Establish groundwater management institutions/structures.
iv. Operationalise the National Water Resources Authority (NWRA).
v. Include policy direction on how institutions should collaborate.

Legislation

i. Regulate abstraction and recharge to protect groundwater quantity and quality.
ii. Enforce existing legislation on environmental management.
iii. Enforce legislation on groundwater data collection and management.
iv. Develop groundwater-specific management legislation.
v. Include groundwater management users in WUAs and Catchment Management Committees (CMCs).
vi. Create a groundwater management association or national body.
vii. Allow for zoning or similar mechanisms to protect overused/vulnerable aquifers.

Institutional

i. Strengthen and harmonise legal and institutional framework in water sector and other related sectors.
ii. Create a National groundwater management association.
iii. Operationalise NWRA.

Strategy/Guidelines

i. Raise awareness on groundwater management issues, through the dissemination of guidelines.
ii. Develop guidelines to explicitly integrate groundwater for catchment-level or basin-level planning.
iii. Undertake comprehensive research on groundwater for proper management.
iv. Build capacity in groundwater development (e.g. drillers, people involved in groundwater exploration, and managers of groundwater projects).
v. Develop guidelines to include guidelines for; Groundwater – PLI development, building groundwater management resilience, Strategic approach to financing groundwater management, and guidelines for O&M of groundwater schemes.
vi. Include transboundary basins in the formal protocols and standards on data collection and storage.

12.4.2 Should haves

Policy

i. Policy to be gender-sensitive in its terminology.
ii. Include the mandatory installation of monitoring infrastructure of boreholes especially for large-scale users.

Legislative

i. Strengthen and harmonise legal and institutional framework in the water sector and other related sectors.
ii. Legislation to be gender-sensitive in its terminology.

Institutional

i. Operationalise the National Water Resources Authority (NWRA).
ii. Community awareness-raising on environmental issues such as catchment protection.
iii. Procurement of state of the technologies for groundwater management.
iv. Stakeholder awareness on groundwater management.

Strategy/Guidelines

i. Establish clear mechanisms for enforcing strategies and guidelines.

12.4.3 Could haves

Policy

i. Develop a National groundwater management policy with clear objectives.
ii. Provide for groundwater management leadership.

Legislative

i. Upgrade existing legislation to clearly promote conjunctive use which refers to the sustainable utilisation of all water sources including groundwater.
ii. Recognise and legalise affordable, small-scale and indigenous solutions to groundwater use.

Institutional

i. Experience sharing nationally and regionally.

Strategy/Guidelines

i. include clear definitions of groundwater services functionality in guidelines.

Moving forward, it is desirable that a road map is developed to close the gaps identified. The ultimate aim of the roadmap is to develop the strategic response required to further advance sustainable groundwater development and management and to address gaps and challenges in a participatory manner.

12.5 CONCLUSIONS

The analysis of the groundwater management PLI framework for the Republic of Malawi presented in this chapter demonstrates that country has made significant strides towards providing an enabling environment for sustainable groundwater management. While the policy and the associated water resources regulations are comprehensive (covering most of the expected dimensions from the DSF), the main challenges in the Republic of Malawi are weak enforcement of these instruments on groundwater protection from pollution and overuse and lack of hydrogeological expertise at the district level. Staffing for some functions such as Groundwater Division is not there in addition to a need to advocate for an inclusive interpretation of the term watercourse amongst policymakers. We conclude that based on the hydrogeological setting of the Republic of Malawi, there is a huge potential for groundwater to contribute to the water security of Malawi, however, this needs to be supported by implementing the provisions of the existing, as well as seeking improvements in the enabling PLI framework in the country. The success of this endeavour to elevate the role of groundwater in the country will very much depend on the convening capacity of the Ministry of Agriculture, Irrigation and Water Development to involve and engage other stakeholders from relevant sectors to sustain cross-sectoral partnerships to increase water security.

REFERENCES

BGS/WaterAid. (2008) Groundwater Quality: Malawi. British Geological Survey/WaterAid. Natural Environment Research Council.Chavula, G.M.S. (2012) *Malawi in.* Pavelic, P.; Giordano, M.; Keraita, B.; Ramesh, V; Rao, T. (Eds.). 2012. *Groundwater availability and use in Sub-Saharan Africa: A review of 15 countries.* Colombo: International Water Management Institute (IWMI), 274 p. doi: 10.5337/2012.213.

Chavula, G.M.S. (2018) *Malawi Country Report on the Water, Energy and Food (WEF) Nexus.* A Malawi Country Report submitted to SADC as part of the SADC - EU Project on "Fostering Water, Energy and Food Security Nexus Dialogue and Multi-Sector Investment in the SADC Region." Lilongwe.

Chilton, P.J., Smith-Carington, A.K. (1984) *Characteristics of the weathered basement aquifer in Malawi in relation to rural water supplies.* Challenges in African Hydrology and Water Resources (Proc. Harare Symp., July 1984) 235–248. IAHS publ. no.144.

Cobbing, J. (2020) Groundwater and the discourse of shortage in Sub-Saharan Africa. *Hydrogeology Journal* 28, 1143–1154.

FAO. (2016) Global Diagnostic on Groundwater Governance.

Global water partnership. (2009). Iwrm Survey and Status Report, p. 29.

GoM. (2017) *Environmental Management Act.* http://www.sdnp.org.mw/enviro/act/contents. html. Accessed: 18 August 2018.

GoM. (2018) *The Action Plan.* Government of Malawi. http://www.sdnp.org.mw/enviro/action_ plan/chap_4.html. Accessed: 4 April, 2020.

IMF. (2017) Malawi: economic development document. *The World Bank.* IMF Country Report No. 17/184.

Laisi, E. (2009) *IWRM Survey and Status Report: Malawi.* Global Water Partnership – Southern Africa.

MBS (Malawi Bureau of Standards). (2005) Malawi standard; drinking water – specification. Malawi Standards Board, MS 214:2005, ICS 13.030.40 (first revision).

MoAIWD. (2017) *National Water Resources Master Plan.* Annexe 2: Groundwater Resources. Ministry of Agriculture, Irrigation and Water Development. Republic of Malawi.

MoAWDI. (2015) *Malawi Rural Water Supply Investment Plan: 2014–2020.* National Water Development Programme.

MoIWD. (2008) *Water and Sanitation Sector: Joint Sector Review Report.* Ministry of Irrigation and Water Development.

MoWD (Ministry of Water Development). (2003) Government of Malawi; Devolution of Functions of Assemblies: Guidelines and Standards.

Pietersen, K., Beekman, H. (2016) Groundwater management in the Southern African Development Community.

Pritchard, M., Mkandawire, T., O'Neill, J.G. (2008) Assessment of groundwater quality in shallow wells within the southern districts of Malawi. *Journal of Physics and Chemistry of the Earth* 33, 812–823.

SADC. (2000) Revised Protocol on Share Watercourses in the Southern African Development Community. https://www.sadc.int/documents-publications/show/Revised_Protocol_on_ Shared_Watercourses_-_2000_-_English.pdf. Accessed: 30 March 2020.

SADC. (2016) Regional Strategic Action Plan on Integrated Water Resources Development and Management Phase IV, RSAP IV, Gaborone, Botswana, 2016.

SADC-GMI. (2019) *Gap Analysis and Action Plan – Scoping Report: Malawi.* SADC GMI report: Bloemfontein, South Africa.

SDNP. (1998) *State of Environment Report for Malawi.* Chapter 6. http://www.sdnp.org.mw/ enviro/soe_report/chapter_6.html. Accessed: 30 March 2020.

Velis, M., Conti, K.I., Frank Biermann, F. (2017) Groundwater and human development: synergies and trade-offs within the context of the sustainable development goals. *Sustainability Science and Implementing the Sustainable Development Goals* 12, 1007–1017.

Wijnen, M., Augeard, B., Hiller, B., Ward, C., Huntjens, P. (2012) *Managing the Invisible Understanding and Improving Groundwater Governance.* Water Partnership Programs, The World Bank. http://www.worldbank.org/water. Accessed: 30 March 2020.

World Bank. (2018) Assessment of Groundwater Challenges & Opportunities in Support of Sustainable Development in Sub-Saharan Africa.

Chapter 13

Groundwater

A juggernaut of socio-economic development and stability in the arid region of Kachchh

P.M. Patel
International Water Management Institute

D. Saha
Central Ground Water Board, Ministry of Jal Shakti, Government of India

CONTENTS

13.1 Introduction... 231
    13.1.1 Kachchh: a true oasis 232
    13.1.2 Water resources of Kachchh: the concealed treasures 234
    13.1.3 2001: the year of change for Kachchh 236
13.2 Comparing the development story of Kachchh (pre-2001 and post-2001) 237
    13.2.1 Achieving new milestones in the agriculture sector and animal husbandry 237
    13.2.2 Attracting large industries, extracting minerals and creating shipping hub 239
    13.2.3 Opening new gateways through the tourism 241
13.3 Groundwater in the story of Kachchh: experience from the field explorations ... 241
13.4 Analysing the factors driving the groundwater depletion in Kachchh 243
13.5 Reviving from the stage of peril 246
    13.5.1 Institutionalising groundwater 246
    13.5.2 Preventive measures (managing supply) 247
    13.5.3 Adaptive measures (managing demand) 248
13.6 Conclusion 249
Acknowledgments 249
Notes 250
References 250

13.1 INTRODUCTION

A belt of land, 160 miles from east to west and from thirty-five to seventy from north to south, Cutch (read Kachchh) is almost cut off from the continent of India, on the north and east by the Arabian sea and the eastern or Kori mouth of the river Indus. From its isolated position, the special character of its people, their peculiar dialect, and their strong feeling of personal loyalty to their ruler, the peninsula of Cutch has more of elements of distinct nationality than any other dependencies of the Bombay Government.

(Campbell, 1880)

DOI: 10.1201/9781003024101-13

13.1.1 Kachchh: a true oasis

The opening paragraph from the earliest accounts of the erstwhile Bombay Presidency Gazetteer prepared during the late 19th century neatly captures the essence of the westernmost peninsula of India, Kachchh. The current administrative district of Kachchh has preserved unique geology, culture, and stories of desert communities' tryst with the destiny. In Sanskrit, Kachchh means an area surrounded by water, a name derived from 'Kachhua' or the tortoise. Ironically, the region surrounded by water is one of the most water-scarce regions of India. Kachchh is surrounded on two sides by sea and by the desert on the other two sides. The Gulf of Kachchh lies on the southern side of Kachchh. As shown in Figure 13.1, Kachchh has 45,652 km^2 of the total landmass, most of which is barren or desert land (54%). The region is isolated from mainland India through Rann of Kachchh and Little Rann[1] of Kachchh, a salt desert spread over 24,617 km^2 and are partially inundated under seawater during high tide. Only 20.7% of the total land in Kachchh mainly known as Mainland Kachchh (see Figure 13.1) is available for agriculture and settlements after the exclusion of forest, wasteland and grasslands. Kachchh earns its oasis status through the greenbelt of Mainland Kachchh that is surrounded by salt beds and sea, and provides to nomadic tribal communities of surrounding desert areas (Table 13.1). The geological formations in Kachchh represent a wide variety of sedimentary and igneous suite of rocks ranging in age from Pre-Cambrian to Recent. The central part of the Kachchh region is marked as Kachchh mainland (Figure 13.1) (Saha and Gor, 2019). The Kachchh mainland is thriving agriculture and economically vibrant oasis in the otherwise dry region of Kachchh (Figure 13.2).

The nature of the land can be introduced from the Köppen Climate Classification that puts Kachchh under subtype 'Bwh' (Tropical and Subtropical Desert Climate) (Beck, et al., 2018). The region also experiences high spatial and temporal temperature variation within its boundaries. The peak summer temperature range is close to 45°C, whereas the cold winter nights can be 2°C to 3°C in some parts. With a daily average of around 35°C, May is the warmest month. January has the lowest average temperature of the year, around 20°C. It follows a similar seasonal cycle like the rest of India with three distinct seasons: Monsoon (June/July–September/October), Winter (November–February) and Summer (March–June). Most of Kachchh receives approx. 350 mm of rainfall within 15–20 rainy days but with very high variability (45% coefficient of variation). The region frequently experienced droughts and below-normal rainfall. Rainfall analysis of 1901–2017 indicates that of every 4 years, 1 year is dry (<250 mm), 2 years are around normal (350 ± 100 mm) and 1 year receives above-normal (>450 mm) rainfall (Patel and Khan, n.d.). Rivers or rivulets are mainly drainages for the rainwater and have very little availability of water beyond the monsoon season. They flow from the central highlands to either north or south to sea or salt beds (Figure 13.1b). To make the living conditions further harsh, Kachchh is one of the most tectonically active zones (Antolik and Dreger, 2003). Records cited by Hathi (2019) show that between 1819 and 1947, Kachchh was struck by seventy earthquakes of varying intensities. The frequency of the jolts has continued consistently since 1947 with frequent tremors and colossus earthquakes in 2001.

With 2.09 million residents (34% urbanisation), Kachchh is one of the least populated regions of India. Historically agriculture and animal husbandry are

Groundwater in the arid region of Kachchh 233

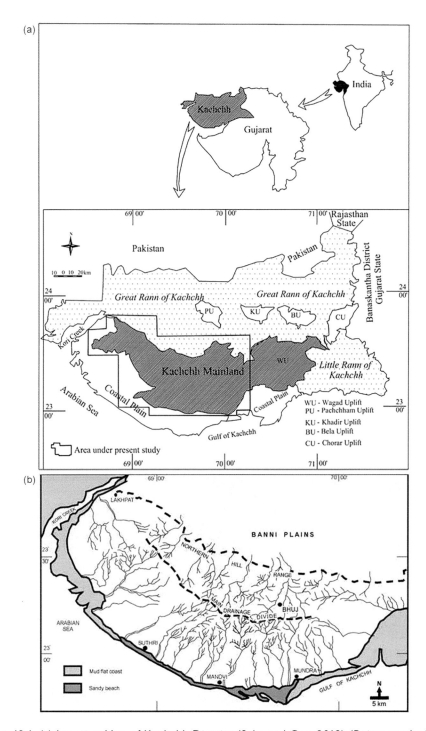

Figure 13.1 (a) Location Map of Kachchh District (Saha and Gor, 2019) (Points marked with arrows are the locations of the surveyed villages) (b) Drainage Map of Kutch.

Table 13.1 Land utilisation pattern

Land use	Area (km^2)	Percent of total area
Forest	1,836	4.02
Irrigated	710	1.56
Unirrigated	6,626	14.51
Culturable wasteland	2,117	4.64
Area not available for cultivation	5,746	12.59
Banni grasslands	4,000	8.76
Area covered by Rann	24,617	53.92
Total area	45,652	

Source: *Census Handbook*, GoG (2001).

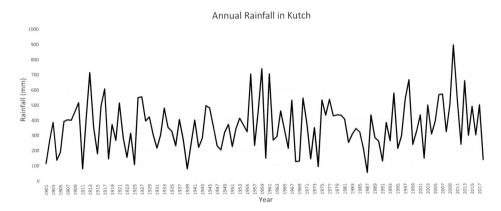

Figure 13.2 Annual rainfall in Kachchh district 1901–2018 (Indian Meteorological Department (IMD)).

predominant activities with almost half of the population associated with the sectors here. Salt processing, lignite mining, bentonite and China clay processing are other natural resources-based economic activities prevalent in the district. Handicraft and handloom, traditional industries are another major employer as the region has its unique traditional heritage for creative appeal (MSME, 2016). In his letter to historian Ramsinh Rathod dated February 20, 1949, then Home Minister and Deputy Prime Minister of India Sardar Vallabhbhai Patel accepted that the Kachchh region was untouched by the modern world. It has a unique culture, heritage and creative acumen of artisans. He also acknowledges the parched land's water woes and the need for plans to revive the geographically important border region for Independent India (Patel, 1949).

13.1.2 Water resources of Kachchh: the concealed treasures

Located in the water-scarce arid region of India, Kachchh suffers from acute shortage of water. More than 197 rivers and streams located here are all ephemeral and do not hold water beyond the months of Monsoon. The scarcity of water led to many unique solutions for the Kachchh region. Kachchh traditionally had water harvesting structures such as shallow wells, tanks and stepwells (also known as *Vav*) as traditional rainwater harvesting to sustain the extreme summer seasons. Apart from these widespread structures in Mainland Kachchh and southern Kachchh, the Banni Uplift (see Figure 13.1) which was mostly inhabited by the *Maldharis* (nomadic tribes) used to build *virdas*[2] for harvesting rainwater. The Banni region has brackish groundwater and saline land. To store the freshwater neither wells nor surface tanks are viable options. These *virdas* are seen as a symbol of the adaptive capacities of the *Maldharis* (Agarwal and Narain, 1997). The water harvesting structures are often created as drought-proofing mechanisms either by the community chiefs or erstwhile rulers. Most of these structures predate the Indian Independence era. The scanty monsoon season often rendered very few spells of good rainfall in Kachchh. The erstwhile structures were good to provide to drinking water needs but not sufficient enough to irrigate the large land available in Kachchh.

After Independence, Kachchh has witnessed the construction of several small and medium dams. By 1965, Kachchh has now 182 surface water minor and medium irrigation projects with an irrigation potential of about 52,000 ha, but actually it caters to about 15,000–20,000 ha (Bharwada and Mahajan, 2002). Currently out of 204 minor and medium irrigation schemes, rarely any scheme was started after 2000. Because of the pertinent water shortage and scanty rainfall, the government has worked on plans for the development of the Narmada canal for drinking and irrigation water demands. (A 400 km long large canal structure to provide approx. 600 MCM water for irrigation and other usages.)

Mainland Kutch was cultivating some horticulture crops such as Mangoes and Date Palm and had irrigated agriculture as recorded in 1880 (Campbell, 1880). The well-off farmers used to irrigate their land with groundwater using *Persian Wheel* and *Charas*. This indicates the groundwater utilisation for agriculture in the middle of arid regions. The advent of new efficient Water Extraction Mechanism (WEM) and affordable deep boring technologies experienced a massive boom in groundwater extraction since the 1970s, the era of atomistic irrigation (Shah, 2009). The key drivers of the massive increase in the small groundwater extraction structures and machines (Diesel Engine and Electric Pumps) are considered to be the private ownership and availability of resources on the owner's will. From the 1970s, majority of Kachchh started using groundwater initially for drinking and domestic supply. Boring and WEM were still not affordable for the farmers to use the groundwater for irrigation requirements. Also, the supply of electricity in the region was limited hence irrigation using diesel-powered engines was not affordable.

Groundwater utilisation soon overpowered the traditional water harvesting systems in Kachchh villages and towns. Bhuj, the capital city of the region, which was known for its marvellous water harvesting system bringing water from nearby hills through the 7-mile-long channel, had to start supplying domestic water needs through groundwater. The harvesting structure, *Hamirsar Lake* wasn not sufficient enough to meet the requirements. Similarly, Anjar, another major city in Kachchh experienced the use of private borewells to meet domestic water requirements, as from 1971 the dug wells and big water tank known as *Sidhasar tank* were not able to sufficiently supply the domestic water needs of the city. Slowly, groundwater became an important source of domestic water supply in the region. In the early 1970s, the groundwater level was mostly 15–20 m below ground and around 30 m below ground in urban locations. The

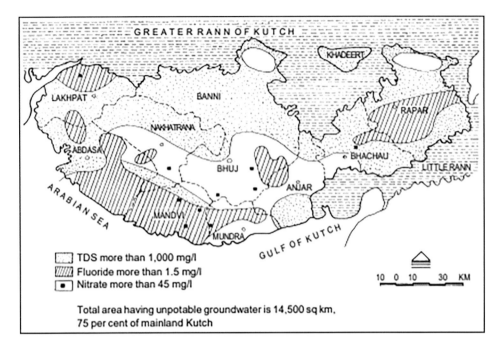

Figure 13.3 Groundwater condition in Mainland Kachchh (Greater Rann of Kachchh and Little Rann of Kachchh are partially desert and partially salt planes that are inhospitable) (Patel, 1996).

newfound wealth in the form of groundwater helped Kachchh with fulfilling the long waited thirst for water (Hathi, 2019).

The region experienced a major drought in 1987 when providing drinking water became a huge challenge. The government focused to cater to drinking water needs through deep tubewells during weak monsoon years and reduce the vulnerability of the region against natural calamities such as droughts. Figure 13.3 suggests the majority of Kutch accept the central part of the mainland had some groundwater quality-related issues but still groundwater developed as the major source of water needs there. Over the years, the parched land of Kachchh experienced a sense of revival after groundwater availability increased gradually through increased deep tubewells and subsidised electricity for pumping. The region also experienced the use of groundwater for agriculture needs in the 1980s and 1990s.

13.1.3 2001: the year of change for Kachchh

Kachchh experienced a major redevelopment after the 2001 earthquake. The earthquake measured 7.7 intensity on the Richter scale and caused huge casualties and destructions. Kachchh was shattered completely with more than 20,000 lives lost just in Kachchh. Of 1,055 villages, 178 were 100% destroyed and 165 villages had more than 70% property destroyed in the earthquake. The major towns of Bhuj, Rapar, Anjar and Bhachau lost more than 70% of property in damages (Hathi, 2019).

To implement the reconstruction and rehabilitation programme, the government formed the Gujarat State Disaster Management Authority (GSDMA) in 2001 (GSDMA, 2001; Garoda, 2019). An estimated outlay of US$ 1,770 million for reconstruction and rehabilitation work was developed. The financial support was channelised through World Bank (US$ 687.5 million), Asia Development Bank (ADB) (US$ 350 million) and other state and central Governments. The region also attracted private investment and philanthropic support from communities attached to the region. Strong community feeling among the people was the reason behind successful rehabilitation that witnessed more than 76% of households destroyed rebuilt without the help of Government or International Agencies (Rediff, 2006). The policy support, international aids and community partnership for the development of Kachchh garnered the unprecedented redevelopment story post-colossal calamity as the most neglected sparsely populated region got national and international attention. The region also experienced infrastructure development (roads, schools, hospitals, etc.) and banking access which led to the mainstreaming of the isolated region. The redevelopment propelled tri-sectoral growth since 2001: (a) agriculture and allied activities, (b) manufacturing and mining activities and (c) services and tourism sector.

A direct impact of the development story could be gauged during the drought year of 2018. The meteorological conditions were considered much worse than the 1987 drought when the region had witnessed a catastrophic water shortage and led to agrarian and livestock crises. In 2018, loss of human life or livestock due to drought could be mitigated as Kachchh and other parts of the Gujarat state have progressed a long way since the 1980s, and extracting deeper groundwater, transporting fodder, food, etc. from other parts are now a viable option for mitigating the impact with added financial stability and access to assistance for most of the population in the region.

13.2 COMPARING THE DEVELOPMENT STORY OF KACHCHH (PRE-2001 AND POST-2001)

13.2.1 Achieving new milestones in the agriculture sector and animal husbandry

Now, agriculture accounts for more than 90% of the groundwater consumption in Kachchh (CGWB, 2017). The increasing urbanisation puts an additional burden on water resources along with irrigated agriculture. A peculiar development in the case of Kachchh was a long spell of high rainfall lasting from 2003 to 2013. But agriculture in the region did not get affected during the low rainfall years from 2014 to 2018. Much of the credit for this goes to be groundwater that has sustained agriculture in the region (Figure 13.4; Table 13.2).

Apart from this, the region has seen an increase in high-value horticulture crops which can be considered as an indicator of progressive agriculture. The area under horticulture crops doubled from around 25,000 ha in 2005–2006 to more than 52,000 ha 2017–2018 (DoH, 2019). This was possible as the traditional farmers who were very much dependent on the rainfed farming fetched fresh investments to upgrade their farming through bank landings and developmental work. The proxy to this claim is the total number of bank branches increased from 187 in 2003 to 453 in 2016–2017 and

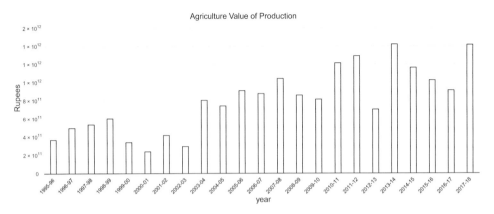

Figure 13.4 Value of agriculture production at constant prices (Patel, 2020a).

Table 13.2 Average area production and yield for major cereals, pulses and oilseeds

Type of crops	Period	Average area under crop ('00 ha)	Avg production ('00 MT)	Avg yield (kg/ha)	Increase (%)
Cereals	2001–2002 to 2005–2006	945	1,064	1,163	80%
	2011–2012 to 2015–2016	693	13,89	2,096	
Pulses	2001–2002 to 2005–2006	662	276	359	15%
	2011–2012 to 2015–2016	764	322	413	
Oil seeds	2001–2002 to 2005–2006	1,520	1,810	1,218	47%
	2011–2012 to 2015–2016	2,266	4,088	1,793	

Source: DoA (2016).

more than half of the lending going to agriculture sector with a steady rise of the total bank deposits over the years across the region (Garoda, 2019) (Figures 13.5–13.7).

Traditionally, Kachchh was not a place of dairy and commercial animal husbandry, as feeding the animals during harsh summertime was a challenge. It is interesting to note that most of the cattle herders in the region were from the nomadic tribes known as *Maladhari* who used to spend monsoon season in Kachchh breeding their cattle. After the monsoon season, the water shortage used to force the herders to walk their caravans towards water-rich central Gujarat. Their chief livelihood used to be the money earned from selling the livestock such as the cattle, horses, and camels bred by them to the farmers and other buyers in central Gujarat. Selling milk or animal and dairy products had never been a viable venture for them as water scarcity often caused a severe shortage of fodder and water requirements for their animals (Bharwada and Mahajan, 2002). Thus, the overall population of the livestock used to be stagnant and dairy development was meagre here.

Recent livestock censuses show a steady rise in livestock population post-2003. The total livestock has increased 21.2% from 1.57 million to 1.91 million between 2003 and 2012 (DCO, 2001, 2011). Estimates from the National Dairy Development Board (NDDB, 2013) show promising milk production development in Kachchh district since

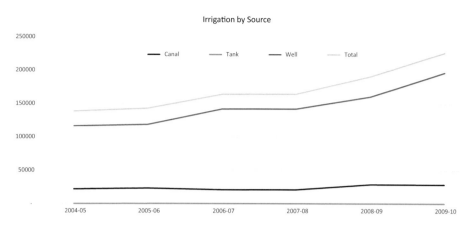

Figure 13.5 Irrigation by source in the district of Kachchh (Patel, 2019).

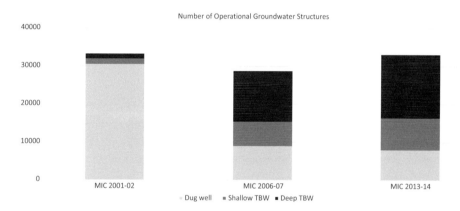

Figure 13.6 Number of operational groundwater structures in Kachchh region (MoJL, 2019).

2007 showing an annual growth rate of 4%. The development of dairy and animal husbandry is supported by the increased area under fodder crops in Kachchh Districts. Agriculture Census of 2010–2011 shows the area under fodder crops has reached 227,918 ha from a mere 39,170 ha in 2000–2001 (MoA, 2015). This suggests the newfound wealth for the dairy farmers and cattle herders in the Kachchh region but the water-intensive summer fodder cultivation has to be fuelled using the groundwater resources that add to the water burden for the region.

13.2.2 Attracting large industries, extracting minerals and creating shipping hub

Kachchh was known for mines and minerals, small industries, artworks and handicrafts, and sea salt extraction. Much of the small and micro industries in Anjar and Bhachau were hard hit by the 2001 earthquake. To revive the industrial stability

240 Groundwater for Sustainable Livelihoods

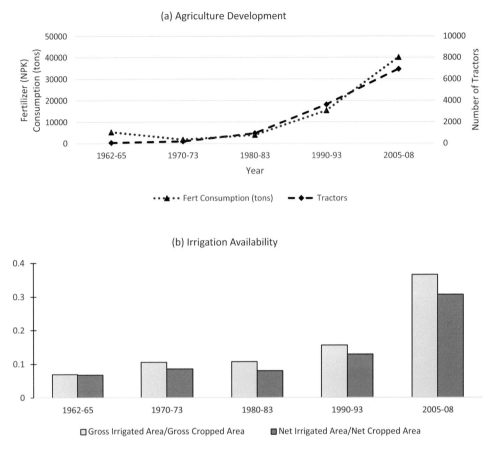

Figure 13.7 Modernising farm practices (a) Use of fertilizers and tractors (Bhalla and Singh, 2012). (b) Status of irrigation (Bhalla and Singh, 2012).

and attract new industrial vibrancy, Gujarat became the first state in India to enact the Special Economic Zone Act 2004 to provide a hassle-free operation regime and encompassing state-of-the-art infrastructure and support services. To help with water requirements, industries were allowed to use untapped groundwater. The total proposed investment by SEZ Developers is around 2,673.73 Billion INR. Kachchh is obliged to have the longest coastal area (406 km coastline) that is suitable for the shipping industry. The region houses India's busiest shipping hubs Kandla and Mundra. Mundra port has registered the fastest Cumulative Average Growth Rate (CAGR) of 35%, and it has become the largest port by cargo handling capacity in India (Garoda, 2019).

The major minerals of the Kachchh region are Gypsum, Clay (White, Natural, Ball and China), Silica Sand, Lime Stones, Granite, Bauxite and Lignite (Coal). Fresh resource mapping helped with more than a 5-fold increase in the extraction of the minerals in Kachchh (Figure 13.8).

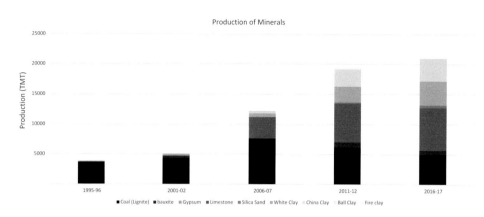

Figure 13.8 Production of minerals (CGM, 2019). TMT, Thousand Metric Ton.

13.2.3 Opening new gateways through the tourism

Although known for recently explored salt beds (White Desert), traditional culture, unique geography, culinary culture and historical monuments, Kachchh was never an attractive getaway for tourists until the redevelopment post-2001 introduced the region to a wider population and made it accessible for tourists. Tourism has certainly helped the district in increasing the economy of the region, but it has put further stress on the limited water resources that the region has. It is difficult to get the number of visitors, but the scale of tourism development can be estimated from the number of visitors at major tourist places. 'Aina Mahel' (Glass Palace) of Bhuj city is one of the most visited tourist destinations. The number of visitors to the Palace increased from 10,340 in 2003 to 157,376 in 2017 (Garoda, 2019). The 15-fold increase has created a demand for luxury and increased urban water demands on top of stress put on urban water demands by rapid urbanisation.

13.3 GROUNDWATER IN THE STORY OF KACHCHH: EXPERIENCE FROM THE FIELD EXPLORATIONS

As 90% groundwater is extracted for agriculture, field exploration to understand the different dynamics has been carried out. It is important to note that investing in groundwater irrigation normally costs 15%–20% of returns the farmers get through improved agriculture. This leads to further investment in the groundwater irrigation infrastructure (Shah, 2009). Also, the usage of groundwater irrigation is highly underreported. Figure 13.6 shows the area under irrigation to be around 40% of the total cropped area, whereas the satellite estimates produced by International Water Management Institute's Global Irrigated Area Map (IWMI-GIAM) programme show barely any cultivated land without basic irrigation (Thenkabail et al., 2009). Similarly, the village census of Bidada village shows 5% of the total area being irrigated. But the ground proofing through field visits and *Talati* (village accountant) records find that each field in the village had at least a dug well or a tubewell.

242 Groundwater for Sustainable Livelihoods

To fulfil the ever-increasing water requirements, groundwater in Kachchh has suf-fered an extreme decline. The Stage of Development[3] (SOD) of Groundwater in the Kachchh region which was around 28.8% in 1984 soon crossed the safe[5] groundwater utilisation mark in1990s and currently most of the blocks of Kachchh are either critical or overexploited (CGWB, 2017). Minor irrigation censuses (MIC) conducted by the Ministry of Water Resources, Government of India show the number of groundwater structures for irrigation purposes is shifting towards deeper structures. This suggests declining water levels in the region (Figure 13.6).

The village survey conducted in five villages of Kutch reveals darker side of the shiny development story of Kachchh. The agriculture dominant villages have seen groundwater depletion to the tune of 2–10 m annually and salinity ingression to a great extent. The condition is further exasperated during the relatively dry years between 2014 and 2018 (Table 13.3). Further, the survey reflected the rampant groundwater overexploitation through multiple cropping seasons. Village Bidada has 8,000 ha total agricultural land under the jurisdiction. It was observed that the village has more than 256 defunct borewells with more than 100 m depth each. Similar is the fate of most other villages of Kachchh that have experienced tubewell boom but groundwater scar-city. Also, the long-term water level data for pre- and post-monsoon periods for the monitoring well at Sanyara indicates a decline of 3.17 m/year (Figure 13.9) (locations marked in Figure 13.1). The increased livestock population and fodder demand force farmers to grow fodder even during the scorching summer of the desert climate. This causes much of the water to be lost in evapotranspiration. Further expansion in the perennial horticulture crops has increased water demand during the summer months that leading to many fold burden on the groundwater. Although having huge water scarcity, many irrigated fields are yet not adopting micro-irrigation technologies such as drip irrigation and sprinkler. Less than 50% of farmers with access to irrigation and capital are yet to adopt the water-saving technologies.

Another gloomy evidence comes from the field experiments to estimate the ground-water usage in the Mainland Kachchh that is underlain by the potential Bhuj Sandstone aquifer system (Saha and Gor, 2020). Their estimates show groundwater extraction for irrigation and domestic purpose are about 841.27 MCM/year (MCM = Million Cubic Metre) and 153.96 MCM/year, respectively, which is more than ten times the estimated

Table 13.3 Groundwater situation in five agriculturally active villages of Kachchh

Village	Mankuwa	Sukhpar	Dhavda	Sanyara	Bidada
Groundwater level (m) (below ground)	90–110	75–90	90–120	90–120	180–200
Total diluted salts (mg/L)	1,000–1,200	1,000–1,500	800–1,000	700–800	3,500–4,000
Long-term groundwater depletion	25 m in 5 years	30 m in 15 years	30 m in 10 years	30 m in 10 years	60 m in 5–7 years

Note: Compiled by authors based on the survey in five villages in Kachchh. Location of Villages are shown in Figure 13.1.

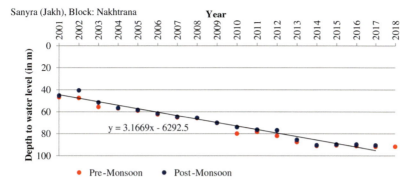

Figure 13.9 Long-term water level trend, Sanyara village, Nakhtarana Block (2001–2018).

recharge of around 103.61 MCM/year. Apart from the Bhuj Sandstone aquifers, good fresh groundwater is stored in coastal aquifers and in paleochannels in the Rann area (Figure 13.1). But these places are already suffering a heavy decline in groundwater availability and quality. The groundwater quality in most parts of the Kachchh region has seen a sharp decline over the years. Rarely any deep tubewell extract water that has TDS less than 1,000 mg/L. More than ninty-nine villages have the nonpotable quality of groundwater in Kachchh and coastal salinity and salinity ingress have seen a rise in the coastal region and Rann of Kachchh (Gujarat Ecology Commission [GEC], 2019).

13.4 ANALYSING THE FACTORS DRIVING THE GROUNDWATER DEPLETION IN KACHCHH

Overdraft of groundwater is the indisputable factor behind the groundwater decline and increased groundwater salinity. The economic development in Kachchh provided alternate income sources for the farming communities. It is experienced that over the years much of the wealth earned outside agriculture has been invested in further developing agriculture and primary investment goes to ensure irrigation which is mostly sourced from groundwater here. Also, the banking sector has significantly increased the liquidity for agriculture activities. The Indian farmers hold stickyness towards farming. Even though having alternate livelihood opportunities, vacating agriculture activities is not considered viable by the farmers.

Figure 13.10b indicates that the large proportion of farmers in Kachchh are not marginal or small farmers and have sufficient resources unlike other parts of India dominated by resource-scarce marginal farmers. The resource available with the larger farmers for experimentation helps in frequent changes in the cropping patterns in Kachchh. The major crop had shifted from *Bajra* (Millet) in the 1960s to cash crops such as cotton, castor, and groundnut in the 2000s that have higher irrigation requirements. Also, wheat production began with increased winter season sowing (Patel, 2019). The shift to water-intensive crops added pressure on the water resources. The very recent shift towards drip irrigation and horticulture crops is doing more harm than good as the rainfed area is converted into irrigated area using subsidised drip irrigation and perennial crops such as pomegranates, mangoes and date palms are

(a) Distribution of Land Holdings

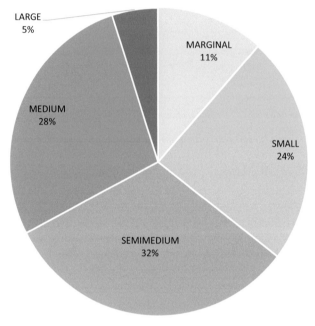

(b) Irrigation Status

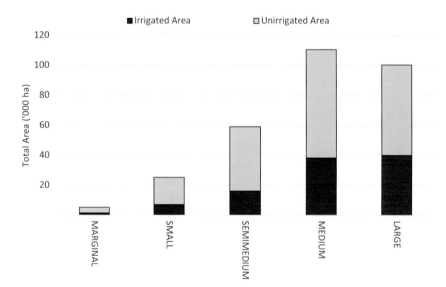

Figure 13.10 (a) Distribution of land holdings by the size group of holdings (Marginal <1 ha, Small:1–2 ha, Semi-medium: 2–4 ha, Medium: 4–10 ha, Large >10 ha) (b) Area under different food grain production by size group of holdings and irrigation status.

cultivated by the orchard owners. These crops require rather round the year water to sustain the extreme heat they experience during the prolonged summer season from February to July in this arid region.

In the case of Kachchh, the initial push for agriculture development was also supported by the obliged rainfall spell during 2003–2013. But the dry spell starting from 2014 has built up the major water stress. 2018 had been declared the worst drought year with most of Kachchh receiving minuscule rainfall not enough to meet domestic drinking water requirements. Still, the farmers with groundwater resources continued to irrigate their crops even in the summer season and hardly any gross cropped area. As documented by Shah (2009), groundwater development is characterised by four stages identified in Figure 13.11. Kachchh has certainly crossed the peak of groundwater development and it is moving towards the fourth stage since 2013. The peculiar case of Kachchh is the rate at which it has surpassed the second and third stages. It did not take even two decades for Kachchh to traverse from sufficient groundwater availability to rapid decline since the agrarian development was accelerated after 2001.

Although farmers may know the long-term consequences of these declining water resources, they discount their future well-being for their present needs. This is an example of the behavioural economics phenomenon of 'Time Inconsistent Behaviour' and 'Present Bias' (Frederick et al., 2002). This explains that the almost free farm power for irrigation provided by the state government has fuelled the rampant overexploitation of groundwater.

It is also expected by the GEC (2019), the proposed industrial development in Saurashtra, Kachchh and North Gujarat will create further pressure on groundwater resources of these regions. The increased industrial zones which are making Kachchh a production hub and power generation hub are also causing a significant negative impact on the water resources. Recent dispute amongst the farmers and the thermal power station guzzling groundwater through multiple large deep tubewells near farm fields reflects on the industrial groundwater extraction. The water resources need to support large-scale cement, machines, tyres, vehicles and automobiles, etc. manufacturing

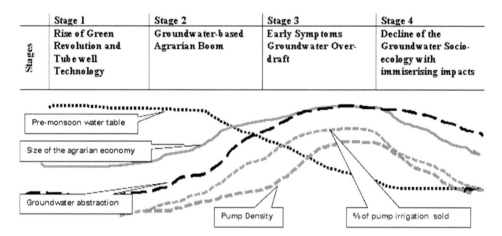

Figure 13.11 Rise and fall of groundwater socio-ecologies (Shah, 2009).

companies having huge water footprints which are mostly unaccounted for in the region. From time to time conflicts between villagers and industries are being witnessed over the groundwater resources.

Apart from agriculture and industrial usage, the increased population and growing urbanisation have also increased domestic water requirements. Currently, just Bhuj, the major city in Kachchh requires more than 40 MLD (Million Litre per Day) water supplied through the municipality, of which 12.5 MLD is provided by municipal borewells, and around 20 MLD is imported from the large dam, *Sardar Sarovar*. The rest is met with the private borewells and tankers abstrac groundwater from the aquifer beneath the city area or in the nearby villages and provide to the city's domestic water requirements. The Indian land laws provide the full rights to the assets below the ground hence there are numerous unreported private borewells constructed throughout the city areas of Kachchh and even in the rural households.

13.5 REVIVING FROM THE STAGE OF PERIL

Ever-increasing water demand in Kachchh is challenging, particuarly with climate change impacting the already erratic rainfall. Historically Kachchh experiencesd 10–15 rainy days in a year. The further shrinking of the rainy season in recent years has increased surface runoff and lessened groundwater recharged. These rainfall anomalies have increased the *Kharif* (Monsoon Sowing Period) crop irrigation requirements as even the crops in the monsoon season require more water for irrigation.

Also, the long-promised water from 'Sardar Sarovar Dam' (A Large Dam on River Narmada) with a dedicated canal for the Kachchh region being constructed (Length of the canal = 385.81 km) that promised to provide around 600 MCM water for irrigation and 60 MCM for drinking purpose is still inadequate for the ever-increasing water demands of the region. To revive the groundwater from the stage of peril there is a need to address demand, supply and the institutionalisation of groundwater.

13.5.1 Institutionalising groundwater

Currently, the rampant use of groundwater is encouraged through the atomistic irrigation regime that provides full control of the groundwater to the owners of the borewell and WEMs. It does not put any liability to conserve the resource and leads to inefficient usage of groundwater and no room to adopt a water-conserving attitude among the farmers. This condition is what is traditionally known as the 'tragedy of commons'. The condition is further worsened due to the availability of almost free electricity for agriculture pumping. Although, the groundwater is virtually owned by everyone who resides on the surface of the resource and everyone shall have equal rights as well as responsibilities for the same, individuals act in the rational self-interest of their gain and neglect the sustenance of common resources. Whereas the resource-rich farmers with capital to finance deeper borewells and higher capacity pumps have continued to extract more and more water that depleted the common resource beneath the surface and forced the resource-poor farmers to fend without water in their shallow wells. The resources are also shared with urban bodies and industries that can put far more capital to extract the groundwater for meeting their water needs.

The rampant overexploitation can be tapped only if the perspective about the groundwater is changed. To balance the demand and supply of groundwater resources, there is a need to recognise the importance of groundwater from an economic sustainability perspective. In the case of groundwater and overall water demand, putting a community-based regulatory framework to ensure water is used judiciously is essential in sustainable groundwater consumption for developing democratic country like India. Efforts are already put up by a few non-governmental organisations (NGOs) in assembling individual farmers to collectively manage their borewells and WEMs. Provisions shall be imposed to penalise the inefficient and prodigal users. Budgeting water judiciously for agricultural needs, domestic needs as well as for the needs of flora and fauna shall be done collectively. Metering of water is the first step towards water use budgeting and creating credibility for water resources among the users. With the advent of cheap metering devices, the cost excuses and implementation issues shall be addressed. Even if the water is not rationed or taxed through the metre, it can help in creating awareness of water usage among the users and sense of responsibility amongst the stakeholders.

13.5.2 Preventive measures (managing supply)

The peninsula of Kachchh has been far-reaching from perennial water sources and nearby regions of Saurashtra and North Gujarat are also experiencing increasing water demand and groundwater stress in recent years. Transfer of water through canals from another place without free riders is a daunting task for Governments and Engineers. Historically, Kachchh survived on surface rainwater harvesting structures. Rainwater harvesting is the most feasible way of ensuring the supply of water. The natural recharge is difficult due to scanty and erratic rainfall but Kachchh experiences a season every 4–5 years with above-normal rainfall when much of the rainwater is not stored and discarded in the form of surface runoff. If this runoff can be diverted to groundwater recharge, it will certainly help with increased groundwater availability. Kachchh is suitable to adopt artificial recharge and rainwater harvesting for groundwater revival. The freshwater stored through water harvesting can be conjunctively used with non-usable brackish water for water multiplying and expanding the water supply for irrigation purposes.

Kachchh experiences very heavy evaporation losses as it faces extreme heat for most of the time in a year. Thus, artificially recharging sandstone and alluvial aquifers through defunct deep tubewell can yield better groundwater revival opportunities than on-surface water harvesting structures. Also, the earthen structure on flashy rivulets can restrict water runoff during rainfall and provide surface storage as well as increase groundwater recharge to some extent by reducing surface runoff to sea. Recent fieldwork in Kachchh reveals that good rainfall in 2020 monsoon has provided with inter-annual storage of groundwater by diverting the surface runoff to groundwater through the defunct borewell in *Bidada* and *Kotada (Jadodar)* villages. The community is taking up the cause through collective fundraising. If they can be provided institutional or financial support from the government, the region can come up with an exemplar story of reviving from the stage of perils through managing their groundwater as experienced with the previous groundwater recharge moments in India (Patel et al., 2020).

248 Groundwater for Sustainable Livelihoods

The per capita water consumption for domestic use rising with urbanisation and the wastewater generated after this use follows the same trend. In the water-scarce region, there shall not be any wastage of water and no wastewater too. Recycling and repurposing of the wastewater after treatment can be used for agriculture in peri-urban areas. This can create approximately 25 MCM in a year from the major urban habitations with piped water supply in Kachchh.

13.5.3 Adaptive measures (managing demand)

Agriculture and allied activities demand the highest water. Cases of other water-scarce, arid agriculture hotspots (such as California and Israel) are good references for Kachchh farmers to develop their agriculture strategy in terms of suitable crops and irrigation methods. For irrigated agriculture, water-saving micro-irrigation such as sprinkler and drip irrigation can give promising outcomes if the entire irrigated agriculture cultivation can be brought under micro-irrigation. The extreme heat and long exposure to sunlight in the arid region of Kachchh can cause high evaporation and evapotranspiration loss of water. On top of this, most of the cultivated land has sandy characteristics that cannot retain much water. Artificial mulching and improved soil water retention capacity through added soil organic matters (manure and compost) and bio inputs will have promising results in reducing irrigation water requirements.

A few specific cases for Kachchh's story are booming dairy, expanding horticulture and adoption of cash crops. These are good strategies for optimising land utility but require higher water input. Kachchh has larger land available as the population is far less than other regions of India but water is scarcer. Thus, unlike the rest of India that adopted varieties of crops with high land productivity, Kachchh needs to solve the equation optimising the water productivity.[4]

A peculiar case of Kachchh horticulture development indicates that crops with lower shelf-life need a special strategy to sell them and realise the anticipated benefits. The high-yield variety of water-saving crops such as date palm has increased water productivity. But the farmers are moving away from it as they are unable to market the product. The horticulture development can help if good market linkages for the products are available and adequate infrastructure for storage, transport and process this horticulture produce is also in place (Patel, 2020b). Similar is the fate of many other horticulture crops such as pomegranates, watermelons and musk melons that are not fetching desired success due to inadequate market access for the sparsely populated distant region of Kachchh. Unless the market linkages are strengthened for these crops, farmers may not adopt capital-intensive but water-efficient horticulture crops here.

The recent increase in area under fodder crops and increased dairy may lead to far more detrimental impacts as growing green fodder needs far more water as compared to any other crops. As per an estimate, 650 tankers[5] are required for irrigating 1 acre of alfalfa.[6] Dairy development and increased fodder crops have been proven detrimental to groundwater resources in many parts of India. The Banaskantha district that shares its border with Kachchh is the best example of dairy development's negative impact on groundwater resources (Singh et al., 2004). Kachchh may follow the same fate as the nomadic communities are settling in the region for animal husbandry and dairy farming.

An important aspect of the arid region of Kachchh is, it receives uninterrupted solar radiation throughout the year. The current 'Surya Shakti Kisaan Yojana' scheme by the government of Gujarat to promote energy-grid connected Solar Pumps[7] can achieve good results in Kachchh. The energy generated during hours of non-operation can be evacuated to the grid. The owners can get monetary compensation for this evacuation. This puts the price for pumping water which is virtually nil as of now due to the free farm power (Shah et al., 2018).

Currently, domestic and industrial water needs are far less than agriculture, but they are growing exponentially with rapid urbanisation and industrial expansion. It will threaten the availability of water for comparatively resource-poor farmers. The industrial propagation in the region needs to attract industries requiring less water and needs to cap and monitor groundwater usage. Even the urban water supplies need to restrict the rampant mismanagement of municipal water supplies that leaves streets flooded with leaking water but the aquifers running dry.

13.6 CONCLUSION

Groundwater has been a juggernaut behind the recent socio-economic development of Kachchh. Groundwater is also an important aspect of increasing drought resilience and rural-agrarian sustainability of the drought-prone arid region. Once known as an oasis of the Western Indian Deserts, Kachchh is now famed as the world's fastest depleting prolific aquifer system. Groundwater depletion is leading to a catastrophic end to the currently glittering development of Kachchh. It needs collective action for the revival of common resources and judicious allocation of the resource for better management and sustenance of livelihood of the people. The three-armed strategy to institutionalise the groundwater, manage demand and augment resources is essential for sustaining the groundwater-fuelled success story of Kutch. Further, the region does not have scope for mishaps to grow unsuitable crops that are not well-market linked or to establish resource mongering industries that are not water efficient. It is needless to point out that, unless the depleting hidden treasure beneath the surface is revived, Kachchh shall again return to the secluded and parched status that it adorned for long before the successful socio-economic development it experienced since 2001.

ACKNOWLEDGMENTS

The authors thank the graceful support of Dr. Tushaar Shah (Emeritus Scientist, IWMI), Dr. Mohan Patel (Secretary, Gujarat Economic Association), Dr. Yogesh Jadeja (Director, Arid Communities and Technologies) and Dr. Tushar Hathi (Retd. Associate Professor, Syamaji Krushna Verma (Kachchh) University) for their essential and valuable suggestions. We also thank Mrs. Jaya Patel for translating references available in the local languages for the authors. The financial and literary support provided by *International Water Management Institute*(IWMI) and *Tata Trust* has been crucial. The on-field support from Mr. Jignesh Senghani, Mr. Jinesh Patel and local NGO, *Prayaas: The Movement for Grassroot Changes* helped the authors with better field insights and experience.

NOTES

1 Rann: Inhabitable Waste Land.
2 Shallow wells dug manually in low depressions normally called *Jheel* (tanks).
3 Stage of Development (SOD): Estimated ground water utilisation out of total dynamic groundwater availability. Safe (SOD <60%), Critical (SOD > 60%, < 100%), Over-Exploited (SOD > 100%).
4 Water Productivity: Production (kg)/Water Required (L).
5 One tanker has 5,000 L water.
6 Most used variety of Fodder in Kachchh.
7 Solar Pumps: Irrigation Pumps that are powered by Photo-Voltaic Solar Panels.

REFERENCES

ACT (Arid Communities and Technologies). (2015). *IWM & PGWM – Piloting Groundwater Management and Governance through the "Neeranchal Programme": Situational Analysis of Kachchh District, Gujarat.* Bhuj: Arid Communities and Technologies.

Agarwal, A., & Narain, N. (1997). *Dying Wisdom: Rise fall and potential of India's traditional water harvesting systems.* New Delhi: Centre for Science and Environment.

Antolik, M., & Dreger, D. (2003). Rupture process of the 26 January 2001 Mw 7.6 Bhuj, India, earthquake from teleseismic broadband data. *Bulletin of Seismological Society of America*, 93(3), 1235–1248. doi:10.1785/0120020142.

Beck, H., Zimmermann, N., McVicar, T., Vergopolan, N., Berg, A., & Wood, E. F. (2018). Present and future Köppen-Geiger climate classification maps at 1-km resolution. *Science Data*, 5. doi:10.1038/sdata.2018.214.

Bhalla, G. S., & Singh, G. (2012). *Economic Liberalisation and India Agriculture.* New Delhi: SAGE Publication India Limited.

Bharwada, C., & Mahajan, V. (2002). Drinking water crisis in Kachchh: a natural phenomenon? *Economic and Political Weekly*, 37(48), 4859–4866.

Campbell, J. M. (1880). *1880- Gazetteer of Bombay Presidency - Vol 5- Kachchh, Palanpur and Mahikantha.* Bombay: Government Central Press.

CGM (Commissioner of Geology and Mining: Geologist). (2019). *Major Minerals' Production.* Bhuj: Government of Gujurat.

CGWB (Central Ground Water Board). (2017). *Dynamic Ground Water Resources of India 2013.* New Delhi: Central Ground Water Board. Retrieved from http://cgwb.gov.in/Documents/Dynamic%20GWRE-2013.pdf.

DCO (Directorate of Census Operations). (2001). *District Census Handbook 2001.* Ahmedabad: Directorate of Census Operations, Gujarat.

DCO (Directorate of Census Operations). (2011). *District Census Handbook Kachchh.* Gandhinagar: Directorate of Census Operations.

DoA (Directorate of Agriculture). (2016). *Area, Production and Yield of Major Crops in Gujarat.* Gandhinagar: Directorate of Agriculture, Government of Gujarat.

DoH (Department of Horticulture). (2019, January 05). *Horticulture Census.* Department of Horticulture, Agriculture, Farmers Welfare and Cooperation Department, Government of Gujarat. Retrieved from https://doh.gujarat.gov.in/horticulture-census.htm.

Frederick, S., Loewenstein, G., & O'Donoghue, T. (2002). Time discounting and time preference: a critical review. *Journal of Economic Literature*, 40, 351–401.

Garoda, M. (2019). "Bhukamp Pachhi nu Kachchh 2001–2017" (Kachchh after the Earthquake 2001–2017). In T. Hathi (Ed.), *"Gujharat no Arthik Itihas" (Economic History of Kachchh)* (pp. 191–358). Anand: Gujarat Economics Association.

GEC (Gujarat Ecological Commission). (2019, December 1). *State of Environment Report Gujarat: Water.* Gandhinagar: Gujarat Ecological Commission. Retrieved from Gujarat Ecological Commission: http://gujenvis.nic.in/PDF/soe-water.pdf.

GSDMA. (2001). *Gujarat Earthquake Reconstruction and Rehabilitation Policy.* Gandhinagar: Gujarat State Disaster Management Authority.

Hathi, T. (2019). *Kachchh no Arthik Itihaas (Economic History of Kachchh).* Anand: Indian Economic Association.

MoA. (2015). *9th Agricultural Census of India 2010–11.* New Delhi: Agriculture Census Division, Department of Agriculture, Cooperation & Farmers Welfare, Government of India. Retrieved from http://agcensus.dacnet.nic.in/.

MoJL (Ministry of Jal Shakti). (2019, December 1). *State Wise Reports.* Retrieved from Minor Irrigation Census: http://164.100.229.38/state-wise-reports.

MSME (Ministry of Micro, Small & Medium Enterprises). (2016). *District Industrial Potential Survey of Kachchh District (2016–17).* Ahmedabad: MSME-Development Institute.

NDDB. (2013). *Milk Production Statistics.* Anand: National Dairy Development Board.

Patel, G.D. (1971). *Gujarat State Gazetteer: District of Kachchh.* Ahmedabad: Government of Gujarat.

Patel, P.M. (2019). Agricultural revisions in drought prone arid region of Kachchh: people led, market oriented growth under adverse climatic conditions. *3rd World Irrigation Forum* (Article No. 149). Bali: Internation Commission on Irrigation and Drainage.

Patel, P.M. (2020a). Community's response to revive depleting groundwater in arid region of Kachchh: a self-propelled movement for the managed aquifer recharge through private defunct borewells. *Addressing Groundwater Resilience under Climate Change.* Online: International Water Resources Association. 29th October, 2020.

Patel, P.M. (2020b). Horticulture pile-up, yet farmers stare at losses in Kachchh. *India Water Portal.* Retrieved from https://www.indiawaterportal.org/articles/horticulture-pile-yet-farmers-stare-losses.

Patel, P.M., & Khan, A. (n.d.). Changing rainfall patterns in an era of climate change: a multiparameter spatiotemporal analysis of trends & impacts for India, *pre-print.* doi:10.21203/rs.3.rs-95659/v2

Patel, P.M., Saha, D., & Shah, T. (2020). Sustainability of groundwater through community-driven distributed recharge: an analysis of arguments for water scarce regions of semi-arid India. *Journal of Hydrology: Regional Studies*, 29, 100680. doi:10.1016/j.ejrh.2020.100680.

Patel, P.P. (1996). *Kachchh Ni Jal Samasya: Swaroop, Kaaran Ane Nivaran.* Bhuj: Kachchh Taari Asmita, Kachchh Mitra.

Rediff. (2006, Jan 31). Kachchh has set a precedent for rehabilitation. *Rediff.com.* Retrieved from https://www.rediff.com/news/2006/jan/31inter11.htm.

Saha, D., & Gor, N. (2020). A prolific aquifer system is in peril in arid Kachchh region of India. *Groundwater for Sustainable Development*, 11, 100394.

Shah, T. (2009). *Taming the Anarchy: Groundwater Governance in South Asia.* Washington, DC: Resources for the Future.

Shah, T. (2014). Towards a managed aquifer recharge strategy for Gujarat, India: an economist's dialogues with hydro-geologists. *Journal of Hydrology*, 518, 94–107. doi:10.1016/j.jhydrol.2013.12.022.

Shah, T., Gulati, A., Hemant, P., Shreedhar, G., & Jain, R. (2009, December 26). Secret of Gujarat's agrarian miracle after 2000. *Economic and Political Weekly*, 44(52), 45–55.

Shah, T., Rajan, A., Rai, G. P., Verma, S., & Durga, N. (2018). Solar pumps and South Asia's energy-groundwater nexus: exploring implications and reimagining its future. *Environmental Research Letters*, 13(11), 115003.

Singh, O.P., Sharma, A., & Singh, R.A. (2004, August 6). Virtual water trade in dairy economy: irrigation water productivity in Gujarat. *Economic and Political Weekly*, 39(31), 3492–3497.

Thenkabail, P.S., Biradar, C.M., Noojipady, P., Dheeravath, V., Gumma, M. K., Li, Y. J., ... Gangalakunta, O.R. (2009). Global irrigated area maps (GIAM) and statistics using remote sensing. In P.S. Thenkabail, J.G. Lyon, H. Turral, & C.M. Biradar, *Remote Sensing of Global Croplands for Food Security* (pp. 41–117). Boca Raton, FL: Taylor & Francis Series in Remote Sensing Applications, CRC Press. Retrieved from https://hdl.handle.net/10568/37342.

Chapter 14

The role of groundwater in economic and social development of Mato Grosso do Sul State, Midwest of Brazil

S.G. Gabas
Federal University of Mato Grosso do Sul

G.F. Dourado
University of California

D.A. Uechi
Federal University of Mato Grosso do Sul

G.H. Cavazzana
Dom Bosco Catholic University

G. Lastoria
Federal University of Mato Grosso do Sul

CONTENTS

14.1 Introduction 254
14.2 Location 254
14.3 Aquifers 255
14.4 Main economic activities 260
    14.4.1 Agriculture 260
    14.4.2 Livestock 261
    14.4.3 Pig farming 261
    14.4.4 Silviculture 262
    14.4.5 Slaughterhouses and tannery 262
    14.4.6 Mining 263
    14.4.7 Tourism 263
14.5 Social aspects 264
    14.5.1 Indigenous communities 264
    14.5.2 Quilombolas 265
    14.5.3 Rural settlements 265
14.6 Environmental aspects 266
14.7 Challenges 267
14.8 Summary 269
Acknowledgements 269
References 269

DOI: 10.1201/9781003024101-14

14.1 INTRODUCTION

The current social and environmental situation of the world has raised some concerns about water availability and its distribution, at global, regional and local scales. The access to water in quantity and quality that meet the minimum parameters considered by the United Nations has been unequal for different populations, even in developed countries (Schaider et al., 2019; Zheng and Flanagan, 2019). In this context, there is the concern of water management bodies regarding the recommendation that no one be left behind (WWAP, 2019).

Social and environmental scientists have been considering the importance of water resources management in the different realms of society, affecting the culture and health of individuals, especially vulnerable communities, as well as political and economic systems as a whole (Canter et al., 1992; Johnson et al., 2001; Otero et al., 2011; Wutch and Brewis, 2014; Hoogesteger, 2017). Despite innumerable papers and important discussions raised by the scientific community, there is still a great number of people facing water scarcity (WWAP, 2019). Recently, some geoscientists referred to the importance of considering the social dimensions and their relevance for achieving sustainable development goals (López-Gunn, 2012; Re, 2015; Velis et al., 2017).

This study presents the importance of groundwater in the social and economic development of Mato Grosso do Sul (MS), in the southern portion of Midwest Brazil. The Cerrado (Brazilian savanna) and Pantanal wetlands occupy most of their territories, which also incorporate forest formations from the Atlantic Forest, all three considered part of the world's biodiversity hotspots. MS encompasses two important watersheds on a continental scale, the Paraná and Paraguay watersheds that form complex river systems and underlying groundwater systems. The state is one of the largest agricultural producing states of Brazil, with a small total population, but with relevant vulnerable communities. Furthermore, the state has a major part of the largest wetland ecosystem in the world, the Pantanal. We aim to broaden the debate about groundwater management in the state and the proposal of a good groundwater governance agenda, contributing to the state's sustainable development. This chapter is organised into six sections: the first introduces the study area; the second presents its location; the third presents the aquifer systems in the region; the fourth discusses the main economic activities; the fifth and the sixth sections address some social aspects and environmental issues, respectively; the final section contains the authors' opinion on the challenges faced by the state to pursue the sustainable development and the improvement of groundwater governance.

14.2 LOCATION

The state of MS has an area of $357,145.535\,km^2$ and an estimated population of 2,778,986 (IBGEa, 2019). From the state's seventy-nine municipalities, sixty-nine are either partially or completely supplied by groundwater. Among them, fifty-eight of these municipalities are located in the east, and twenty-one are located in the west, being supplied by the Paraná and Paraguay hydrographic basins (Figure 14.1), respectively.

Groundwater in Mato Grosso do Sul State 255

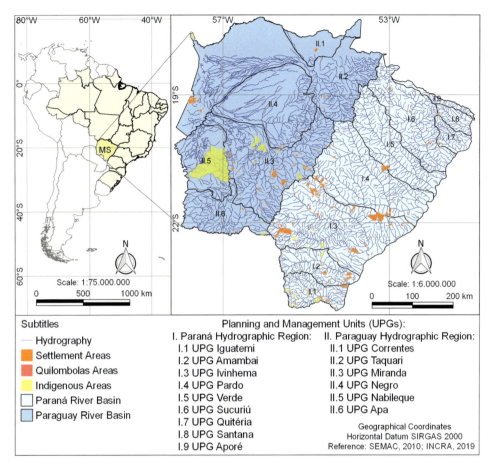

Figure 14.1 Location of Mato Grosso do Sul, its hydrographic basins and vulnerable communities.

14.3 AQUIFERS

MS has rocks from three tectonic contexts: the Precambrian basement formed by meta-igneous and metasedimentary rocks, the sedimentary rocks of a Paleo-Mesozoic sedimentary basin that include intercalated volcanic rocks (Paraná Basin), and the Cenozoic sedimentary basin (Pantanal Basin). According to the surface lithology (CPRM, 2006) and their hydraulic properties, MS has nine hydrostratigraphic units (Figure 14.2), so named: Precambrian Aquifer System (PCAS), Calcareous Precambrian Aquifer System (CPCAS), Furnas Aquifer System (FAS), Aquidauana-Ponta Grossa Aquifer System (APGAS), Guarani Aquifer System (GAS), Serra Geral Aquifer System (SGAS), Bauru Aquifer System (BAS), Tertiary Aquifer System (TAS) and Pantanal Aquifer System (PAS).

There are no regional hydrochemistry studies or groundwater quality monitoring data. The only hydrogeological survey of the entire state provides hydraulic and

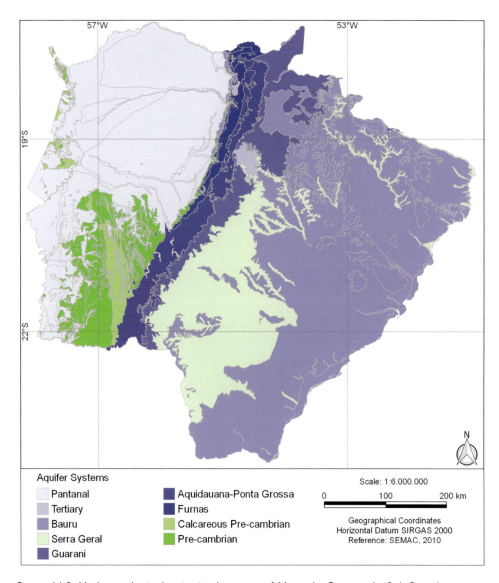

Figure 14.2 Hydrogeological units in the state of Mato de Grosso do Sul, Brazil.

physicochemical data of some of the aquifer systems, collected from the 200 wells owned by the state sanitation agency (Sanesul) (Sanesul/Tahal, 1998). However, because the analyses of all physicochemical parameters were done for less than 10% of the wells, the hydrochemical classification of these aquifers was not possible; only pH values were considered for all. In addition, some graduate studies focused on some aquifers were described by Lastoria (2002), Castelo Branco Filho (2005) and Gastmans (2007) for the Serra Geral, Cenozoic and Guarani aquifers, respectively. In terms of the percentage of water volume exploited, the municipalities in the Paraná (Paraguay) Basin account for 56.6% (34.8%) of all groundwater extracted. In respect to the total volume of water extracted for public supply in the state, 52.7% comes from groundwater (SEMAC, 2010). The principal characteristics of the aforementioned aquifer systems are subsequently described.

The PCAS consists of metamorphic rocks from the basement of the Paraná and Pantanal Basins. It encompasses a large variety of rocks including metasedimentary rocks of distinct metamorphic degrees, metavolcanic rocks and granite gneisses. This hydrogeological system is found primarily in the southwestern part of the state and near the city of Corumbá, its outcrop areas.

The CPCAS is important for two municipalities in the state, Bonito and Bodoquena, where tourist activities are concentrated. Although the city of Bonito is a popular attraction, in part due to its water resources, its groundwater has not been sufficiently studied. There are no hydrochemical studies associated with this aquifer system. A pH of 6.9 was reported by Sanesul/Tahal (1998); however, that study included the Precambrian aquifer. Wells were drilled in both carbonated and non-carbonated rocks of this age, and only one well was used to collect hydrochemical data for this system. Enetério (2009) found pH values between 7.1 and 7.8 in water samples from dug wells in Bonito. Dias et al. (2007) reported a pH of 8.1 to be the highest value in groundwater samples in Bodoquena, possibly from the Calcareous Precambrian System. Although the economy in the area is driven by tourism and agriculture, studies regarding regional vulnerability are non-existent. However, Enetério (2009) identified the aquifer's vulnerability to contamination by leachate from cemeteries.

FAS is important for the municipalities of Rio Negro, Rio Verde de Mato Grosso, Sonora and Coxim, and Pedro Gomes, in the northern part of the state. A flowing artesian well is present in the latter municipality, indicating the confined behaviour of the aquifer (Ponta Grossa Formation). The groundwater flow in the region moves towards the Taquari, Negro and Aquidauana Rivers, maintaining the base flow of these river systems that are critical for the region.

Considering storage capacity and hydraulic conductivity, rocks from the APGAS have low water productivity. The hydrogeological survey done by Sanesul/Tahal (1998) has not considered it an aquifer. However, we decided to consider it a hydrogeological unit due to its importance for the local water supply, as it is the only source of water to some rural properties or small districts.

The GAS is a particularly important unit, composed of Mesozoic eolian arenites. The aquifer has outcrop areas in the occidental region that continues in a wide band along a northeast-southwest direction up to the northern Paraguay Hydrographic Region (Gastmans, 2007, Gastmans et al., 2010a). The outcrop area is approximately 10 km wide to the south and 150 km to the north (OAS, 2009), being partially responsible for the direct recharge of the GAS. Regarding hydrochemistry, the water quality from the GAS is classified as Zone I (Type A; OAS, 2009), characterised by a primarily calcium-carbonate composition, followed by calcium-magnesium and calcium-sodium bicarbonates, with little mineralisation. Gastmans (2007) classified the water from GAS as calcium carbonated or calcium magnesium where the aquifer is open or in wells with a thin basalt layer and as sodium-bicarbonate waters that evolve to bicarbonate-sodium chloride/sulphate in areas of greater aquifer confinement. In Campo Grande, the capital of the state, there are twenty deep wells that exploit water from this aquifer for human supply. Gastmans et al. (2010b) reported that there was no contribution from other water sources toward GAS recharge in the western part of the aquifer and that the underground flow velocity, calculated with Darcy's Law, is less than 10 m/year, indicating that the deep-water recharge in the aquifer occurred between 30,000 and 10,000 B.P. in the Last Glacial Maximum. Recharge rate estimates of the confined aquifer are important for its management plan, as the renewal rate of this water resource is very slow.

The SGAS is a fractured and heterogeneous aquifer that is phreatic in southeastern MS and composed of basalt lava flows that occurred in the Jurassic period during the formation of the Paleo-Mesozoic sedimentary basin. It is the most exploited aquifer; Campo Grande alone has more than 100 wells exploiting this system for water supply. The discharge zones are the rivers and springs, which occur where the fractures (particularly horizontal ones) intercept the topography. The entire basaltic outcropping surface is an area of direct recharge. SGAS, in its non-outcropping portion, holds an indirect recharge with contribution from the underlying (Guarani) and overlying (Bauru) aquifer systems (Lastoria, 2002; Lastoria et al., 2007). Regarding water quality, the pH varies between 5.5 and 9.5, showing low mineralisation with the greatest values of electrical conductivity found to be from 150 to 240 µs/cm, in the southern region, between Dourados and Fátima do Sul. The calcium and magnesium cations predominate over those of sodium and potassium, meanwhile, the dominant anion is bicarbonate, followed by chloride, and nitrate, probably with anthropogenic origins (Lastoria, 2002).

The BAS is a porous aquifer composed of sedimentary rocks (mainly arenites) from the Paraná Basin. This aquifer system is one of the most important in the state as 32.9% of the registered wells exploit water from it, and 7.8% collect water from this system combined with the SGAS (SEMAC, 2010). It is responsible for the discharge of regional groundwater into important rivers in MS, such as the Pardo, Verde and Sucuriú Rivers as well as smaller rivers from the Paraná Hydrographic Basin, as demonstrated by Cavazzana et al. (2019) in a small watershed within this area. Regarding pH values, there are variations from 5.5 to 7.2 (Sanesul/Tahal, 1998) that can be explained by the heterogeneous sedimentary deposits in this group, containing calcite cement in some layers.

The TAS is a sedimentary porous aquifer composed of a non-continuous lateritic deposit that was preserved in the highlands to the north. This rock formation forms the plateau that occurs, in the municipalities of São Gabriel do Oeste, Sonora and Chapadão do Sul. Most farms in these counties are supplied by the TAS. This unit was not individualised in the State Water Resources Plan due to its small size in relation to the scale and use of groundwater considered in that study. However, this system is important to the local and state economy. In São Gabriel do Oeste, the aquifer is exploited for irrigated agriculture by most of the rural properties located in the highest part of the municipality. The wells are 30 m deep on average in the region due to the shallower water table. Previous studies on water quality in the cultivated areas show that the groundwater is predominantly of calcium bicarbonate type and, secondly, sodium bicarbonate, with pH values ranging from 4.7 to 6.43, low electrical conductivity, from 1.44 to 20.0 µs/cm (Souza et al., 2014; Ferraro et al., 2014) and with high concentration of coliforms in wastewater-irrigated areas (Pahl et al., 2018).

The PAS is a porous and phreatic aquifer system with an outcrop area of 96,917 km^2 (IBGEa, 2019). It consists of sediments from the Pantanal Basin, which are predominantly fine sandy sediments that are scarcely compacted. Although in Corumbá, the largest municipality in the Pantanal wetlands, the public water supply system uses surface water, the aquifer is the main source of water on lowland farms. These farms host significant beef cattle ranching that is responsible for the primary economic activity in the region. Few properties use water from the bays and freshwater lakes, common in the Pantanal landscape, because many of them dry up during prolonged droughts. These waters are characterised predominantly as sodium bicarbonate, some of which are mixed bicarbonate, and locally, sodium chloride. The general classification of the waters is considered as $Na^{+1} > Ca^{+2} > Mg^{+2}$ and $HCO_3^{2-} > Cl^- > SO_4^{2-}$. Although the waters are generally soft, some wells have hard water (Castelo Branco Filho, 2005). The potentially exploitable groundwater reserve for the total aquifer area (162,199.57km^2) is 188.53 m^3/s (Brazil, 2018). According to SEMAC (2010), its potential exploitable reserve in the state is estimated to be 117.67 m^3/s. Different infiltration coefficients were used in both estimations, 0.15% in Brazil (2018) and 0.10% in SEMAC (2010). It is important to note that the state has approximately 60% of the total Pantanal freshwater wetlands.

There are great differences among the hydraulic characteristics of the aquifer systems (Table 14.1). In terms of exploitation, the most exploited aquifer, considering the number of wells, is Serra Geral Aquifer (49%) followed by Bauru Aquifer (34%). The main users of groundwater are human consumption for private and public supply (35%), and animal watering (25%) and other uses (e.g., recreation) (34%) (Figure 14.3).

Table 14.1 Hydraulic properties of Mato Grosso do Sul's aquifers systems

Aquifer system	Average depth (m)	Average flow (m^3/h)	Specific capacity (m^3/h/m)	Transmissivity (m^2/d)
PCAS	110.0	27.0	1.4	-
CPCAS	115.0	19.5	6.3	-
FAS	150.0	45.0	1.6	91.2
APGAS	150.0	5.8–20.0	0.4–1.10	39.7
GAS	112.0	20.0–320.0	2.4	1.0–650.0
SGAS	150.0	0.3–144.0	0.3–50.0	80.0–400.0
BAS	107.0	5.0–110.0	0.5–2.5	1.0–630.0
TAS	30.0	19.0	1.2	150.0–350.0
PAS	47.0	18.0	0.13–30.0	0.1–864.0

Note: PCAS, Precambrian Aquifer System; CPCAS, Calcareous Precambrian Aquifer System; FAS, Furnas Aquifer System; APGAS, Aquidauna-Ponta Grossa Aquifer System; GAS, Guarani Aquifer System; SGAS, Serra Geral Aquifer System; BAS, Bauru Aquifer System; TAS, Tertiary Aquifer System; PAS, Pantanal Aquifer System.

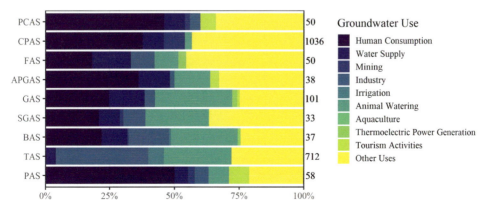

Figure 14.3 Groundwater uses (%) and the number of wells per aquifer system in Mato de Grosso do Sul, Brazil. (Data from: Imasul, 2020.)

14.4 MAIN ECONOMIC ACTIVITIES

The economic and population growth of a region is highly dependent on the availability of water resources. The economic development of MS relies heavily on the primary sector of the economy, such as agriculture and agribusiness (Lima et al., 2016), whose productivity is directly linked to the regional water supply. The overall abundance of water resources and availability of land formed a positive scenario mainly for agriculture (both subsistence and commercial), farming, forestry, grazing and industry consolidation. These factors have increased the competitiveness of the state in the Brazilian economy. Brazil has 26 states with great economic characteristics and differences. The current position of MS in the Brazilian economic context of agricultural production is presented in Table 14.2. Recently, the state has ranked the sixteenth position in the national Gross Domestic Product (Castelão et al., 2020). Further descriptions of the main economic activities are described in the following section.

14.4.1 Agriculture

MS occupies the fifth position in crop production among the other states (Table 14.2). It is one of the main producers of soybean and corn, standing out among the five largest grain producers in the country (IBGEa, 2019). Agriculture also contributes to the production of renewable energy sources (biodiesel and ethanol) for the state and nation (Castelão et al., 2020). The total grain production in 2019 was 19.7 million tons (Conab, 2020). The municipalities with the largest agricultural production are Dourados, Maracaju, Ponta Pora and Rio Brilhante, in the south-central region, where basaltic residual soils occur (outcrop area of the SGAS). In this region, there is also a large sugarcane production, that in 2018 reached 49,784,753 tons in 680,611 ha of the planted area. Therefore, sugarcane bagasse is also used as a renewable energy source, encompassing 42% of the total renewables in the state (Castelão et al., 2020).

Table 14.2 Crop production, beef cattle production and total revenue from agricultural exports (2019/2020) of the largest producers among the Brazilian states

State	Crop production		Cattle (slaughtered)		Total exports	
	Million Tons	%	Head	%	US$	%
MT	73,754.8	29.3	5,649,896	17.6	16,850,406,465	17.4
PR	39,853.5	15.9	1,452,174	4.51	12,778,774,665	13.2
GO	26,712.8	10.6	3,013,431	9.37	5,421,934,951	5.60
RS	26,266.6	10.4	1,966,444	6.11	11,988,866,649	12.4
MS	**19,717.8**	**7.8**	**3,585,036**	**11.2**	**4,999,250,392**	**5.16**
MG	15,110.0	6.0	2,846,455	8.85	7,888,113,873	8.14
BA	9,637.2	3.8	1,196,050	3.72	3,944,312,968	4.07
SP	9,193.5	3.7	3,326,168	10.3	15,259,244,785	15.8
SC	6,507.4	2.6	536,299	1.67	6,196,664,583	6.40
TO	5,679.9	2.3	1,032,557	3.21	1,107,215,157	1.14
Others	18,991.4	7.6	4,326,662	13.5	10,415,839,872	10.8

Data from Conab (2020).

The sugarcane mills use groundwater from the SGAS, GAS and BAS for processing sugarcane into sugar and ethanol; the wastewater effluent generated (vinasse) is used to fertirrigate sugarcane fields. As there is no groundwater monitoring network, it is not possible to know whether this practice is impacting the water quality of the SGAS; one of the most exploited aquifer systems in the state and the most exploited in the two largest cities, Campo Grande and Dourados. Although about 68.4% of the total water used in Brazil is destined for irrigation (ANA, 2018), in MS this is still an uncommon practice (Figure 14.3) as most of the properties adopt extensive agricultural systems. Irrigation is predominantly practiced for rice production using surface water, usually in Ivinhema, Pardo, Amambai and Miranda (SEMAC, 2010). However, family farming is a growing trend in the state; of the 71,164 thousand economically active properties registered in the state, 61% account for family farming. Most of these properties rely on groundwater to produce mainly temporary and permanent crops, horticulture, floriculture, certified seeds and seedlings, and livestock.

14.4.2 Livestock

MS stands out for its beef cattle ranching in an extensive system supplying the national meat production chain, holding the second position among the producing states (Table 14.2). The grazing area of natural and cultivated pasturelands accounted for 18,439,835 ha in 2017, with a herd of 20,896,700 head of cattle in 2018 (IBGEa, 2019). About 10.8% of the total water consumed in Brazil is destined for animal water supply (ANA, 2018). The estimated use of water for animal watering in the state is 25.33 m^3/s (Imasul, 2020), a demand almost equivalent to human supply; although, in general, the extensive livestock systems use rainwater harvesting systems for animal watering. Pasturelands are mostly rainfed in extensive systems, as irrigation of pastures is still poorly practiced in the state.

The use of intensive production systems has been increasing recently, due to the decrease in the availability of land with agricultural use capacity (Lima et al., 2016). Despite requiring greater technological investments, intensive beef production systems have obtained better results with improved productivity and production control. Intensive livestock requires the use of larger volumes of water for cleaning animal shelters and irrigating pastures to maintain a constant supply of forage. The intensive system has been growing throughout the state and country but with limited adoption in drier regions with irregular rainfall (Obeed Al-Azawi et al., 2017). Therefore, the use of groundwater resources is a promising strategy to support livestock production in the state, not only in the extensive system but also in the intensive system.

14.4.3 Pig farming

Pig farming is a growing activity in MS, currently raking the state in the seventh position in the country, with the slaughter of 2 million head of pig in 2019, but expected to reach the fourth place soon (Asumas, 2019). São Gabriel do Oeste is the largest producer, accounting for an increase of 119.57% in production between 2007 and 2017 (IBGEa, 2019), followed by Glória de Dourados and Itapora. The activity is highly dependent on water, especially in the intensive confined rearing system (Guerini Filho et al., 2015), which is the most employed in the state and country (Asumas, 2019). One farm in São Gabriel do Oeste, for example, has a capacity to produce 4,300 pigs per cycle, that consume 76,000 L of water per day, for drinking and cleansing the 8 tons of waste per day in the swine facility.

Unlike cattle ranching, groundwater is the major source of water used in the swine production process. In São Gabriel do Oeste, smaller farms exploit water mostly from the TAS, in shallow wells up to 80 m deep, and larger farms exploit water from wells deeper than 100 m, also reaching the GAS. In Glória de Dourados and Itapora, the SGAS is exploited.

The pig farming industry has reached high environmental standards at a level of excellence in the state. With the installation of anaerobic biodigesters, farms treat swine manure while capturing methane emissions produced in this process to generate electricity. That becomes an extra income due to the sales of carbon credits to the carbon offset market. Finally, the liquid effluent resulting from the treatment is used for fertirrigation.

14.4.4 Silviculture

Silviculture of pine and mainly eucalyptus plantations have been stimulated by the demand for coal and pulp production by the steel and paper industries in Três Lagoas. The area occupied by eucalyptus plantations increased 7.8-fold between 2006 and 2012 in the state, with a main 19% increase between 2011 and 2012 (Fernandes et al., 2016). Between 2014 and 2018, there was another increase of 25%, reaching the number of 1,133,218 ha, ranking MS as the third-largest producer of eucalyptus in Brazil (IBGEa, 2019).

Water availability is an important factor for eucalyptus seedling development, especially during the exponential growth phase, which demands the highest water consumption, in particular in the first weeks (Silva et al., 2015). The irrigation of eucalyptus has been considered by companies as a strategy to achieve greater production in smaller areas to face the international market competition. Especially in the eastern portion of the state, where drought periods can last 7–8 months. In this context, the use of groundwater can become a key contributor to economic growth in this changing scenario. Most of the cultivated areas are located near the city of Três Lagoas, in the outcrop area of the BAS.

14.4.5 Slaughterhouses and tannery

With the movement of large cattle herds to the Midwest region of the country, many slaughterhouses and tanneries began to settle near the supply centres. In 2017, there were forty-nine cattle, nine pig and five poultry slaughterhouses in the state (Mato Grosso do Sul, 2018). In 2019, around 3.3 million head of cattle, 1.9 million head of pig and 145.3 million head of poultry were slaughtered in these facilities (MAPA, 2019). Most water used in the slaughtering, cooling process and general operations comes from underground water bodies (SEMAC, 2010). Generally, these facilities are installed in places outside the urban areas with limited access to the public water supply service, due either to geographical distance or demand. In this case, the facilities can have grant concessions issued by the state regulator to install tubular wells, drilled to exploit groundwater.

14.4.6 Mining

There are mineral extraction sites in the municipalities of Corumbá and Ladário, on the border with Bolivia. With hills sustained by Precambrian sedimentary rocks of the Urucum Massif, the area of 131,105.5 ha contains mineral deposits of iron and manganese. The annual production of the Urucum mine is approximately 2.5 million tons of iron ore in an open pit and 800,000 tons of manganese in an underground mine (Vale do Rio Doce Mineração, 2019). The selective extraction and separation of manganese and iron from the iron–manganese ore is done in aqueous environment. The source of water for this activity is the groundwater pumped from the underground mine. With the advancement of manganese mining operations, a drastic decrease in streamflows of rivers that drain the slopes of the massif occurred. The discharge from springs that maintained the base flow was affected, causing a conflict with the local population, which is discussed in item 6.

In some outcrop areas of the Precambrian karst formations, such as in Corumbá and in the southwestern portion of the state, there are limestone deposits (MME, 1982; CPRM, 2006) that form the second largest reserve in the country. Limestone has been extracted and processed for its extensive use as agricultural limestone to increase the low pH of soils throughout the country as well as for the cement industry. Groundwater is used in the process of spraying the internal roads for dust suppression.

14.4.7 Tourism

The main tourist attractions are concentrated in the municipalities of Bonito, Bodoquena and Jardim, in the Serra da Bodoquena (plateau) (Moretti et al, 2016; Ribeiro, 2018), where the PCAS outcrops, and in the Pantanal region of Corumbá, where the PAS occurs in the lowlands and the PAS and PCAS occur in the highlands (Dos Santos and Abid Mrecante, 2010; Pacheco et al, 2017). Corumbá is famous for recreational fishing, birdwatching and numerous tours into the Pantanal wetlands to explore its impressive biodiversity, activities that tourists can engage while staying in farms, hotels, inns or still fishing lodges ('boat-hotels').

In the Serra da Bodoquena, there is a national park that contains fascinating natural water recreation areas intrinsically related to groundwater derived from the most extensive continuous karst areas in Brazil. Ecotourism is the main economic activity, driven by the presence of pit caves, potholes, waterfalls, freshwater spas, canyons, lakes, crystal-clear rivers, wetlands and forests with unique biodiversity and wildlife. Cave systems are formed by the circulation and accumulation of water that drive the chemical dissolution of limestone rocks. The high concentration of underground limestone acts as a natural filter that makes the rivers exceptionally clear. Groundwater discharge directly into the river channels and from numerous spring flows form rivers with vivid blue-green tones. This region is a major destination for domestic and international ecotourism driven by its natural beauties that provide activities such as

264 Groundwater for Sustainable Livelihoods

hiking, tree climbing, cycling, horseback riding, boat rides, tubing, stand-up paddle-boarding, rappelling, ziplines, snorkelling and diving.

Although the Serra da Bodoquena has been an active tourist spot for a few decades, there is no study of the integration of surface and groundwater in the region. The hydrogeological study in the region was recommended in the State Water Resources Plan, aiming at the management of some conflicts already existing at the time, between the tourism businessmen and other local stakeholders, as well as to evaluate the risk of karst subsidence in the region. Hotels, inns and other facilities rely entirely on groundwater for water supply.

14.5 SOCIAL ASPECTS

MS has one of lowest population densities in Brazil (6.86 inhabitants/km^2), with an estimated population of 2,778,986, mostly concentrated in urban areas (85.6%) (IBGEa, 2019). It occupies the tenth position in quality of life among the other states, with a Human Development Index (HDI) of 0.729 (IBGEa, 2019). 63.3% of the seventy-nine municipalities have between 10,000 and 50,000 inhabitants; 30.4% have less than 10,000 inhabitants. Only four municipalities have a population of over 100,000 inhabitants, being one of them the capital of the state, Campo Grande, with 786,797 inhabitants (IBGEa, 2019). Among the state's population, some vulnerable communities should be given more attention due to their socio-cultural and economic aspects. Noteworthy are indigenous communities, quilombolas and rural settlements. Some features of such communities are described in subsequent sections.

14.5.1 Indigenous communities

MS has the second largest indigenous population in Brazil. According to the last census of 2010, the indigenous communities were of 73,295 inhabitants (IBGEb, 2019). In 2019, these populations are estimated to account for 83,241 people (IBGEb, 2019). There are eleven indigenous ethnicities that currently live in the state: Terena, Kinikinau, Kaiowa, Guarani, Kadiweu, Ofaie, Guato, Chamacoco, Ayoreo, Atikum and Camba (Chamorro and Combès, 2018). They live in forty-five traditional territories (Figure 14.1), distributed among thirty-two of the seventy-nine municipalities. Amambai is the county with the largest indigenous population in the state, 7,158 people that representing approximately 10% of the total population (IBGEb, 2019).

Although the traditional indigenous lifestyle is dependent on rivers and lakes, all the indigenous communities are mainly supplied by groundwater due to the reduction and delimitation of their lands. In addition, some non-traditional land uses surrounding the indigenous territories, such as land leasing for intensive farming, have led to water quality-related problems in surface water bodies. The changes affecting the indigenous sacred and traditional lands and their respective consequences may be the reason for the several cases of suicide among young people in these populations (Staliano and Modardo, 2019). The communities are supplied by wells installed and monitored by the Special Secretariat of Indigenous Health (SESAI), part of the Ministry of Health. Data on the catalogued wells and water quality are not available.

Some countries, such as New Zealand and Australia, have been considering indigenous ecological knowledge and integrating it to hydrogeological understanding undertaken in some areas in order to better inform water management decisions (Stepheson and Moller, 2009; Bohensky and Maru, 2011; Liedloff et al, 2013). The use of indigenous knowledge could be valuable in MS, considering the lack of hydrogeological studies, their respective costs and the presence of large and complex systems, such as the Pantanal lowlands. The implementation of an integrated water resources management strategy that considers groundwater use, indigenous knowledge (Gabas et al., 2020) and practices and tribal water rights to Native Brazilians can provide reliability and possibly diminish conflicts.

14.5.2 Quilombolas

Quilombolas are rural communities of the descendants of former African slaves who live on subsistence agriculture and the exploitation of natural resources on lands that have been donated, occupied or acquired a long time ago. The quilombolas settlements were first established by fugitive slaves that escaped from slave quarters of the farms where they were exploited, taking refuge in nearby forests (Conde et al., 2017). The inhabitants are mainly self-sustained, selling or exchanging their excess production in local markets. The traditional agricultural activities developed in their land allow the production of cassava and maize flours, milk, honey, rapadura (unrefined whole cane sugar), brown sugar, molasses and other homegrown or handmade agricultural goods.

Quilombolas communities are recognised, registered and demarcated by the National Institute of Colonization and Agrarian Reform (INCRA). In Brazil, there are 1,747 quilombola areas distributed throughout the country (INCRA, 2019). In MS, it is estimated that 1,721 families are distributed in 35 quilombola communities present in 19 municipalities (INCRA, 2019). There are few studies in the literature about this type of community. Most studies are related to public health (Souto et al., 2012; Barroso et al., 2015). Conde et al. (2017) described their social organisation, relationships with the environment and traditional knowledge of local natural resources.

As they are essentially rural communities, most communities are supplied by groundwater from tubular wells built under the responsibility of INCRA and operated and monitored by the National Health Foundation (Funasa). The water is used for general domestic use, irrigation and the processing and manufacturing of agricultural products. However, there are no specific studies that evaluate the quality of these water resources in these communities. Funasa assesses water quality by analysing some physical–chemical parameters, such as pH, turbidity, dissolved oxygen, oxi-reduction potential, electrical conductivity, total dissolved solids, salinity, temperature and free residual chlorine, and microbiological parameters, faecal coliforms and *Escherichia coli*.

14.5.3 Rural settlements

The rural settlements are areas where farmers (settlers) are participating in an Agrarian Reform project, which can be set up on expropriated or public land, distributed and established by the regional superintendencies of INCRA. According to INCRA (2019), there are 204 rural settlements in MS, where 27,647 families are settled in a total area of 712,358.0876 km², with a total capacity for 32,013 families, in 54 municipalities. Most of the settlements are supplied by groundwater from deep wells built by INCRA in the community centre for human water supply. For example, in the Meneguel and Nova Alvorada Settlements, in the municipality of Nova Alvorada do Sul, 91.1% of the households rely on water wells for their water supply (Alvarenga and Rodrigues, 2004). Each of the settled families is responsible for the water supply for their own economic activities. However, as there was no study directed to the determination of water availability before implementing the settlements, water scarcity problems are quite common in most of them.

14.6 ENVIRONMENTAL ASPECTS

There are natural and human-influenced environmental issues of water resources use, allocation and conservation that should be considered in the socioeconomic development of MS. In this context, here we discuss three representative examples: a conflict between agriculture and tourism that occur in the outcrop area of the PCAS, a conflict between mining and rural settlers in the region of the PAS and the conflict of wells installed close to septic tanks, a conventional practice scattered throughout the state.

One of the classic conflicting linkages of tourism with agriculture is occurring in Bonito. In 2008, the soybean production in Bonito was 38,700 tons in a total of 15,000 ha of planted area, with an average yield of 2,580 kg/ha, occupying the twentieth position in the state. The production has been increasing since then. In a decade, soybean reached 160,080 tons of grain (MAPA, 2019). It is important to emphasise that this expansion occurred even though the region was classified as a critical area in the state's Ecological-Economic Zoning plan and indicated as one of the priority areas for biome conservation (Mato Grosso do Sul, 2015). The expansion of agricultural frontiers has been indicated to negatively impact the water quality of the main watercourses of the municipality, consequently affecting the environment and harming tourism. During the rainy season between 2018 and 2019, some tours were cancelled due to the increased water turbidity (Folha de São Paulo, 2019), an unusual phenomenon in the area. As there are no ongoing water-related studies and no groundwater monitoring, a technical evaluation is made difficult, although the problem is probably driven by land use and land cover change.

Several concerns have also risen regarding mining activities, mostly in the Urucum Massif in Corumbá, where mining operations occur since the early 20th century. With the advancement of manganese mining at the end of the century, there was an increase in groundwater pumping for the ore treatment process, reaching a consumption up to three times higher than the city's demand. Consequently, the flow of the Urucum Stream, the main drainage of the massif, gradually decreased until it dried up in 2004. The managing bodies, the Federal Public Prosecution Service, the State Public Prosecution Service and in particular the Brazilian Institute of Environment and Renewable Natural Resources (IBAMA), demanded the recovery of streamflow in order to guarantee the water supply and the ecological services that matter to the rural settlement on the slope of the massif. This stream restoration and rehabilitation process began in 2005 by the mining company, with the progressive increase in groundwater discharge recovering the streamflow. Also, water wells were installed to provide another source of water to the community. The stakeholders agreed to the delivery of an average flow of 1,057.09 m³/month, reached only in the first half of 2019. Restriction of streamflow shall be carried out until the mine has been exhausted and the natural streamflow recovered (Dos Santos, 2019).

One of the major environmental and public health problems in Brazil is the lack of basic sanitation. Many Brazilian cities do not have sewage collection and treatment systems. In MS, the state's sanitation agency currently operates 46 sanitary sewage systems and 62 sewage treatment stations; however, more than 50% of the population still uses septic tanks. This reality becomes even more relevant in rural communities and properties. In rural areas, shallow tubular wells are usually installed close to houses or facilities such as pens, pigsties and chicken coops, therefore, close to septic tanks and other sources of contamination. This is observed both in small, low-income properties and in large rural properties. The main accountable factor for that is the absence of effective public policies to address this issue. Additionally, there are some technical factors, such as the absence of specialised professionals and highly technological apparatuses that help water well drilling companies locate and drill tubular wells. Furthermore, education is another limiting factor as, in general, the Brazilian population is unaware of the basic functioning of the hydrological cycle and its relationship with groundwater, thus taking little account of its importance.

14.7 CHALLENGES

Brazil is a country used to the abundance of natural resources, therefore it is not prepared for water scarcity periods, as demonstrated by São Paulo, the richest city in Brazil, that nearly ran out of water in 2014, during the region's worst drought in recorded history (Gupta, 2019). Considering population growth, environmental problems and climate change, as pointed out by recent studies (Cullet et al., 2017; Smerdon, 2017; Aslam et al., 2018), groundwater is going to be the main source of water in the future. In MS, this is the ongoing situation as almost 80% of the cities are supplied by groundwater; although the actual total abstraction is largely unquantified. The capital of the state, Campo Grande, faced water scarcity problems in 2019 that required the adoption of drought contingency measures. As a result, the company responsible for the city's water supply decided to drill new deep wells. However, the state lacks a groundwater quality monitoring programme as well as knowledge about its aquifer systems and their total exploitation and users. The improvement of the grant system and the implementation of proper water and food security protocols depend on this information. These actions could help not only local and regional water supply authorities but also highly water-dependent economic sectors, such as agriculture and industry, in planning their future.

Multsch et al. (2020) simulated potential water demands of irrigated agricultural areas in Brazil, to predict negative impacts on streamflow, caused by the replacement of rainfed by surface-water irrigated agriculture. The authors mention that irrigation should be discontinued in 54% of the currently irrigated areas to avoid water scarcity in that case. Their results point out critical scarcity in the Pantanal and southern MS. Projections for future climate scenarios in MS state indicate an increase in extreme events, demanding more water for agriculture (Berezuck et al., 2017). This fact means a huge potential economic impact on the state's economy; therefore, an increase in groundwater exploitation is expected, the 'agriculture groundwater revolution' (Campana, 2007).

Considering climate change impacts, positive rainfall anomalies are mostly expected in the west (Pantanal); low to medium vulnerability is expected in most of the state, with high vulnerability in the regions of Dourados, Paranhos, Paranaiba and parts of Corumbá and the lowest adaptive capacity and resilience in the south (MMA, 2017). Corn and soybean are predicted to have an area reduction mostly in southern MS (Zilli et al., 2020). Pasture yield will also be affected by climate change, affecting the livestock sector through losses in productivity and losses in soybeans and corn production, as they are used as livestock feed (Zilli et al., 2020). Meanwhile, restrictions imposed by the sugar cane agro-ecological zoning (AEZ) tend to favour its expansion in eastern MS (Zilli et al., 2020). In that scenario, the use of groundwater for irrigation will be an asset for countering the restrictions brought by climate change.

Even being prepared for a nonstationary climate, MS ought to face its social inequality. Nowadays, there is a great difference in water security among the population, leaving the vulnerable communities behind (mainly the indigenous, quilombolas and rural settlement communities). Climate change might reinforce and worsen this scenario, increasing poverty due to their lower resilience to environmental changes. As pointed out by some authors, better water allocation planning (Foster et al., 2015), governance (López-Gunn, 2012) and justice (Otero et al., 2011) can support resilience and equitable growth for all users.

The implementation of adequate management practices depends not only on the institutional arrangements, but also on the engagement with local communities and the highly water-dependent economic sectors (Re, 2015). In our perspective, all these requirements for good management practices are only attainable if a collaborative approach is delineated and pursued by local and regional agencies and stakeholders, based on reliable and public data. The effort to reach this collaborative approach is even more important when considering the limited knowledge of the aquifer systems in proper scale. Therefore, successful management practices rely on adequate governance arrangements that could only be achievable by an integrated work of identification, planning, decision-making and implementation of actions related to groundwater management and directives associated with urban, industrial and agricultural water use efficiency and drought resiliency.

Population awareness and commitment to water resources issues are limited to the access and quality of their education. Despite the low performance of the Brazilian education system, groundwater is a hidden resource usually not included as a subject for primary and secondary education. For instance, the frequent installation of tubular wells near potential sources of contamination is understandable, as we only protect and care about what we understand. In this context, we can clearly see the need to involve stakeholders in water resources issues, making them more accountable to responsive groundwater management. That requires a change in the attitudes and perceptions among individuals, institutions, professionals, and decision-makers. Investments in education and social awareness could raise the communities' social and political willingness and understanding towards water resources management problems.

14.8 SUMMARY

In general, MS state has abundant surface and groundwater resources. However, there are some regions that suffer from water deficits. The state is nowadays a major consumer of groundwater for domestic, agricultural and industrial use. The most exploited aquifers are the Serra Geral and BASs. Although not all aquifer systems are greatly exploited, they are all essential for specific parts, activities and populations. Although the state has recently implemented a grant system, there is a lack of information about its aquifer systems due to shortage of hydrogeological surveys and groundwater quality monitoring. Agriculture and livestock are the main economic activities and are not yet dependent on groundwater. However, considering climate change perspectives these activities may become major users of groundwater in the future. Other activities, such as tourism and pig farming are totally supplied by groundwater resources. Despite the low population, the state has relevant indigenous groups, rural settlements and quilombola populations that are vulnerable communities mostly supplied by groundwater. Some challenges for the future are to enhance hydrogeological knowledge of the region, to create a groundwater monitoring programme, to include the vulnerable communities in the water resources management system, to encourage the participation of all users in water resources discussions and to educate the population, especially the children, about groundwater-related issues.

ACKNOWLEDGEMENTS

The authors express their thanks to Josciane Simplicio and Jaime Verruck, secretary and the head, respectively, of the Secretaria de Estado de Meio Ambiente, Desenvolvimento Econômico, Produção e Agricultura Familiar-SEMAGRO, for allowing the use of some figures from the state Water Resources Plan, and to Leonardo Sampaio Costa e Kelson Santos, from the State Environmental Agency (Imasul); to Anna Chistina Menno dos Santos, environmental analyst of the Instituto Brasileiro do Meio Ambiente e dos Recursos Naturais Renováveis-IBAMA, for providing many information about the mining conflict and to Giovanni Gabas Coelho from the Secretariat for Trade and International Relations, Ministery of Agriculture, for providing some agricultural production data.

REFERENCES

Agência Nacional de Águas – ANA. (2018) *Conjuntura dos recursos hídricos no Brasil 2018: informe anual/Agência Nacional de Águas.* Brasília: ANA, 72p. [Online] Available from: https://www.ana.gov.br/noticias/ana-lanca-conjuntura-dos-recursos-hidricos-no-brasil-2018 [Acessed 12th February 2019].

Alvarenga, M.R., Rodrigues, F.P. (2004) Indicadores socioeconômicos e demográficos de famílias assentadas no Mato Grosso do Sul. *Revista de Enfermagem.* UERJ, 12(3), 286–291.

Aslam, R.A., Shrestha, S., Pandey, V.P. (2018) Groundwater vulnerability to climate change: a review of the assessment methodology. *Science of The Total Environment* [Online], 15 January, 612, 853–875.

Associação Sul Matogrossense de Suinocultores-Asumas. (2019) *Mapeamento da suinocultura brasileira*. In: https://asumas.com.br/. [Accessed 8th October 2019].

Barroso, S.M.; Melo, A.P.S.; Silva, M.A.; Guimarães, M.D.C. (2015) Efficacy of the patient health questionnaire (PHQ-9) for screening depression in the Quilombola population of Bahia State, Brazil. *International Journal of Epidemiology*, 44(suppl 1), i65–i165.

Berezuk, A.G., Silva, C.A., Lamoso, L.P., Schneider, H. (2017) Climate and production: the case of the administrative region of grande dourados, Mato Grosso do Sul, Brazil. *Climate*, 5(49). [Acessed 08th July 2020].

Bohensky, E.L., Maru, Y. (2011) Indigenous knowledge, science, and resilience: what have we learned from a decade of international literature on "integration"? *Ecology and Society*, 16(4), 6.

Brazil. Ministério do Meio Ambiente (2018) Plano de recursos hídricos da Região Hidrográfica do Rio Paraguai. Relatório final, Março/2018. 401p.

Campana, M. (2007) The agricultural groundwater revolution: opportunities and threats to development. *Ground Water*, November, 45(6), 656–666.

Canter, L., Nelson, DI., Evrett, J.C. (1992) Public perception of water quality risks – influencing factors and enhancement opportunities. *Journal of Environmental Systems*, 22(2), 163–187.

Castelão, R.A., de Souza, C.C., Frainer, D.M. (2020) Southern Mato Grosso state (Brazil) productive system and its impact on emissions of carbon dioxide (CO_2). *Environment, Development and Sustainability*, 1–15.

Castelo Branco Filho, H. (2005) Distribuição espacial e temporal das características hidroquímicas das águas subterrâneas do Pantanal do Rio Negro. Rio de Janeiro: Master Dissertation, Universidade Federal do Rio de Janeiro, 198p.

Cavazzana, G.H.; Lastoria, G., Gabas, S.G. (2019) Surface-groundwater interaction in unconfined sedimentary aquifer system in the Brazil's tropical wet region. *Brazilian Journal of Water Resources, RBRH, Porto Alegre*, 24(8), 15.

Chamorro, G., Combès, I. (2018) *Povos indígenas de Mato Grosso do Sul: história, cultura e transformações sociais*. 1st edn. Dourados: UFGD.

Companhia de Pesquisa e Recursos Minerais – CPRM. (2006) *Mapa Geológico do Estado de Mato Grosso do Sul*. São Paulo: DAEE/IG/IPT/CPRM. Escala 1:1.000.000.

Companhia Nacional de Abastecimento – Conab (2020) Available at https://www.conab.gov.br/. [Accessed 15th July de 2020].

Conde, B.E.; Ticktin, T.; Fonseca, A.S.; Macedo, A.L.; Orsi, T.O.; Chedier, L.M.; Pimenta, D.S. (2017) Local ecological knowledge and its relationships with biodiversity conservation among two *Quilombola* groups living in the Atlantic rainforest, Brazil. *PLoS One*, 12(11), e0187599.

Cullet, P., Stephan, R.M. (2017) Introduction to groundwater and climate change: multi level law and policy perspectives. *Water International*, 18 August, 42(6), 641–645.

Dias, C.A., Oliveira, D.M., Freire, D.C.T., Vianna, S.A.C. (2007) Caracterização das águas subterrâneas em municípios do Estado de Mato Grosso do Sul (CD-ROM). *I Simpósio de Recursos Hídricos do Norte e do Centro-Oeste*, Cuiabá,

Dos Santos, A.C.M. (2019) Personal communication at Ibama office in Campo Grande on September 2019.

Dos Santos, E., Abid Mercante, M. (2010) Turismo en la Cuenca del Alto Paraguay. *Aspectos positivos y negativos Estudios y perspectivas en turismo*, 19(5), 673–687.

Enetério, N.G.P. (2009) Avaliação da suscetibilidade do aqüífero freático à contaminação por necrochorume em Bonito-MS. Campo Grande: Master Dissertation, Universidade Federal de Mato Grosso do Sul, 102p.

Fernandes, G.W., Coelho, M.S., Machado, R.B., Ferreira, M.E., Aguiar, L.M.S., Dirzo, R., Scariot, A., Lopes, C.R. (2016) Afforestation of savannas: an impending ecological disaster. *Natureza & Conservação*, 14(2), 146–151.

Ferraro, A.A., Gabas, S.G., Lastoria, G. (2014) Origem de Metais Pesados em Aquífero Livre De São Gabriel do Oeste, Mato Grosso do Sul. *Geociências*, 34(4), 801–815.

Folha de Sao Paulo (2019) Frias, S. *Águas turvas afetam turismo em Bonito e passeios são cancelados*. https://www1.folha.uol.com.br/cotidiano/2019/04/aguas-turvas-afetam-turismo-em-bonito-e-passeios-sao-cancelados.shtml. [Accessed 19th April 2019].

Foster, S., Evans, R., Escoleto, O. (2015) The groundwater management plan: in praise of neglected "tool of our trade". *Hydrogeology Journal*, 23, 847–850.

Gabas, S.G., Lastoria, G., Uechi, D.A. (2020) Inclusion of indigenous communities in water resources management in the Middle West of Brazil: a proposal. [Lecture] In: Geothics and Groundwater Management, Porto, Portugal, 18th May.

Gastmans, D. (2007) Hidrogeologia e hidroquímica do Sistema Aqüífero Guarani na porção ocidental da Bacia sedimentar do Paraná. Rio Claro: PhD Thesis at Geoscinece and Environment Institute, Universidade Estadual Júlio de Mesquita Filho, 238p.

Gastmans, D., Chang, H.K., Hutcheon, I. (2010a) Groundwater geochemical evolution in the northern portion of the Guarani Aquifer System (Brazil) and its relationship with diagenetic features. *Applied Geochemistry*, 25, 16–33.

Gastmans, D., Chang, H.K., Hutcheon, I. (2010b) Stable isotopes (2H, 18O and 13C) in groundwaters from the northwestern portion of the Guarani Aquifer System (Brazil). *Hydrogeology Journal*, 18(6), 1497–1513.

Gastmans, D., Kiang, C.H. (2005) Avaliação da hidrogeologia e hidroquímica do Sistema Aqüífero Guarani (SAG) no Estado de Mato Grosso do Sul. *Águas Subterrâneas*, 19(1), 35–48.

Guerini Filho, M., Dal Soler, A.L., Reginatto, V.P., Casaril, C.E., Lumi, M., Konrad, O. (2015) Análise do consumo de água e do volume de dejetos na criação de suínos. *Revista Brasileira de Agropecuária Sustentável*, 5(2), 64–69.

Gupta, H. (2019) Assessing water security in the Sao Paulo metropolitan region under projected climate change. *Hydrogeology and Earth System Sciences*, 23(12), 4955–4968.

Hoogesteger, J. (2018) The ostrich politics of groundwater development and neoliberal regulation in Mexico. *Water Alternatives*, 11(3), 552–571.

Instituto Brasileiro de Geografia e Estatística – IBGEa. (2019) *Cidades: Mato Grosso do Sul* [Online]. Available from: https://cidades.ibge.gov.br/brasil/ms/panorama. [Acesso em: 01 dezembro 2019].

Instituto Brasileiro de Geografia e Estatística – IBGEb. (2019) *Indígenas*. [Online]. Available from: https://indigenas.ibge.gov.br/. [Acessed 18th November 2019].

Instituto de Meio Ambiente de Mato Grosso do Sul- Imasul. (2020) Water resources Management Board. Available from: https://www.imasul.ms.gov.br/.

Instituto Nacional de Colonização e Reforma Agrária- INCRA. (2019) *Relação de processos abertos Quilombolas*. [Online] Available from: http://www.incra.gov.br/quilombola/ [Accessed 13th September 2019].

Johnson, N., Revenga, C., Escheverria, J. (2001) Managing water for people and nature. *Science*, May 11, 292(5519), 1071–1072.

Lastoria, G. (2002) Hidrogeologia da Formação Serra Geral no Estado de Mato Grosso do Sul. PhD Thesis at Geoscience and Environment Institute, IGCE– UNESP. Rio Claro. 133p.

Lastoria, G., Chang, H.K., Sinelli, O., Hutcheon, I. (2007) Evidências da conectividade hidráulica entre os sistemas aquíferos Serra Geral e Guarani no Estado de Mato Grosso do Sul e aspectos ambientais correlacionados. *XV Encontro Nacional de Perfuradores de Poços, Gramado*. CD-ROM.

Liedloff, A.C., Woodward, E.L., Harrington, G.A., Jackson, S. (2013) Integrating indigenous ecological and scientific hydro-geological knowledge using a Bayesian Network in the context of water resource development. *Journal of Hydrology*, 499, 177–187.

Lima, J.F. de, Piffer, M., Ostapechen, L.A.P. (2016) O crescimento econômico regional de Mato Grosso do Sul. *Interações*, Campo Grande, 17(n. 4).

López-Gunn, E. (2012) Groudnwater governance and social capital. *Geoforum*, 43, 1150–1151.

Mato Grosso do Sul. (2015) Zoneamento Ecológico Econômico do Estado de Mato Grosso do Sul. [Report] Campo Grande, MS. 199p.

Ministério da Agricultura, Pecuária e Abastecimento-MAPA. (2019) *Agropecuária brasileira em números*. [Online]. Available from: http://www.agricultura.gov.br/assuntos/politica-agri-cola/agropecuaria-brasileira-em-numeros. [Accessed 10th November 2019].

Ministério de Minas e Energia - MME. (1982) Secretaria Geral. *Projeto RADAMBRASIL*. Folha SE. 21 Corumbá: Geologia. Rio de Janeiro.

Ministério do Meio Ambiente – MMA. (2017) Índice de Vulnerabilidade aos desastres natu-rais relacionados as secas no contexto da mudança do clima. Ministério do Meio Ambiente, Brasília.

Ministério Público Federal-MPF (2019) *Mapa Quilombolas*. [Online] Available from: http://www.mpf.mp.br/ms/atuacao/mapa-quilombolas [Accessed 14th September 2019].

Moretti, E.C., Salinas Chávez, E., Do Noascimento Ribeiro, A.F. (2016) El ecoturismo en áreas kársticas tropicales: Parque Nacional Sieera da Bodoquena, Mato Grosso do Sul, Brazil y Parque Nacional Viñales, Pinar Del Rio, Cuba. *Gran Tour*, 13, 82–104.

Multsch, S., Krol, M.S., Pahlow, M., Assunção, A.L.C., Barretto, A.G.O.P., Lier, Q.J.V., Breuer, L. (2020) Assessment of potential implications of agricultural irrigation policy on surface water scarcity in Brazil. *Hydrology and Earth System Sciences*, 24, 307–324.

Obeed Al-Azawi, A.A., Ward, F.A. (2017) Groundwater use and policy options for sustainable management in Southern Iraq. *International Journal of Water Resources Development*, 33(4), 628–648.

Organization of American States– OAS. (2009) *Aquífero Guarani – Síntese hidrogeológica do Sistema Aquífero Guarani*. [Report] Brasília: CD-ROM.

Otero, I., Kallis, G., Aguilar, R., Ruiz, V. (2011) Water scarcity, social power and the production of an elite suburb. *Ecological Economics*, May 15, 70(7), 1297.

Pacheco, A.O.de C., Benini, E.G., Mariani, M.A.P. (2017) La economia reativa en Brasil: el Desarrollo del turismo local en el pantanal sur de Mato Grosso. *Estudios y perspectivas en turismo*, 26(3), 678–697.

Pahl, C.B.C., Lastoria, G., Gabas, S.G. (2018) Microbial contamination of groundwater in a swine fertigation area. *Brazilian Journal of Water Resources, Porto Alegre*, 23(42), 12.

Paula de, F., Chang, H.K., Chang, R.C. (2005) Hidroestratigrafia do Grupo Bauru (K) no Es-tado de São Paulo. *Águas Subterrâneas*, 19(2), 19–36.

Re, V. (2015) Incorporating the social dimension in the hydrogeological investigations for rural development: the Bir-Al-Nas approach for socio-hydrogeology. *Hydrogeology Journal*, 23, 1294–1304.

Ribeiro, A.F. do N. (2018) What Bonito is this one? Territorial disputes on lands agro-eco-tour-ism. *Entre-lugar*, December, 9(18), 37–67.

Sanesul/Tahal. (1998) *Estudos Hidrogeológicos de Mato Grosso do Sul*: [Report] Relatórios v. I a V, 14 mapas esc. 1:500.000; Campo Grande.

Schaider, L.A., Swetschinski, L., Campbell, C., Rudel, R.A. (2019) Environmental justice and drinking water quality: are there socioeconomic disparities in nitrate levels in U.S. drinking water? *Environmental Health*, 18, 3.

Secretaria de Estado de Meio Ambiente e Desenvolvimento Econômico, Produção e Agricul-tura Familiar – SEMAGRO. (2018) *Perfil Estatístico de Mato Grosso do Sul*. [Report] Ano base: 2017. Campo Grande: SEMAGRO, 100p.

Secretaria de Estado de Meio Ambiente, do Planejamento, da Ciência e Tecnologia. Instituto de Meio Ambiente - SEMAC. (2010) *Plano estadual de recursos hídricos de Mato Grosso do Sul*. Campo Grande: Editora UEMS, 196p.

Silva, C.R.A., Ribeiro, A., Oliveira, A.S. de, Klippel, V.H., Barbosa, R.L.P. (2015) Desen-volvimento biométrico de mudas de eucalipto sob diferentes lâminas de irrigação na fase de crescimento. *Pesquisa Florestal Brasileira*, 35(84), 381–390.

Smerdon, B. (2017) A synopsis of climate change effects on groundwater recharge. *Journal of Hydrology*, December, 555, 125–128.

Souto, R.G., Santo, L.R.E., Ribeiro, F., Almeida, J.M., Silveira, M.F. (2012) Evaluation of intestional parasites and hepatic schistosomiasis of the "quilombola" community of Sao Francisco, MG. *Motricidade*, April, 8(52), SS95.

Souza, A.A., Lastoria, G., Gabas. S.G., Machado, C.D. (2014) Avaliação da água subterrânea nos Aquíferos Cenozoico e Guarani em São Gabriel do Oeste-MS: subsídios à gestão integrada. *Ciência e Natura*, 36(2), 169–179.

Staliano, P, Mondardo, M.L. (2019) Onde e como se suicidam os Guarani e Kaiowá em Mato Grosso do Sul: confinamento, Jejuvy e Tekoha. *Psicologia: Ciência e Profissão*, 39, 9–21.

Stepheson, J., Moller, H. (2009) Cross-cultural environmental research and management: challenges and progress. *Journal of the Royal Society of New Zealand*, 39(4), 130–140.

Vale do Rio Doce Mineração. (2019) *Conheça as minas do Sistema Centro-Oeste no Mato Grosso do Sul*. [Online] Available from: http://www.vale.com/brasil/PT/aboutvale/news/Paginas/conheca-minas-sistema-centro-oeste-mato-grosso-sul.aspx [Accessed 16th December 2019].

Velis, M., Conti, K.L., Biermann, F. (2017) Groundwater and human development: synergies and trade-offs within the context of sustainable development goals. *Sustainability Science*, 12, 1007–1017. [Accessed 19th June 2019].

Wutch, A., Brewis, A. (2014) Food, water and scarcity toward a broader anthropology of resource insecurity. *Current Anthropology*, 55(4), 444–468. [Accessed 15th June 2020].

WWAP (UNESCO World Water Assessment Programme). (2019) The United Nations World Water Development Report 2019: Leaving No One Behind. [Report] Paris, UNESCO. Available in: https://en.unesco.org/themes/water-security/wwap/wwdr/2019. [Accessed 15th March 2019].

Zheng, Y., Flanagan, S. (2019) The case for universal screening of private well water quality in the U.S. and testing requirements to achieve it: evidence from arsenic. *Environmental Health Perspectives*, commentary, 1–6.

Zilli, M., Scarabello, M., Soterroni, A.C., Valin, H., Mosnier, A., Leclère, D., Havlík, P., Kraxner, F., Lopes, A.L., Ramos, F.M. (2020) The impact of climate change on Brazil's agriculture. *Science of the Total Environment*, 740, 139384.

Chapter 15

Valuing groundwater use

Resolving the potential of groundwater in the Upper Great Ruaha River Catchment of Tanzania

D.B. Mosha
Sokoine University of Agriculture

J.L. Gudaga
Amani College of Management and Technology

D. Gama and J.J. Kashaigili
Sokoine University of Agriculture

CONTENTS

15.1 Introduction 276
15.2 Methodology 278
    15.2.1 Study area 278
    15.2.2 Research design and data collection methods 279
    15.2.3 Data analysis 280
        15.2.3.1 Descriptive analysis 280
        15.2.3.2 Financial analysis 280
        15.2.3.3 Cost-benefit analysis basic assumptions 281
        15.2.3.4 Binary logistic regression 282
15.3 Results and discussion 283
    15.3.1 Demographic and socio-economic characteristics of respondents 283
    15.3.2 Groundwater source and usage 284
    15.3.3 Groundwater usage at household level 285
    15.3.4 Determinants of quantity of water use per day 286
    15.3.5 Irrigation practices using groundwater source in the study area 287
    15.3.6 Reasons for accessing and using groundwater 288
    15.3.7 Cost-benefit analysis 289
        15.3.7.1 Short-term cost-benefit analysis 289
        15.3.7.2 Financial viability of groundwater 289
    15.3.8 Socio-economic factors influencing the use of groundwater by
        smallholder farmers 291
    15.3.9 Threats and challenges on the groundwater use 291
15.4 Conclusions 292
Acknowledgments 292
References 292

DOI: 10.1201/9781003024101-15

15.1 INTRODUCTION

Globally, the use of groundwater has been growing rapidly, especially in low-income countries, where a range of factors including urbanization, industrialization, land-use changes, and population growth are putting pressure on water provision from surface waters, especially in drylands (Gronwall and Oduro-Kwarteng, 2018). Groundwater is typically viewed as a more resilient water supply option to surface water options for household supply (MacAllister et al., 2020). The population of sub-Saharan Africa (SSA) is projected to increase to between 1.5 and 2 billion with approximately 50% of the population living in urban areas by 2050 (Gronwall and Oduro-Kwarteng, 2018). This rising population will require safe drinking water, and water for irrigation to enhance food security (Osborn et al., 2015).

Changing climates including rising incidences of drought, higher temperatures, and changes in the frequency and intensity of extreme events accentuate water supply challenges in Africa (Mwakalila, 2014; Sappa et al., 2015). In the Upper Great Ruaha River Catchment (UGRRC) of Tanzania, for instance, it is suspected that river flow has been diminishing while the demand for irrigation water has increased (Mwakalila, 2011). Securing access to sufficient surface water is a growing challenge in this semi-arid catchment. Groundwater is increasingly considered as the most viable solution to sustain irrigation (Taylor et al., 2019). Groundwater, in principle, holds particular benefits including drought resilience and ubiquity as its widespread availability provides a buffer against climate variability (MacDonald et al., 2012; Taylor et al., 2013a). Here, we examine the role of groundwater as a freshwater source to sustain livelihoods in the UGRRC (Figure 15.1). Privately owned shallow wells and hand-dug wells have become the fastest-growing source of freshwater in the Usangu Plains.

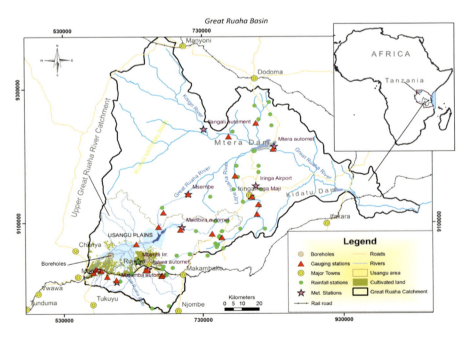

Figure 15.1 Map of the Upper Great Ruaha River Catchment showing major towns, drainage network, and hydrological monitoring stations.

Drawing on existing literature, three major factors are driving the heightened interest in groundwater development and management for future in Tanzania. The first, and arguably the most pressing issue is poverty, reflected in the need to realize the UN Sustainable Development Goals 2 and 6, and the aim of the National Water Policy to provide improved access to safe water supplies to all communities (Osborn et al., 2015; URT, 2002). To reach this goal, the delivery of groundwater through well-placed and appropriately constructed and maintained boreholes is vital. A vivid example is that of the city of Dodoma where the piped water supply derives entirely from a single well-field (Taylor et al., 2013b). Second, increased water use for livestock and small-scale irrigation improves livelihoods and food security, and a pathway out of poverty (Tucker et al., 2014). Groundwater represents an untapped source of water in the UGRRC but technical, socio-economic, and institutional factors have so far constrained its use (Gudaga et al., 2018). Third, climate change intensifies precipitation and amplifies temperatures that strongly impact the availability of surface water. There is increasing evidence that while the availability of rainwater and surface water are projected to become more erratic and less reliable as a result of climate change, groundwater are less affected (Komakech, 2018; MacDonald et al., 2012) and may often be enhanced (Cuthbert et al., 2019). Enhancing water storage capacity, both above and below ground, is widely accepted as a coping strategy against hydrological extremes such as floods and droughts (Taylor, 2009; Damkjaer and Taylor, 2017).

A key question is whether groundwater can transform the productivity and climate resilience of smallholder farming to alleviate poverty and enable inclusive economic development. Many research papers and reports in the UGRRC (e.g. Kashaigili, 2010; Mwakalila, 2011, 2014; Gudaga et al., 2018; Gama, 2018) and across SSA (Howard, 2017) highlight that little is known about the role of groundwater use in supporting agricultural livelihoods or opportunities to expand this role in the future. Drawing evidence from focus-group discussions (FGDs) and ninety household interviews in the Usangu Plains (Figure 15.1), this chapter explores the following questions: (a) what are the uses of groundwater and what are the benefits of using it? (b) what are the costs and benefits of using groundwater for irrigation by smallholder farmers and how can they inform policymakers on how to unlock the potential of groundwater resources? (c) what are the factors that influence the adoption of groundwater-fed irrigation by smallholder farmers and what are the expected challenges? In this chapter, we adopted Abric et al. (2021) definition on smallholder referring smallholders are subsistence farmers, generally with landholdings that are smaller than 2 ha, privately owned, and under the complete control of the household head (Abric et al., 2011).

Field research was conducted under the *GroFutures* (Groundwater Futures in SSA) project under the UPGro (Unlocking the Potential of Groundwater for the Poor) program (2015–2020). Data were collected from three villages: Ubaruku, Mwaluma, and Nyeregete in Mbarali District (Figure 15.2), surveyed from 2017 to 2018. The FGDs and in-depth interviews occurred in 2016, 2017, and 2018. The protocol for interviews was approved by Sokoine University of Agricultural; opinions presented here reflect views on groundwater resources and users from these villages in the Usangu Plains. We show that groundwater is a vital resource for both low- and medium-income households, and in future is a pathway for rice commercialization and medium-scale livestock keeping.

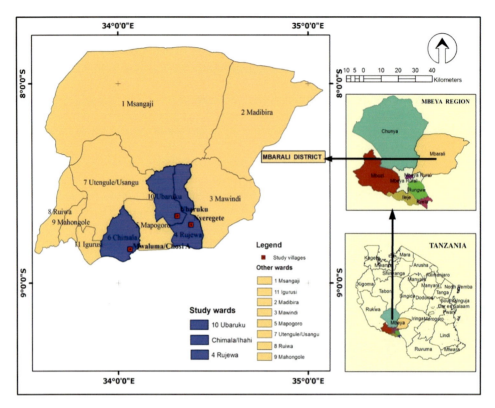

Figure 15.2 Map of Mbarali District showing the locations of the study villages: Ubaruku, Nyeregete, and Mwaluma.

15.2 METHODOLOGY

15.2.1 Study area

The study was conducted in the UGRRC and considered both technical and socio-economic aspects of groundwater resource assessment and management. The UGRRC was selected as a suitable site for investigation of groundwater matters as part of a larger study of allied basin observatories in Tanzania, Ethiopia, Niger, and Nigeria under *Gro-Futures*. The UGRRC is situated in the upper reaches of the Rufiji River Basin (RRB). It lies within a semi-arid belt from North to South through the central portion of Tanzania (Mwakalila, 2014). It encompasses an extensive wetland, comprising seasonally flooded grassland and a much smaller area of a permanent swamp locally known as *Ihefu*, which is supplied by streams draining humid upland catchments in the Uporoto and Kipengere mountains. This area is critical to Tanzania as it provides livelihood for smallholder farmers and agro-pastoralists, contributes to food security, and sustains wetlands and their associated biodiversity that, in turn, flows downstream to the Ruaha National Park as well as the Mtera and Kidatu hydropower plants (Walsh, 2012).

The UGRRC is characterized by two distinct landscapes – central plain (Usangu Plains in Mbarali District) and highlands (Uporoto Mountain ranges). The villages of Ubaruku, Nyeregete, and Mwaluma are located in Mbarali District (Figure 15.2). The district lies between Latitudes 7°41′ and 9°25′ south, and between Longitudes 33°40′ and 35°40′ east at an altitude range of 1,010–1,100 m above sea level (Mwakalila, 2011). The climate of the area is mostly semi-arid with seasonal temperature and rainfall variations. Temperatures range from 20°C to 25°C, whereas the annual rainfall varies between 500 and 700 mm/year. The area has a unimodal type of rainfall which falls from November to May, and which is normally scattered and varies across the Mbarali District. Rainfall is generally unreliable and localized droughts are common (URT, 2010). According to 2012 national census, the Usangu Plains has a total population of ~790,500 and an annual growth rate of 2.7% (National Bureau of Statistics (NBS), 2012). The population is multi-ethnic and multi-cultural in which Sangu is the dominant indigenous ethnic group; other ethnic groups include Bena, Hehe, Maasai, Sukuma, and Nyakyusa. There has been a significant change in ethnic composition with increasing competition in land-use systems (SMUWC, 2001). Figure 15.2 shows the administrative boundaries of the study area.

15.2.2 Research design and data collection methods

The study uses mixed data collection methods that include: (a) review of existing documents on groundwater use and management in Tanzania; (b) FGDs; (c) key informant interviews; (d) household surveys; and (e) direct observation. In September 2015, a stakeholders' policy dialogue workshop involved twenty-eight participants was conducted to gain an understanding on the existing use of groundwater and the institutional framework governing water resources in the UGRRB and Tanzania. Between 2016 and 2018, the FGDs and key informant interviews were conducted to collect qualitative data on the role played by various stakeholders in groundwater development and management.

FGDs involved an average of ten participants per FGD (32 in total) comprising a diversity of gender social groups (men and women; youth and elders; and wealth category – poor, medium, and rich households). The proportion of women and men participants per group were broadly similar. Different categories of social groupings were consulted to get highlights on various aspects of the research including access points and responsible water management institutions as well as challenges and opportunities. Participants' direct observations were also undertaken to gain information on well locations and conditions, water access points, water abstraction technologies, and storage facilities. A total of twenty-six participants were involved in key informant interviews guided by a checklist of questions.

A quantitative survey was also deployed and involved ninty respondents comprising both male and female heads of households selected randomly from three sampled villages in Mbarali District. A semi-structured questionnaire was used to collect quantitative data. The questionnaire captured a number of variables included: household demographic data, groundwater source, uses and management aspects, technology for water abstraction, distribution, and storage as well as water and production costs, factors constant or enhance development and management of groundwater resources.

280 Groundwater for Sustainable Livelihoods

15.2.3 Data analysis

15.2.3.1 Descriptive analysis

Qualitative data collected were subjected to content analysis, reducing, and clustering information into small meaningful units based on themes, trends, frequency, and strongly held opinions. Interpretations were made by researchers and subsequently used in the discussions. Descriptive and econometric analyses were undertaken using the Statistical Package for Social Science (SPSS) (Pallanti, 2007). Descriptive statistical analysis was computed to determine frequency and percentages of socio-economic characteristics of the respondents. An econometric analysis, outlined in the following section, was conducted to examine the relations between dependent and independent variables.

15.2.3.2 Financial analysis

Net present value (NPV) and cost-benefit ratio (CBR) were calculated to evaluate the long-term financial viability of using groundwater for small-scale irrigation. Information on surface water irrigation was included in this analysis in order to compare the profitability with and without groundwater irrigation while other factors such as climate change notwithstanding. NPV and CBR were computed from equations 15.1 and 15.2 following Lin and Nagalingam (2000):

$$\text{NPV} = \sum_{i=0}^{n} \frac{B_t - C_t}{(1+r)^t} \tag{15.1}$$

$$\text{CBR} = \frac{\sum_{t=1}^{T} \frac{B_t}{(1+r)^t}}{\sum_{t=1}^{T} \frac{C_t}{(1+r)^t}} \tag{15.2}$$

where
  B = benefits at year 2016 (market value of yield at year 2016)
  C = Cost at year 2016 (market value of inputs, fees, and other production costs)
  t = the time in years, i.e. 30 years ($t = 30$)
  r = discount rate 12%, 18%, and 20%
  $(1+r)^t$ = discount factor

The cost (C) component included the initial capital cost of the borehole, operation, and maintenance cost, water fee, market prices of inputs, the cost of plowing, planting, weeding, and harvesting. Discounting reflects the time value of money. Benefits and costs are worth more if they are experienced sooner such that all future benefits and costs should be discounted to their present value for the investments with long lifespan. As the discount rate increases, the present value of future benefits and costs reduces. A discounting rate of 12% was used in this analysis as per the Bank of Tanzania (BOT) and outlined in their Monthly Economic Review (February 2017). Apart from the fixed discounting rate used by the Central Bank in Tanzania (BOT), the study also considered interest rates of 18% and 20% used by different microfinance banks in Tanzania as they are the main credit sources for smallholder farmers. There is, however, considerable uncertainty over the correct discount rate, and also high uncertainties are expected in agricultural production that can affect investment financial viability. In order to check how investment in motor pump-based smallholder irrigation is sensitive to cost and income changes, different scenarios were assumed. Scenario 1 posits an increase in production costs and reduced income, whereas scenario 2 posits an increase in production costs and increased income (Gebregziabher et al., 2016). More specifically, scenario 1 proposes a 25% increase in production costs and 10% decrease in income. According to Gebregziabher et al. (2016), the assumptions are that production cost and income may continue in an increased and decreased trend, respectively. On this basis, scenario 1 proposes a 25% increase in production costs and 10% decrease in income, whereas scenario 2 proposes a 100% increase in the production costs and 25% increase in income. Gebregziabher et al. (2013) noted that the size of land for production affects the investment economic viability due to the economies of scale whereby cost per unit of output generally decreases with an increase in the scale of production. Estimating the lifespan of an intervention is uncertain, subjective, and widely debated. Since the use of groundwater involves fixed costs which are capital intensive, lifespan is a critical variable to determine the viability of an investment as this affects the income stream over the proposed lifespan of the investment. It has been argued that boreholes are drilled and function for a lifespan of 20–50 years (Carter et al., 2014). Here, we employed an investment lifespan of 30 years that falls within this range.

15.2.3.3 Cost-benefit analysis basic assumptions

Cost-benefit analysis (CBA) was applied to estimate the direct costs and benefits accrued from investing in groundwater by smallholder farmers. In-line with the CBA framework, the analysis was carried out based on the following considerations:

1. All costs and benefits are estimated in incremental terms and compared against the do-nothing option of no groundwater irrigation.
2. The analysis starts at year 0 when the initial investment costs of the groundwater infrastructure occurred, whereas maintenance and operational costs were assumed to start from the second year after the investment.
3. All production costs and benefits from using groundwater for irrigation were regarded with the assumption that, since it was difficult to forecast the cash flows for the entire lifespan of the investment, the constant value was used in measuring project viability throughout the lifespan of the project. Costs and benefits have been quantified and valued in TZS using November–December 2016 market prices.
4. The depth of hand-dug and motorized wells in the study areas was extracted from reports from the Mbarali District Council, Rufiji Basin Water Board, and well labels. Depths range from 9 to 23 m (mean: 15 m) for dug wells and 14–100 m for machine-drilled wells. This analysis considers three different well depths of 40, 50, and 100 m as initial capital increases as well depths increase. Because shallow

282 Groundwater for Sustainable Livelihoods

(<15 m) wells, whether hand-dug and machine-drilled, are vulnerable to seasonal groundwater-level decline, a minimum borehole depth of 40 m was chosen to support small-scale groundwater irrigation.

5. Two production seasons in a year for groundwater irrigation were assumed where rice could be produced during the wet season and during the dry season the same field would be used to cultivate any other crop. This assumption follows the argument that through groundwater, a farmer has an added advantage of irrigating his/her farm during the dry season. Empirical evidence was observed during data collection, whereby some households that owned wells (mostly dug wells) had irrigated backyard gardens during the dry season. Vegetables and tree fruits were grown in these gardens for their own consumption and for sale in the local market. At Mont Fort Secondary School, rice seedlings, vegetables, onions, and orchard crops were grown in school gardens using groundwater in the dry season.

6. This analysis used onion as the second crop during the dry season. This was due to the argument that rice was reported as both a cash and food crop grown during the wet season, whereas onions, watermelons, and vegetables were reported as cash crops grown in the dry season. Thus, rice and onion were selected in estimating the viability of investing in groundwater irrigation by smallholder farmers. By considering such scenarios, the relative profitability of using groundwater for small-scale irrigation was compared to surface water irrigation.

7. Operation and maintenance were estimated to take 10% of the investment cost per year. This was estimated from the communal deep well supplying water to the villages Ubaruku and Mpakani where hydroelectric power is used as a source of energy.

15.2.3.4 Binary logistic regression

Binary logistic regression was used to determine the relationship between independent variables (age, education level, household's size, gender, occupation, and credit access, and income level) in influencing groundwater use for irrigation. The independent variables are categorized into two distinct groups that are binary and continuous. The variables used in the regression are presented in Table 15.1. The hypothesis was concerned with the influence of household characteristics on groundwater usage. SPSS was employed to compute the Binary logistic regression.

Table 15.1 Description of variables used in the logistic regression model

Variable	Description
Y	Groundwater use for irrigation (1 = yes, 0 = no)
X_1	Gender of household head (1 = female, 0 = male)
X_2	Households size
X_3	Age of the respondent (years)
X_4	Access to financial institutions (1= yes, 0 = no)
X_5	Education level of households head (1= educated, 0 = illiterate)
X_6	Households income (TZS)
X_7	Social network membership (1 = yes, 0 = no)

Valuing groundwater use in Tanzania   283

15.3 RESULTS AND DISCUSSION

15.3.1 Demographic and socio-economic characteristics of respondents

Table 15.2 presents the demographic and socio-economic characteristics of the respondents. Most of the respondents (73.2%) were in the age ranging from 18 to 35 years, which is productive and active working-age group. This age group suggests that the selected respondents are the best representative sample since most of them are engaging in various activities and probably using incentives available for economic development.

According to information presented in Table 15.2, most of the respondents (77%) were male with female constituting 23%; 81% of the respondents were married. The number of male respondents is high probably because, in many rural settings, most households are headed by men, who are normally the key speakers for their households; very few women are designated as household heads in an individual family.

In terms of education level, almost three-quarters of respondents (72%) had primary education, very few (2%) had attained secondary education, and about 25% had no formal education. Education plays a major role in the socio-economic development of many societies through the adoption of technology and innovation of new initiatives in order to alleviate poverty (Gama, 2018). Given prevailing educational backgrounds, there may be a requirement for farmers to receive information and training on groundwater and its use before they can consider investing in this source of water, despite declining surface water and more restrictive rules and by-laws restricting its use for irrigation during the dry season.

Table 15.3 shows that more than half of respondents (62%) are engaged in crop farming, while others earn a living through other means including livestock keeping and petty business such as tailoring, brick making, crop selling, etc. It was observed crop production plays a significant role in income and livelihood support of many smallholder farmers. Crops grown in the study area include rice, maize, vegetables, onions,

Table 15.2 Survey respondent characteristics

Household details		Nyeregete (n = 33)	Ubaruku (n = 34)	Mwaluma (n = 30)	Total (N = 97)
Household head age group (years)	18–38	8 (24)[a]	6 (18)	9 (30)	23 (24)
	39–59	15 (46)	19 (56)	14 (47)	48 (50)
	≥60	10 (30)	9 (27)	7 (23)	26 (27)
Household head gender	Male	22 (67)	29 (85)	24 (80)	75 (77)
	Female	11 (33)	5 (15)	6 (20)	22 (23)
Household heads' marital status	Single	0 (0)	1 (3)	0 (0)	1 (1.0)
	Married	27 (82)	30 (88)	22 (73)	79 (81)
	Divorced	2 (6)	1 (3)	1 (3)	4 (4)
	Widow	4 (12)	2 (6)	7 (23)	13 (13)
Household head education level	Primary	26 (79)	26 (77)	18 (60)	70 (72)
	Secondary	0 (0)	2 (6)	0 (0)	2 (2)
	Illiterate	7 (21)	6 (18)	12(0)	25 (26)
Household size (group)	2–5	16 (49)	14(41)	23 (77)	53 (55)
	6–9	14 (42)	20 (59)	7 (23)	41 (42)
	≥10	3 (9)	0 (0)	0 (0)	(3)

[a] Values given in parantheses are percentages; note that rounding errors may lead totals of 99% or 101%.

284 Groundwater for Sustainable Livelihoods

Table 15.3 Number and proportion of households in income-generating activities

Type of economic activities	Nyeregete	Ubaruku	Mwaluma	Total
Crop production	12 (40)[a]	20 (67)	28 (93)	60 (67)
Crop production & livestock keeping	14 (47)	2 (7)	0 (0.0)	17 (19)
Crop production & petty business	0	7 (23)	1 (3)	9 (10.0)
Employment & crop production	0	1(3)	1 (3)	2 (2)
Crop production, livestock keeping, business	4 (13)	0 (0.0)	0 (0.0)	2 (2)

N = 30 for each village.
[a] Values given in parantheses are percentages; note that rounding errors may lead totals of 99% or 101%.

watermelons, sweet potatoes, and fruits. Participants in FGDs reported that horticultural crops were mainly grown in backyard gardens irrigated by groundwater and play a significant role in improving household income, food security, and nutritious status. These results are consistent with outcomes reported previously by Vilholth (2013).

15.3.2 Groundwater source and usage

Table 15.4 shows the distribution of respondents by groundwater source in the UGRRC. Approximately 66% rely on boreholes (shallow and deep boreholes) and the remainder (34%) rely on hand-dug wells. Twenty-five dug wells and five functioning types of machinery drilled wells were observed during the survey. These statistics corroborate numerous previous studies indicating that hand-dug wells and boreholes are available in rural settings in SSA where surface water supply distribution networks do not exist or are unreliable (Collins, 2000). Hand-dug wells are constructed, excavated, and lined by human labor (Figure 15.2). Most wells are small (80 cm diameter) and range in depth from 5 to 40 m. In the UGRRC and more broadly in Tanzania, hand-dug wells use a rope and a bucket to access groundwater.

Participants of the FGDs reported that hand-dug wells are under the management and ownership of individual households in which neighboring households have access to this water source. In Nyeregete, the majority (>50%) of users reported collecting adequate groundwater from hand-dug wells in contrast to users in the other two study villages (Ubaruku and Mwaluma). Water supplies from the hand-dug wells are used for domestic hygiene, livestock, house construction, and irrigation.

The other form of groundwater supply is boreholes (Figure 15.3). Borehole depths range from 40 to 120 m. These water sources are under the ownership and management of either village government or an organization, for example, in Ubaruku village there is Community Water Supply Organization popularly known as UBAMPA (a collaboration

Table 15.4 Distribution of respondents on accessing groundwater sources

Source	Nyeregete	Ubaruku	Mwaluma	Total
Boreholes	12 (40)	26 (87)	21 (80)	59 (66)
Hand-dug wells	18 (60)	4 (13)	0 (20)	31 (34)
Total	30 (100)	30 (100)	30 (100)	90(100)

Note: Values given in parantheses are percentages.

Figure 15.3 A hand-dug well (a) and a borehole equipped with hand pump (b) in Nyeregete village. (Photos: © Mosha 2016.)

between Ubaruku and Mpakakani villages), and Rujewa Small Urban Water Sanitation Authority. Field observations revealed that boreholes were equipped with either hand-pump or motorized using either solar power, diesel, or electricity. Figure 15.3 depicts a shallow borehole with a hand-pump and a borehole with a hand-pump.

In all study villages, the majority of the groundwater users reported collecting an adequate amount of water throughout the year, and are of the view that water from the borehole is safe and suitable for drinking and cooking. Groundwater users reported being able to collect adequate quantities of water from the boreholes throughout the year, representing a resilience to seasonal climate variability not permitted by surface waters.

15.3.3 Groundwater usage at household level

Findings from household surveys and FGDs reveal that groundwater has uses beyond domestic purposes and includes livestock, brick making, and irrigated gardening. Table 15.5 shows the overall volume of groundwater used per day at the household level. The results showed that the mean daily volume of water used in each household was 221 L. As each household has an average of five people,

Table 15.5 Quantity of groundwater used per household per day in liters (n = 90); standard deviation (σ)

Village	N	Household water usage		Per capita water usage	
		Mean	σ	Mean	σ
Nyeregete	30	261	105	50	37
Ubaruku	30	236	96	40	20
Mwaluma	30	166	64	40	24
Total	**90**	**221**	**98**	**43**	**28**

286 Groundwater for Sustainable Livelihoods

this per capita usage is equivalent to 43 L/day in the households surveyed in three villages. This total exceeds the minimum criterion of 20 L per person per day recommended nationally (URT, 2015) and observed in a similar study in Ethiopia (Tucker et al., 2014).

Generally, the amount of groundwater collected from water points was higher in Nyeregete village than in the other two villages (Table 15.5). The difference can be attributed to a number of factors including the availability of other water sources and distance from the household to the water point or source. For instance, in this study, 100% of the households in Nyeregete village relied on groundwater. In Mwaluma and Ubaruku, groundwater users were not using groundwater to water their gardens because they perceived that groundwater is hard water. In Nyeregete some water user irrigated their home garden using watering cans. Further, users reported variations in per capita water usage as a function of the proximity and availability (operationally) of water points. The relationship between the amount of water used and the distance from the compound has similarly been shown by Tucker et al. (2014).

The other main usage of groundwater is livestock husbandry. Livestock plays an important role in people's livelihood in all three villages, particularly in Nyeregete village where some of the villagers are agro-pastoralists (Mang'ati) who rely on the livestock for most of their income and food. Further, farmers in all three villages in this study reported that they, independent of their economic standing, are generally able to increase the amount of water provided to livestock in the dry season from groundwater sources.

15.3.4 Determinants of quantity of water use per day

Table 15.6 shows groundwater user perceptions of the determinants of quantity of water use per household. In all villages, users reported the key determinants of using a large amount of water are household size, number of under-5 years, and wealth status of the household. Among these variables, wealth status of the household ranks highest 94% (Table 15.6). Findings from FGDs indicate that poor households consistently use less water than the better-off and middle-class groups, due to partly engaging in fewer household activities, lower incomes, and labor shortages. Poorer households collected less water for domestic use (hygiene and cooking) because of less labor and cash. Labor shortages influence users to opt for the nearer hand-dug wells although the water is less safe compared to boreholes that are more distant located.

Table 15.6 Factors perceived to influence water quantity used per households

Factor	Frequency	Percentage	Rank
Household size	29	91	2
Number of under-5 years	27	84	3
Wealth status of the household	30	94	1
Other factors (distance to source; season)	18	56	4

Source: Qualitative data, 2016; reported percentages are not all based out of a total of thirty-two due to variations in responses.
N = 32 FGD participants.

These findings are also consistent with those of Tucker et al. (2014), who reported poorer households to use less water because they have less labor for water collection and fewer storage and transport assets.

Reported differences as a function of economic status/wealth category are more highly pronounced in Ubaruku, and Nyeregete than in Mwaluma. Situational analyses in the two later villages show that people had very few economic development projects or petty business as compared to the former village. Moreover, the extent of dependence on groundwater increases up to 100% in all three villages during the dry season, from May to November each year.

15.3.5 Irrigation practices using groundwater source in the study area

In terms of irrigated agriculture, groundwater was observed and reported to irrigate rice nurseries particularly in November and December, and various horticultural crops (such as green vegetables, tomatoes, fruits tree, banana, and onions) in household gardens. In Mont Fort Secondary School, field observation showed a range of horticultural crops (e.g. vegetables, bananas, and yams, as well as citrus, mangoes, guava, avocado, and pawpaw trees) that have been grown and irrigated by groundwater since 1998. Students use groundwater for all their basic needs. Hand-dug wells are located around the household compound homestead, their construction was reported to be financed by households themselves.

The mean average cost for constructing a dug well was estimated to be TZS 250,000 (USD 114) and this includes the lining of the earth walls with bricks and a top cover (made with timber or aluminum corrugated sheets) ranging in depth from 9 to 23m. Apart from construction costs, it also requires pumps or buckets tied with ropes for fetching water from the wells. Manual-driven pump in the study area costs about TZS 500,000 (USD 227), whereas a bucket with a rope will cost about TZs 20,000 (USD 9). The cost of drilling these wells ranged from TZS 150,000 (USD 68) to 180,000 (USD 82) per meter depth. This cost is relatively cheap when compared to the cost of drilling a borehole.

For boreholes, the costs of the preliminary survey alone range from TZS 1,000,000 (USD 455) to 1,500,000 (USD 682) depending on the distance between the drilling company and the site. Community boreholes were reported to be used mostly for domestic water consumption with an exception of Mont Fort Secondary School where a borehole was found to be used not only for domestic water supply, but also for livestock, fish ponds, and small-scale irrigation activities such as orchards, vegetables, and rice seedlings. Figure 15.4 displays rice seedlings and rice in a farm both irrigated by groundwater. The construction of boreholes was financed by the government of Tanzania through the Ministry of Water and development partners including UNICEF, DANIDA, and WWF.

Groundwater used to provide supplementary irrigation in the UGRRC is increasing but, so far, its use in intensive irrigation activities has yet to be realized. Farmers noted that they had very limited knowledge of groundwater quality, abstraction, and irrigation technologies. These responses suggest more intensive use of groundwater for irrigation may be considered if there was confidence that this magnitude of use could be sustained. This inference is consistent with that drawn by Villholth (2013) who noted that many smallholder farmers in Tanzania lag in their use of groundwater for irrigation because of lack of information and knowledge on groundwater resources and their usage.

288 Groundwater for Sustainable Livelihoods

Figure 15.4 Groundwater irrigation technology and rice farming in Mont Font Secondary School. (Photos: © Gama 2016.)

15.3.6 Reasons for accessing and using groundwater

Table 15.7 summarizes explanations provided in FGDs of factors that influence access to, and use of, groundwater resources. The results show that 40% of the respondents used groundwater because they perceived it to be safe. Other reasons include that it is the only source of water available and near walking distance. For instance, FGDs reported that the majority of the people in Ubaruku village used groundwater rather than water from irrigation canals because it was considered to be safe, whereas people

Table 15.7 Factors influencing access to and use of groundwater source (n = 90)

Reasons	Male	Female	Total
Near walking distance to groundwater source	8 (18)[a]	6 (13)	14 (16)
Adequate groundwater source	2 (4)	4 (9)	6 (7)
Affordability to groundwater charges	2 (4)	6 (13)	8 (9)
Is the only source available	14 (31)	12 (27)	26 (29)
Groundwater is clean and safe	19 (42)	17 (38)	36 (40)
Total	45 (100)	45 (100)	90 (100)

[a] Values in parentheses are percentages.

from Nyeregete village reported that it was the only source available. With regard to the situation of water availability in Nyeregete, one of the key informants reported that "…I thank God for groundwater availability at Nyeregete village. I sometimes ask myself, what would happen in our village if we should have no groundwater resource? Perhaps many people could migrate to other villages to sustain their livelihoods".

15.3.7 Cost-benefit analysis

15.3.7.1 Short-term cost-benefit analysis

Table 15.8 summarizes the results of the economic analysis of the use of both surface water and groundwater for irrigation on annual basis. Smallholder irrigation by both surface water and groundwater small-scale irrigation has a positive gross margin of TZS 630,415 (USD 287) and 4,820,415 (USD 2191), respectively. Computed values are strongly influenced by crop prices, the prices of inputs and outputs, and the prevailing market situation. Positive gross margins imply that the use of both surface water and groundwater for irrigation is able to cover the costs of production. It is worth noting that highest gross revenue was obtained from the use of groundwater for irrigation despite its production costs being double that of surface water. These calculations suggest the use of groundwater by smallholder farmers is economically viable. A study by Shah et al. (2013) concludes that groundwater-fed irrigation is economically viable for smallholder farmers because it supports dry-season irrigation.

15.3.7.2 Financial viability of groundwater

Table 15.9 shows a summary of NPV and CBR calculations for 1 ha of rice and 1 ha of onion. As shown in Table 15.9, the highest NPV was observed for investing in a machine-drilled well of 40 m depth with the value of TZS 38,636,794 (USD 17,562), 23,032,915 (USD 10,470), and 19,807,103 (USD 9,003) at the discounting rate of 12% 18%, and 20%, respectively. Likewise, investing in a well with a depth of 50 and 100 m also had positive NPVs at the same discounting rate albeit less than that observed when investing in 40 m deep well due to the increased costs of drilling.

Investing in groundwater has positive NPVs at a discounting rate of 12%, 18%, and 20% per hectare for all considered well depths. In other words, the present value of the benefits stream is greater than the present value of the cost stream, suggesting

290 Groundwater for Sustainable Livelihoods

Table 15.8 Profitability of using groundwater (GW) and surface water (SW) for irrigation

Operation	Parameter	SW (TZS/ha)	GW (TZS/ha)
Production cost[a]	**Wet season (Rice)**		
	Nursery management	40,000	40,000
	Plowing	162,500	162,500
	Furrowing	162,500	162,500
	Inputs (fertilizer, seeds, and pesticides per acre)	296,250	296,500
	Planting	210,000	210,000
	Weeding	165,000	165,000
	Bird scaring	50,000	50,000
	Harvesting	500,000	500,000
	Total cost of production (rice)	**1,586,250**	**1,586,250**
	Dry season (Onion)		
	Nursery management	NA	60,000
	Plowing and basin preparation	NA	212,500
	Inputs (fertilizer seeds and pesticides)	NA	1,775,000
	Planting	NA	150,000
	Harvesting	NA	212,500
	Total cost of production (onion)		2,410,000
Other cost	Water use fee per year	50,000	150,000
	O and M[b]	0	2,300,000
	Others total cost	**50,000**	**2,450,000**
Benefits	Crop yield Rice (tonne/hectare/year)	4.25	4.25
	Crop yield Onion (tonne/hectare/year)	NA	20
	Output Price Rice (TZS/tonne)	533,333	533,333
	Output Price Onion (TZS/tonne)	NA	450,000
Total revenue	**TZS/tonne/year**	**2,266,665**	**11,266,665**
Gross margin[c]		**630,415**	**4,820,415**

Source: Data represent farm statistics from the harvest of the cropping season 2016.
Note: 1 USD = 2,200 TZS.
[a] Production cost per hectare per season.
[b] Operation and Maintenance Cost per year.
[c] Total revenue from sale of crop — total cost of crop production.

that investment in groundwater by smallholder farmers is financially viable. The CBR was also greater than 1 and according to decision criteria, projects with CBR which is positive and greater than 1 are financially viable because the discounted benefits are higher than the discounted costs. These results are consistent with similar economic analyses of smallholder farmers employing groundwater-fed irrigation by Abric et al. (2011), Namara et al. (2011), and Dittoh et al. (2013). The CBR is, however, strongly linked to the depth of the borehole, with a borehole of 100 m only marginally positive when 1 ha is irrigated.

Table 15.9 Summary of the results of the cost-benefit analysis for irrigation by groundwater (GW) using different well depths and surface water (SW)

Parameter	GW 40 m (TZS/ha)	GW 50 m (TZS/ha)	GW 100 m (TZS/ha)	SW (TZS/ha)
Investment	7,800,000	9,437,500	23,000,000	
Production cost				
Maintenance cost & operation	780,000	943,750	2,300,000	‾
Inputs cost	3,996,250	3,996,250	3,996,250	1,586,250
Water use fee	150,000	150,000	150,000	50,000
Total production cost/season	4,926,250	5,090,000	6,446,250	1,636,250
Crop value	11,266,665	11,266,665	11,266,665	2,266,665
Net benefit	6,340,415	6,176,665	4,820,415	630,415
NPV at 12%	38,636,794	35,997,029	14,133,330	4,534,025
NPV at 18%	23,032,915	20,879,629	3,045,165	2,947,353
NPV at 20%	19,807,103	17,763,101	833,783	2,615,663
CBR at 12%	6.55	5.27	1.69	
CBR at 18%	4.48	3.61	1.16	-
CBR at 20%	4.05	3.26	1.04	-

15.3.8 Socio-economic factors influencing the use of groundwater by smallholder farmers

Logistic regression was used to analyze socio-economic factors that influence the use of groundwater by smallholder farmers. The inferential test for goodness-of-fit (Hosmer & Leme statistic) indicates that the model fits the data well at $p < 0.05$. The descriptive measures of goodness-of-fit also support that the model fits the data weakly (Cox & Snell $R^2 = 0.19$) and moderately (Nagelkerke $R^2 = 0.39$). Of the considered variables (Table 15.1) including gender of household head, household size, respondent age, education level of household head, access to financial institutions, household income, and social network membership, households size ($p < 0.05$) was the only variable (beta value of 0.38) that was statistically significant as the determinant of groundwater use.

The outcome of the logistic regression implies that, when, the household size increases by one unit, there is an increase in the probability that the households will use groundwater for irrigation by approximately 38% (based on the computed beta value). One plausible explanation for this outcome is the availability of adequate labor to be deployed in groundwater smallholder irrigation. Furthermore, this finding indicates that an increase in the number of households (to an average of 6) leads to an increase in the ability and desire to diversify the available groundwater resource for irrigation that enhance food security and livelihoods support. Similar findings have been found in Ethiopia where higher household sizes were related to greater groundwater use (Tucker et al., 2014).

15.3.9 Threats and challenges on the groundwater use

Major threats to groundwater resources in Usangu Plains include pollution, overexploitation, and management of abandoned water wells. There is also inadequate data and information on the safe yield of wells not only in the UGRRC but also more widely in Tanzania (Sappa et al., 2015; Mjemah et al., 2010). In addition, farmers reported some challenges facing them on how to utilize groundwater effectively that include (a) lack of awareness, (b) lack of capital, (c) long distance to community boreholes which was said to provide quality water; (d) suitability of groundwater for irrigation – in terms of salinity, alkalinity, and acidity; (e) well and aquifer yields; and (f) competition for use with pastoralists.

15.4 CONCLUSIONS

The analysis of groundwater use in three villages of the UGRRC in southern Tanzania shows that groundwater offers important benefits to these communities. The most significant economic benefit is improved agricultural production, and in particular, the ability to produce a second crop during the dry season. CBR reduces with depth of boreholes as deeper boreholes have higher capital costs but are likely to provide more benefits. Irrigation using groundwater is more reliable and more profitable than using surface water due to the ability to produce a dry-season crop. It is concluded that wider adoption of groundwater-fed irrigation by smallholder farmers will provide opportunities to increase crop production and ensure food security and has the potential to increase income through the sale of surplus. The broad lessons of our analysis are expected to be relevant to other parts of semi-arid Tanzania, where groundwater-fed irrigation can address surface water shortages associated with climate variability and change.

ACKNOWLEDGMENTS

The authors acknowledge the support of a grant, *GroFutures*, under the UK government's NERC-ESRC DFID UPGro program (ref. NE008592/1), and assistance provided by Richard G. Taylor (UCL) in the preparation of this chapter.

REFERENCES

Abric, S., Sonou, M., Augegard, B., Onimus, F., Durlin, D., Soumaila, A. & Gadelle, F. (2011) *Lessons Learned in the Development of Smallholder Private Irrigation for High-Value Crops in West Africa*. World Bank, Washington, DC.

Carter, R., Chilton, J., Danert, K. & Olschewski, A. (2014) *Siting of Drilled Water Wells - A Guide for Project Managers*. Rural Water Supply Network, SKAT Foundation, St Gallen. 348p.

Collins, S. (2000) *Hand-dug Shallow Wells*. Series of Manuals on Drinking Water Supply Volume 5. SKAT Foundation, St. Gallen.

Cuthbert, M.O., Taylor, R.G., Favreau, G., Todd, M.C., Shamsudduha, M., Villholth, K.G., MacDonald, A.M., Scanlon, B.R., Kotchoni, D.O.V., Vouillamoz, J.-M., Lawson, F.M.A., Adjomayi, P.A., Kashaigili, J., Seddon, D., Sorensen, J.P.R., Ebrahim, G.Y., Owor, M., Nyenje, P.M., Nazoumou, Y., Goni, I., Ousmane, B.I., Sibanda, T., Ascott, M.J., Macdonald, D.M.J., Agyekum, W., Koussoubé, Y., Wanke, H., Kim, H., Wada, Y., Lo, M.-H., Oki, T. & Kukuric, N. (2019) Observed controls on resilience of groundwater to climate variability in sub-Saharan Africa. *Nature* 572, 230–234.

Damkjaer, S. & Taylor, R.G. (2017) The measurement of water scarcity: defining a meaningful indicator. *Ambio* 46, 513–531.

Dittoh, S., Awuni, J.A. & Akuriba, M.A. (2013) Small pumps and the poor: A field survey in the Upper East Region of Ghana. *Journal of Water International* 38(4), 449–464.

Foster, S., Tuinhof, A., & van Steenbergen, F. (2012) Managed groundwater development for water-supply security in sub-Saharan Africa: investment priorities. *Water SA* 38, 359–366. https://doi.org/10.4314/wsa.v38i3.1.

Gama, D.G. (2018) *Financial Viability of Groundwater Use for Irrigation by Smallholder Farmers in the Usangu Plains, Tanzania.* A Dissertation Submitted in Partial Fulfilment of the Requirements for the Degree of Master of Science in Environmental and Natural Resource Economics of Sokoine University of Agriculture, Morogoro, Tanzania. 94p.

Gebregziabher, G., Hagos, F., Haileslassie, A., Getnet, K., Hoekstra, D., Gebremedhin, B., Bogale, A. & Getahun, G. (2016) Does investment in motor pump-based smallholder irrigation lead to financially viable input intensification and production? An economic assessment. LIVES Working Paper 13. Nairobi, Kenya: International Livestock Research Institute (ILRI).

Gebregziabher, G., Villholth, G., Hanjrab, A., Yirgac, M. & Namara, E. (2013) Cost-benefit analysis and ideas for cost sharing of groundwater irrigation: evidence from north-eastern Ethiopia. *Water International Journal* 38(6), 852–863.

Gronwall, J. & Oduro-Kwarteng, S. (2018) Groundwater as a strategic resource for improved resilience: a case study from peri-urban Accra. *Environmental Earth Sciences* 77(6), 1–13.

Gudaga, J.L., Kabote, J.S, Tarimo, A.K.P.R., Mosha, D.B. & Kashaigili J.S (2018) Effectiveness of groundwater governance structures and institutions in Tanzania. *Journal of Applied Water Science* 8, 77.

Howard, G. (2017) Groundwater research into policy within the context of Africa & the SDGs. DFID. [Online] Available from: http://iahbritish.wpengine.com/wp-content/uploads/2017/11/Howard_Ineson2017.pdf [Accessed 19th March 2019].

Kashaigili, J.J. (2010) *Assessment of groundwater availability and its current and potential use and impacts in Tanzania.* Report for the International Water Management Institute, Colombo, Sri Lanka.

Komakech, H.C. & de Bont, C. (2018) Differentiated access: challenges of equitable and sustainable groundwater exploitation in Tanzania. *Water Alternatives* 11(3), 623–637.

Lin, G.C.I. & Nagalingam, S.V. (2000) *CIM Justification and Optimization.* Taylor and Francis, London. 36p.

MacAllister, D.J., MacDonald, A.M., Kebede, S., Godfrey, S. & Calow, R. (2020) Comparative performance of rural water supplies during drought. *Nature Communications*, 11, 1099.

MacDonald, A.M., Bonsor, H.C., Ó Dochartaigh, B.É. & Taylor, R.G. (2012) Quantitative maps of groundwater resources in Africa. *Environmental Research Letters* 7, 021003.

Mjemah, I.C., Van Camp, M., Martens, K. & Walraevens, K. (2010) Groundwater exploitation and recharge rate estimation of a quaternary sand aquifer in Dar-es-Salaam area, Tanzania. *Environmental Earth Sciences* 63(3), 559–569.

Mwakalila, S. (2011) Assessing the hydrological conditions of the Usangu wetlands in Tanzania. *Journal of Water Resource and Protection* 3, 876–882.

Mwakalila, S. (2014) Climate variability, impacts and adaptation strategies: the case of mbeya and Makete districts in great Ruaha catchment in Tanzania Department of Geography, University of Dar es Salaam, Dar es Salaam, Tanzania. *Journal of Water Resource and Protection* 6, 43–48.

Namara, R., Awuni, J., Barry, B., Giordano, M., Hope, L., Owusu, E. & Forkuor, G. (2011) *Smallholder Shallow Groundwater Irrigation Development in the Upper East Region of Ghana.* International Water Management Institute, Research Report Colombo, Sri Lanka.

National Bureau of Statistics (NBS) (2013) 2012 Tanzania population and housing census distributed by administrative areas, National Bureau of Statistic, Dar Es salaam, Tanzania. 244p.

Osborn, D., Cutter, A. & Ullah, F. (2015) *Universal Sustainable Development Goals: Understanding the Transformational Challenge for Developed Countries.* Stakeholders Forum. [Online] Available from: http://www.stakeholderforum.org [Accessed 12th October 2020].

Pallanti, J. (2007) *SPSS Survival Manual. A Step by Step Guide to Data Analysis Using SPSS for Window* (3rd Edition). Open University Press. England SL 2QL.

Sappa, G., Ergul, S., Ferranti, F., Sweya, L.N. & Luciani, G. (2015) Effects of seasonal change and seawater intrusion on water quality for drinking and irrigation purposes, in coastal aquifers of Dar es Salaam, Tanzania. *Journal of African Earth Sciences*, 105, 64–84. doi:10.1016/j.jafrearsci.2015.02.007.

Shah, T., Verma, S. & Pavelic, P. (2013) Understanding smallholder irrigation in Sub-Saharan Africa: results of a sample survey from nine countries. *Water International* 38(6), 809–826.

Sustainable Management of the Usangu Wetland and its Catchment project (SMUWC) (2001) *Groundwater in the Usangu Catchment.* Final Report. Sustainable Management of the Usangu Wetland and its Catchment project, Government of Tanzania.

Taylor, R.G. (2009) Rethinking water scarcity: role of storage. *EOS, Transactions, American Geophysical Union* 90(28), 237–238.

Taylor, R.G., Favreau, G., Scanlon, B.R. & Villholth, K.G. (2019) Topical collection: determining groundwater sustainability from long-term piezometry in Sub-Saharan Africa. *Hydrogeology Journal* 27, 443–446.

Taylor, R.G., Scanlon, B.R., Doell, P., Rodell, M., van Beek, L., Wada, Y., Longuevergne, L., LeBlanc, M., Famiglietti, J.S., Edmunds, M., Konikow, L., Green, T., Chen, J., Taniguchi, M., Bierkens, M.F.P., MacDonald, A., Fan Y., Maxwell, R., Yechieli, Y., Gurdak, J., Allen, D., Shamsudduha, M., Hiscock, K., Yeh, P., Holman, I. & Treidel, H. (2013a) Groundwater and climate change. *Nature Climate Change* 3, 322–329.

Taylor, R.G., Todd, M., Kongola, L., Nahozya, E., Maurice, L., Sanga, H. & MacDonald, A. (2013b) Evidence of the dependence of groundwater resources on extreme rainfall in East Africa. *Nature Climate Change* 3, 374–378.

Tucker, J., MacDonald, A.M., Coulter, L., & Calow, R. C. (2014) Household water use, poverty and seasonality: wealth effects, labour constraints, and minimal consumption in Ethiopia. *Water Resources and Rural Development* 3, 27–47.

United Republic of Tanzania (URT) (2002) *National Water Policy. The Ministry of Water and Livestock Development.* Tanzania: Dodoma.

United Republic of Tanzania (URT) (2010) Official website of Mbeya Region. [Online] Available from: http://www.mbeya.go.tz [Accessed on 20th July 2019].

United Republic of Tanzania (URT) (2015) *Water Sector Status Report 2015.* Dar es Salaam: Ministry of Water, United Republic of Tanzania.

Villholth, K.G. (2013) Groundwater irrigation for smallholders in Sub-Saharan Africa: a synthesis of current knowledge to guide sustainable outcomes. *Water International Journal* 38(4), 369–391.

Walsh, M. (2012) The not-so-great Ruaha and hidden histories of an environmental panic in Tanzania. *Journal of Eastern African Studies* 6(2), 303–335.

Chapter 16

Conjunctive use of surface and groundwater

Operational and water management strategies to build resilience, water security, and adaptation

G.F. Marques
Federal University of Rio Grande do Sul (IPH/UFRGS)

C.D.P. Mattiuzi
Brazilian Geological Survey (CPRM/SGB)

S.D. Cota
Nuclear Technology Development Center (CDTN/CNEN)

M. Pulido-Velazquez
Universitat Politècnica de València

CONTENTS

16.1 Introduction.. 296
16.2 Main challenges faced by small cities, rural settlements, and irrigated
        agriculture in South America and Brazil... 296
        16.2.1 The role of surface and groundwater: theory, practice gaps, and
                  reality... 297
        16.2.2 Status on vulnerability and its impact on people, the environment,
                  and the economy.. 298
16.3 Conjunctive use – concept, definitions, and potential 300
16.4 Potential applications in Brazil... 301
        16.4.1 The advantages of conjunctive use under hydrological uncertainty:
                  economic benefits and operational strategies 301
                  16.4.1.1 Operational strategies... 302
        16.4.2 How to integrate existing (and new) water resources management
                  framework and instruments.. 307
16.5 Conclusions 309
References.. 310

DOI: 10.1201/9781003024101-16

16.1 INTRODUCTION

Water systems, from urban centers to rural and agricultural regions have increasingly complex challenges as water becomes scarce, infrastructure ages and climate change affects hydrological variability and uncertainty. Under this context, the systems need strategies to adapt, not only improving rational use, but also infrastructure investment and operations. While the role of groundwater as a key water supply source is widely recognized, and more in a context of decreasing available freshwater resources and growing demands under global and climate change (Taylor et al., 2013; Kløve et al., 2014; Wada and Bierkens, 2014), its sustainable development to meet future challenges also depends on the integration with surface water supplies. Such integration has been conducted in other countries as conjunctive use operations, which have contributed to mitigating aquifer overdrafts, improving supply reliability, reducing operational costs, and increasing the water system's flexibility.

This chapter explores how conjunctive use operations of groundwater and surface water can be implemented to improve water supply reliability to urban and irrigated agriculture demands in Southern Brazil. We discuss the application of conjunctive use strategies and water management instruments under a comprehensive and integrated approach to improve water supply reliability and flexibility in a sustainable way. The discussion is followed by field examples where such strategies should build upon. We believe such combination is necessary to allow groundwater resources to fully contribute to promoting economic growth in the long run, going beyond the aquifer and incorporating its operation to broader and integrated water management required under the challenges ahead.

16.2 MAIN CHALLENGES FACED BY SMALL CITIES, RURAL SETTLEMENTS, AND IRRIGATED AGRICULTURE IN SOUTH AMERICA AND BRAZIL

Poverty in rural areas is one of the targets related to the UN 2030 Agenda for Sustainable Development Goals (SDGs). It is the focus of the 1.1 Target ("eradicate extreme poverty") for Goal 1 – No Poverty, 2.3 Target ("double the agricultural productivity and incomes of small-scale food producers") for Goal 2 – Zero Hunger, and others. According to Castañeda et al. (2018), a study based on household data for the year 2013 from eighty-nine countries found out that 80% of extremely poor and 76% of the moderate poor population (considering the sampled population) dwell in rural areas. Workers who are considered extremely poor are four times more abundant in the agriculture sector than in the non-agriculture sectors of the economy. In the context of Latin America and Caribbean areas, 53% of the extremely poor population lives in rural areas and only 18% of the rural residents can be considered as non-poor. Among the working adults in the same geographic area, 68% can be considered as extremely poor (Castañeda et al., 2018).

The extremely poor population in rural areas can be also concentrated in not-so-productive regions, such as arid environments, and have to face additional challenges related to the lack of infrastructure and poor allocation of funds for production incentives (Campos et al., 2018). Poverty in the context of the rural environment is strongly related to the population's access to natural resources, including land and water; therefore, assuring access to these resources is key for achieving the SDGs. This is especially important when considering the vulnerability of these populations

Groundwater management strategies **297**

to climate-related events, such as droughts and floods. People in such conditions are dependent on the few resources they possess and usually do not have alternative means of income for coping with the effects of natural disasters.

Water security is already at risk for many and the situation will become worse in the next few decades. According to the United Nations World Water Development Report (2018), nearly 6 billion people will suffer from clean water scarcity by 2050, as a result of increasing demand for water, reduction of water resources, and increasing pollution of water, driven by population and economic growth (WWDR, 2018).

In Brazil, regions such as the semiarid in the northeast are experiencing conflicts among different users and consumptive uses, which are commonly associated with irrigation, due to high water demand or low availability (ANA, 2017). At the same time, over 60% of the population of this region is considered to be extremely poor (Aquino et al., 2018). Irrigation is the main use of water in Brazil; according to the National Water Agency, 52% of all withdrawal was directed to irrigation in a total of $1,083\,m^3/s$ in the year 2017 (ANA, 2018).

In this context, groundwater plays a significant role. In a recent study, Hirata et al. (2020) point out that while the real numbers are unknown, it is estimated in more than 2.5 million wells, pumping over 17.58 million m^3/year ($557\,m^3/s$), which serve 30%–40% of the population. At the average Brazilian utility water rate ($0.62 USD/m^3$), this would represent $10.8 USD billion/year in value. According to Hirata and Conicelli (2012), while groundwater is the main source of water in small- and medium-size cities, contributing to the socioeconomic development of the country, in some cases it is the only water supply source for poor population, and with questionable quality. If the UN 2030 Agenda goals are to be met, increasing the income of familiar, small-scale producers of food and reducing poverty and inequality, it is essential to find new methods to increase local water availability, improving the robustness and resilience of the water supply systems that rely on local sources.

16.2.1 The role of surface and groundwater: theory, practice gaps, and reality

Groundwater is part of the hydrological cycle and its behavior is well understood in academia, especially regarding the interaction between surface and sub-surface processes in the hydrological cycle (Freeze et al., 1979). Hydrologically, surface water bodies can set up the following three configurations with groundwater flow: (a) groundwater flows toward the surface water body, (b) surface waters contribute to the underground flow, and (c) rivers contribute or receive groundwater during sometimes the year, or even at one or the other course; there are also cases of disconnection of the two sources, in which there is no direct influence between the river and the aquifer (Larkin et al., 1992).

Depending on the configuration of the groundwater flow, the extraction of water from watercourses can decrease the surface contribution to the aquifer and contributes to the depletion of the water table. In other cases, withdrawal of groundwater implies the reduction of its contribution to surface water bodies, which might affect downstream surface water uses and groundwater-dependent ecosystems (Kløve et al., 2011). Due to the interrelated effects of these systems, surface and groundwater allocations should be performed in a conjunctive way (Silva, 2007).

Given this well-known behavior, aquifers play important roles in integrated surface and groundwater management, acting as strategic reserves and promoting greater assurance on long-term water availability, as they are less subject to temporal variability (Companhia de Pesquisa de Recursos Minerais (CPRM), 2008). While a deep understanding of river and aquifer system characteristics and their behavior over time is deemed essential, in practice, the water policies and water management instruments designed for its governance have often failed to recognize aquifer boundaries and fluxes, resulting in disputes, impacts, and litigation. The serious groundwater mining and conflicts in the High Plains aquifer (US) provide an example, which prompted following legislation reform to recognize interconnection with surface water and implement regulatory control (Giordano and Villholth, 2007). In Brazil, surface water permits can be either state or federal domain, whereas groundwater is always state domain. This adds complexity to aquifer management, as a close coordination between state and federal water managing bodies is required to evaluate withdrawal limits, safe yields and issue pumping permits, which is rarely the case (OECD, 2015).

Another critical issue that compromises the necessary knowledge base is the limited data gathering system. In South America, only a few countries have long-term monitoring network for groundwater. A summary of livelihoods to monitoring practices in Latin American has shown that, in 2013, only Argentina, Brazil, and Colombia have engaged in continuous monitoring on a national level. The level of coverage of these networks is variable. For example, up to November 2019, over 400 wells are integrated into the RIMAS (acronym in Portuguese for Integrated Network of Ground Water Monitoring), operated by CPRM/SGB – Brazilian Geological Survey. Established in 2009, the network priories sedimentary aquifers of high social-economic importance, covering thirty-one aquifers. While it indicates that positive efforts are underway to better know and manage groundwater in Brazil, this context also highlights the existing gap in the awareness about the current stage of exploitation of groundwater as well as the potential for future expansion. Hirata and Conicelli (2012) point out further potential impacts to the aquifers, by climate change, which may reduce recharge in the north and northeast regions of Brazil and increase the pressure over available resources. To cope with those challenges, the authors highlight the importance of better knowledge of hydraulic and chemical aquifer characteristics, improving society awareness, and effective management actions.

In this context, the major gaps and challenges to implementing integrated aquifer management are (a) to evaluate the actual availability (stock), fluxes, and interactions of the underground resource; (b) to monitor and control withdrawals from both surface and groundwater bodies; and (c) to combine (a) and (b) to design effective water management institutional, economic and operational instruments that would bring integrated aquifer management to reality.

16.2.2 Status on vulnerability and its impact on people, the environment, and the economy

Inequity in water access is a problem with multiple factors. Demographic, socioeconomics, political and cultural aspects can be involved in the development of environmental injustice (Aleixo et al., 2016). According to Arsky and Santana (2017), the Brazilian 2010 census indicated that 72.2% of its rural population rely on wells, small reservoirs, and other local water supply sources with unreliable quality and high risk of disease. For Carlos (2020), even if Brazil fulfills its 2033 goal for sanitation services, reaching full water coverage in urban areas, its rural population would still lag behind with 23% of the rural inhabitants without access to potable water. These differences reflect the difficulties in extending reliable services for those areas and this context is also reflected in other countries of Latin America.

Silva et al. (2012) listed different challenges related to this unbalance between meeting water supply demand in rural and urban areas, such as the lack of economies of scale and limited capacity to finance infrastructure for water abstraction, treatment, distribution, and sanitation; lack of expertise in maintaining water infrastructure; and additional difficulties related to the water supply in arid and semiarid environments, which requires installing and operating more energy-intensive systems.

Some of these results were posteriorly confirmed by using the experience and perceptions of Brazilian experts in the rural water supply. Machado et al. (2019) were able to prioritize six factors considered critical for the success of the water supply in rural areas, reflecting the challenges faced to improve good supply in rural communities. In order of importance, these are:

- Existence of policies and plans on the national level to address the issue;
- Commitment to delivering water accordingly to the quality standards;
- Existence of managerial and institutional support for local providers;
- Adaptation of the tariff structure to the local context to ensure the continuity of the water supply;
- Existence of post-construction technical support for the communities and providers; and
- Existence of local service providers that are responsible for managing the system on a routine basis.

Besides these more general issues, the communities located in arid or semiarid areas also have to address problems of availability and quality of local water sources, both superficial and groundwater, related to hydrological, pedological, and climatological issues. This will require a good level of diversification and integration of available water sources, to take advantage of each other's particularities including seasonal variability, storage capacity, long-term trends, and quality. In small rural communities, decentralized water distribution systems are less expensive, often more viable, alternative to water access (Massoud et al., 2009). Carlevaro et al. (2011) further discuss how local options (e.g. surface water, groundwater, and rainwater) should be combined to result in a cost-effective water supply system. This integration is a key element to boost resilience, as the ability of the system to face changes and maintain well-being (Boltz et al., 2019).

In terms of quantitative hydrology, these regions are characterized by irregular temporal distribution of precipitations and high evapotranspiration rates. The average annual precipitation in these areas varies between 500 and 850 mm, with 70% of the rain volumes concentrated in January to April, and the real evaporation is between 450 and 700 mm/year (potential evapotranspiration reaching 2,600 mm/year) (Gheyi et al., 2012). In the northeast of Brazil, most of the main rivers are intermittent and remain dry for several months of the year.

Also, the geological framework in these regions of Brazil is mostly associated with shallow soils and crystalline bedrocks with very low infiltration and storage capacities. These characteristics, combined with insufficient precipitation and the strong dependence on irrigation, contribute to the concentration of minerals in the soil, promoting soil salinization. Depending on the local geological settings, groundwater can also have moderate to high salinity (Gheyi et al., 2012), requiring treatment to ensure adequate quality standards. In semiarid regions with highly productive sedimentary aquifers; however, intensive abstraction requires proper water management to guarantee long-term sustainability.

The combination of all the elements presented leads to vulnerability scenarios, especially for rural small communities, which have led to a recurrent water supply crisis, as discussed in Villar (2016). Conjunctive use should provide a valuable and much needed contribution in this aspect. The following section presents an in-depth presentation of conjunctive use concepts, definitions, and studies in Brazilian regions and discussions about the main benefits, challenges, and opportunities, highlighting how conjunctive use strategies can contribute to improving water supply resilience.

16.3 CONJUNCTIVE USE – CONCEPT, DEFINITIONS, AND POTENTIAL

The use of groundwater has increased rapidly and intensely to meet several demands, but sometimes without adequate technical support (Sahuquillo, 2009). Groundwater resources are decreasing: around 20% of the world's aquifers are estimated to be over-exploited, leading to serious consequences such as subsidence, decreased runoff, saline intrusion, among others (Gleeson et al., 2012; Bierkens and Wada, 2019). Recently, conjunctive management of water resources has been presented as a key principle by the Global Groundwater Governance project (GGGFA, 2015).

The United Nations Food and Agriculture Organization has defined conjunctive use as "the combined use of surface and groundwater resources to minimize unwanted physical, environmental and economic effects and optimize the balance of water demand and supply" (FAO, 1995). Conjunctive use has also been defined as "the situation in which groundwater and surface water sources are developed for supply, but not necessarily using both sources continuously over time or providing each user with water from both sources" (Foster et al., 2010). The main purpose of conjunctive use of groundwater and surface water is to maximize the inherent benefits of using each source, and that integrated management of both can complement the water system and optimize productivity and water use efficiency.

In summary, conjunctive use consists of a management strategy in which the available surface and groundwater resources are used in a coordinated and integrated manner, in order to complement the properties of both sources (Pulido-Velazquez et al., 2003; Ross, 2017). Conjunctive use strategies must be developed from management plans and methods based on technical, scientific, and economic studies on demands, availability, and operational decisions, reducing the effect of variability and uncertainty on water availability and improving economic returns. Coordination on the use of water sources can bring several benefits, including increased water security and supply reliability, improved water quality, decreased loss, increased water system resilience to climate change, and adaptation to uncertainty in availability and changes in water supply demands (Marques et al., 2010; Ross, 2017). Conjunctive use management strategies aim mainly to obtain the greatest benefit from the use of water from various sources. (GGGFA, 2015; Alley, 2016), and their application is already a reality in many places; for example, to reduce impacts from annual and seasonal imbalances and to meet agricultural and industrial demands (Raul and Panda, 2013; Rezaei et al., 2017); reducing scarcity (Dai et al., 2016; Draper et al., 2003; Abdolvandi et al., 2014).

The advantages promoted by conjunctive use are not restricted to solving supply problems, but also to increasing the economic efficiency of the system (Sahuquillo, 1985; Singh, 2016). One of the direct benefits is the optimization of water allocation in relation to the availability and distribution of this resource, enabling the development of a use strategy over time. In addition, this type of management approach also increases the security of water supply, for human and agricultural uses, reduces evaporative losses, minimizes irrigation and treatment costs, avoids excessive depletion of surface and groundwater sources, and consequently reduces impacts on the environment (Evans and Evans, 2011).

As an example, Jenkins et al. (2004) analyzed the optimization of a large-scale water system operation in California (including more than 90% of statewide demands) indicating that greater economic gains would be possible given the increased operational flexibility provided by conjunctive use strategies. In this case, the availability of underground (much higher than surface) storage has allowed surface reservoirs to be operated less conservatively without, however, increasing the risk of system failure as a whole. For Marques et al. (2006), the availability and price associated with the exploration of surface and groundwater directly affect the operation of conjunctive use strategies, which may serve as inducers of effective solutions or render unfeasible some operations of alternate use, when the differences in exploitation costs are very large. Riegels et al. (2013) show that it is possible to increase welfare while meeting ecological and groundwater sustainability goals by using water pricing to support a conjunctive management strategy in which price signals encourage surface water use during wet years and groundwater use during dry years. Further investigation of economic benefits associated with conjunctive use can be found in Pulido-Velazquez et al. (2016) who revised the state-of-the-art in conjunctive use models with emphasis on hydroeconomic approaches.

16.4 POTENTIAL APPLICATIONS IN BRAZIL

16.4.1 The advantages of conjunctive use under hydrological uncertainty: economic benefits and operational strategies

Diversification in the use of water sources to meet water demands is a recurring situation in many countries, and in Brazil it is no different. According to the Atlas of Urban Water Supply, 14% of Brazilian municipalities are supplied with a mixed system, with the potential to increase this percentage (ANA, 2010). The São Paulo Metropolitan Region, which is supplied by a reservoir system, has recently experienced an unprecedented water crisis and the conjunctive use with groundwater has emerged as an alternative to minimize water shortages (Bertolo et al., 2014). However, it must be highlighted the importance of carrying out studies of feasibility and impacts, otherwise it could aggravate the conditions of water resources systems, as occurred in the Metropolitan Region of Recife, in which overexploitation of the underground source caused a serious problem of saline intrusion (Hirata et al., 2012). Diversified supply systems are already a reality, for example, in the São Luís Island region, in Maranhão State, with tubular wells, water reservoirs, and water transposition from the mainland to the island (CAEMA, 2012).

Groundwater has also been used for agriculture, such as fruit production in the Chapada do Apodi, a semiarid region in the state of Ceará, where groundwater availability has subsidized the region's agricultural and economic development (Medeiros et al., 2003). In the western region of Bahia state, water from the Urucuia Aquifer has been used for irrigation and expansion of the agricultural frontier. However, it is important to note the importance of this source, which is responsible for the base flow of the tributaries of the left bank of the São Francisco River and the tributary springs of the right bank of the Tocantins River (Gaspar, 2006). Despite the clear benefits, groundwater use in agriculture demands proper planning (and integration with surface water when possible) in order to avoid adverse consequences such as lowering of the water level and salinization of the soil, common when groundwater is used extensively and exclusively. Regions where these problems already occur, as in the Jandaíra and Beberibe Aquifer Systems in the Brazilian Northeast (ANA, 2005), are examples worthy of further analysis for conjunctive use strategies.

The combined use of surface and groundwater is already observed in several locations in Brazil; however, in many cases, there are no studies of impacts on water sources or operating rules. There are several opportunities for the application of conjunctive use strategies in the country, mainly due to the diversification of water sources and the high groundwater and surface water potential throughout the Brazilian territory.

16.4.1.1 Operational strategies

Operational strategies for small urban and agricultural systems in Brazil will require coordination between state and municipal governments. Groundwater rights exist under a permitting system, run by the state (the state is responsible for managing the aquifers and issuing the water permits), while urban water supply systems are part of sanitation plans, the responsibility of municipalities. The Brazilian National Water Law has the watershed as a management unit, and the watershed plans are the management instruments that should bring together all necessary actions from the sanitation side (e.g. urban water supply systems for potable water) and actions for broader management of raw water systems (e.g. protection of watersheds, surface and groundwater permits). Hence, any sound operational strategy for conjunctive use needs a well-established coordination framework, capable of:

* Integrating hydrologic and hydrogeological databases between different states overlying a common aquifer;
* Integrating the water user database between different states;
* Establishing common rules and procedures for issuing both surface and groundwater permits across different states;

Groundwater management strategies 303

- Establishing common rules and procedures for environmental licensing for different municipalities to protect vulnerable aquifer recharge areas against contamination; and
- Defining necessary actions in the watershed plan (e.g. new water infrastructure) to function as either passive or active conjunctive use elements.

Besides allowing decisions that integrate groundwater and surfaces water use, the strategic actions above are necessary to prevent externalities and third-party effects from water and land use in one municipality, or state, into the other, sharing the same aquifer. These include aquifer overdraft, contamination, and ensuing litigation.

An important advantage of conjunctive use is to bring the groundwater storage availability to complement surface water (which is often more variable spatially and temporally) contributing to addressing problems such as drought and shortages (Foster et al., 2010). There are several possible strategies for conjunctive use, which may involve different time arrangements and that depend on the type of problem faced. Marques et al. (2010) presented a breakdown of possible strategies (Table 16.1).

The following sections explore general operational strategies, followed by two examples where operational strategies were investigated.

Local-Scale Operations – Contaminant Management

An example of contaminant management through continuous use and groundwater pumping relocation (Table 16.1) can be found in a study conducted in a small rural district in Brazil, close to the city of São Francisco, state of Minas Gerais. In 1993/1995, several cases of dental fluorosis were diagnosed in this locality. In multidisciplinary research, Velasquez et al. (2003) identified over thirteen wells with fluoride concentration above 0.8 mg/L and three small rural communities in the district presented endemic and dental fluorosis, with severe cases among children (Figure 16.1). In these three small rural communities, the concentration of fluoride in the water ranged from 1.18 to 3.9 mg/L.

According to Velasquez et al. (2003), the presence of fluoride in the water was a consequence of fluorite in the limestone karst aquifer (natural source contamination). Local municipality government installed reverse osmosis (RO) treatment plant (Figure 16.2),

Table 16.1 Strategies for conjunctive use (Marques et al., 2010)

Problem	Strategy	Time frame	Operation
Drought, inter-annual imbalances	Drought cycling	Annual to 10-year period	Store surface water in wet years, and use groundwater in dry years
Seasonal imbalances	Seasonal cycling	Seasonal	Use groundwater in dry months and surface water in rainy months
Early stage of quick development of a region	Initial intensive exploration	Years to decades	Initial intensive groundwater use to support development and delay infrastructure investments
Saline intrusion, contaminant dispersion	Continuous use	Continuous, as long needed	Surface and underground relocation, recharge management, treated water injection

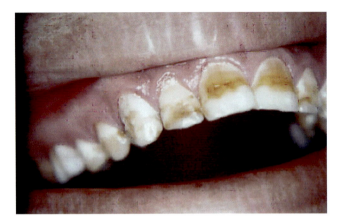

Figure 16.1 Dental fluorosis occurrence in children – São Francisco municipality, Minas Gerais state, Brazil. (Picture credit: E.F. Ferreira, 2002. Velasquez et al., 2003.)

but its treatment capacity was limited to 64 m³/month in treating water from a well with 1.92 mg/L fluoride concentration. After treatment, the water was distributed to the three small rural communities which combined demand (households and schools) totals 3,440 m³/month. The remaining demand was met with water from other wells with high fluoride concentrations.

In order to investigate alternative operational arrangements to mitigate the problem, Marinho et al. (2007) simulated the operation of this small system under different conditions, including connection with a neighbor well and a rainwater harvesting system. The neighbor well has a high capacity (264 m³/hour) and has no fluoride in the water. This well also supplies to other rural communities. The simulation model included the three small communities (as demand nodes) the RO treatment plant, the wells (as water supply nodes), and the pipelines (as links). The simulation was performed with WEAP software (Water Evaluation and Planning) (Sieber et al., 2005) and it explored four different scenarios. The first (base) scenario had no limits in the fluoride concentration in the water delivered to the communities, which was similar to real conditions but in violation of Brazilian Ministry of Health Standards No. 635/75. The latter limits fluoride concentration in the water to a minimum of 0.6 and a maximum of 0.8 mg/L for regions with average max temperature in the 26.4°C–32.5°C range, which is the case of São Francisco. The next three scenarios imposed the 0.8 mg/L limit to the water delivered and bring in other arrangements: Scenario 2: Fluoride limited to 0.8 mg/L, RO treatment online; Scenario 3: Fluoride limited to 0.8 mg/L, RO treatment online, rainwater harvesting system; and Scenario 4: Fluoride limited to 0.8 mg/L, RO treatment offline, integration with neighbor well. When a 0.8 mg/L limit was imposed in the simulation, the system model rearranged the use of the available water supply sources so that their combined flow and concentration would always meet the limit.

The base scenario resulted in a water supply coverage of 90.3%, but with only 2.1% of the demand being met with treated water. The remaining 88.3% received water with 1.92 mg/L fluoride concentration. As expected, when the 0.8 mg/L limit on the

Figure 16.2 Reverse osmosis treatment plant, São Francisco municipality, Minas Gerais state, Brazil. (Picture credit: Fernando Marinho, 2005. Oliveira et al., 2007.)

fluoride concentration was imposed in scenario 2, the supply coverage was reduced to only 3.17% (from 90.3% in the base scenario). Even though the water from different wells could be mixed to optimize the use of the RO plant capacity (the RO plant produces water with no fluoride), it was still severely limited. At this point, both scenarios reflected what was already known: the inhabitants of the three communities were not supplied enough water to meet their water demand in quantity and quality.

Scenario 3 brought in the use of a rainwater harvesting system integrated with groundwater use. The region has long-term average precipitation of 965.7 mm/year, with significant seasonal variation (the climate is in the transition from subhumid to semiarid). The rainwater harvesting system is based on household roof water collection and storage in ferro-cement tanks. Under this scenario, the supply coverage varied according to the monthly rainfall, ranging from 27.3% to 45.2% (up from 3.17% in Scenario 2). Despite its relatively small contribution to water supply even during the wet (most favorable) months, rainwater provided a significant improvement in both water quality (the 0.8 mg/L standard was always met) and in quantity by integrating the use of the available supply sources.

Finally, scenario 4 integrated the neighbor well into the system, boosting the supply coverage to 100%, but maintaining the use of local wells with just enough pumping to

306 Groundwater for Sustainable Livelihoods

meet the health standard concentration. Under this arrangement, the local wells still provided up to 1,433.5 m³/month (41.6% of the 3,440 m³/month total demand), while the neighbor well provided the remaining 2,006 m³/month. By combining the operation of local and neighbor wells with the RO treatment, this operational arrangement increased the flexibility of the local rural water supply system with the following three major points:

- It reduced the costs of importing water from another neighbor system;
- It reduced the reliance on external neighbor wells, which contributes to minimizing potential conflicts with other users; and
- It avoided pumping exclusively from low fluoride concentration wells, which contributes to minimizing the possibility of contaminant migration due to changes in the potentiometric level and hydraulic gradients (Datta, 2005; Das Gupta and Onta, 1997).

Regional-Scale Operations – Hydroeconomic Water Allocation

Another recent study addressed a different economic context. Mattiuzi et al. (2019) assessed the economic water allocation and the potential of conjunctive use of surface water and groundwater operations at an irrigated agricultural region in Southern Brazil. The authors applied a hydroeconomic model to calculate water scarcity and scarcity costs, with and without conjunctive use strategies in the Santa Maria River basin system, state of Rio Grande do Sul, Brazil. This region has a high production of rice and soybean crops and high agricultural water demands. Recent drought events lead to scarcity scenarios and conflicts among users.

Hydroeconomic models combine economic concepts in water resources management models, and represent, at a regional hydrological scale, the technical, environmental, and economic aspects of the water systems. This approach provides decision support tools for conjunctive management of surface and groundwater, contributing to improving management strategies that promote efficiency and transparency in water use, as discussed in detail in Jackeman et al. (2016). In the Santa Maria River basin study (Mattiuzi et al., 2019), the hydroeconomic model resolved water allocation by following linear piece-wise economic penalty functions, developed from the marginal economic benefit functions calculated by the agricultural production model (Howitt et al., 2012).

The results of the hydroeconomic model simulations for water allocation from 2001 to 2015 showed that, under the current permit scheme allocation adopted in the basin, scarcity costs figures were nearly R$ 2 billion (accumulated over 15 years), while under an economic water allocation the figure was reduced to R$ 0.47 billion over the 15-year period, and water scarcity would be down to one third. When groundwater operations were integrated into surface water, scarcity costs were reduced to R$ 0.11 billion over 15 years, and under more economically efficient water allocation there were no scarcity costs (Mattiuzi et al., 2019).

A subsequent analysis in Mattiuzi and Marques (2019) further identified conjunctive use operating strategies to reduce scarcity and scarcity costs in this same region, based on previous calculations in Mattiuzi et al. (2019). Conjunctive use strategies included:

- In months of lower water demand, land use and soil management activities and practices should prioritize groundwater recharge;

Groundwater management strategies 307

- All year long, implement strategies for permanent recharge including green infrastructure in urbanized areas (e.g. permeable pavements, rain gardens, and other natural drainage system practices);
- Water credits: implementation of a water accounting system based on estimates of the amount of water recharged according to the management practices adopted by each producer;
- In months of higher water demand, groundwater sources would be used to supplement surface water sources (depending on the region);
- Assuming that water charges had been already implemented in the basin, the amount charged for groundwater use would be reduced due to the amount of water credits each user has accounted for. Water charges are also management instruments in the National Water Resources Policy;
- In case of no implementation of water charges, the pumped groundwater volumes would be calculated according to the amount of water credits accounted for; and
- Between years, alternate surface and groundwater use for irrigation, so that higher surface water use in wet years increases groundwater recharge to replenish the stock, while higher groundwater use in dry years (when less surface water is available) depletes groundwater reserves, keeping storage space for next wet years to replenish groundwater.

The authors concluded that the application of conjunctive use strategies, through technical studies and integration between management teams, decision-makers, and users, can contribute to the effectiveness of water management, avoid conflicts, and increase economic benefits. The authors also mentioned the importance of carrying out studies of detailed hydrogeological, land use, and vegetation cover mapping of the region for the development and implementation of conjunctive use strategies.

16.4.2 How to integrate existing (and new) water resources management framework and instruments

The examples presented in the previous section highlighted the opportunities for the development of conjunctive use programs. The Brazilian Water Law (Brazil, 1997) defines several water management instruments that can and should be used to encourage and enforce the use of surface water or groundwater sources according to seasonality and distribution within predetermined operational strategies. Three instruments are discussed here: water permits, water charges, and water resources plans.

Water permits aim to ensure quantitative and qualitative control of water use and the effective exercise of water access rights, through the preservation of multiple uses (Brazil, 1997). Permits can be used in conjunctive use strategies to drive water allocation toward higher priority uses in times of scarcity. This mechanism could be defined when granting the permit, to raise awareness of users and managers to the prioritization of water in a watershed. The suggested water credits could be accounted with the water permits and the alternating surface and groundwater operations could be enforced through variable water permits, as an example.

Water charges aim to signal that water is an economic good and give the user an indication of its real value, to encourage rational use of water and obtain financial resources to finance studies and projects within the river basin (Brazil, 1997). All water uses that are subject to permits are also subject to charge. Water charging can be explored in conjunctive use strategies to encourage more economically beneficial water uses, and also to subsidize studies and to finance works to improve the water system. Currently, just a few Brazilian watersheds have implemented water charges, and their effectiveness has been heavily questioned (OECD, 2017).

Water resource plans, as proposed on the Brazilian Water Law, aim to support and guide water management. The plans have a long-term horizon (usually about 10 years), and besides providing a diagnosis of the water resources situation within a river basin, it also analyzes demographic growth alternatives, predicts productive activities evolution, changes in land occupation patterns, hydrological balance between availability and future demands of water resources, identifies potential conflicts, proposes actions and projects to be developed to meet standards and goals, among others (Brazil, 1997). Water resources plans can define, in an integrated fashion, the required hydrology, geology, and hydrogeology studies to map the surface water–groundwater interaction, the necessary monitoring network (what to measure, where and the density of the data collection) and the required infrastructure to make the operations feasible and finally the directives to the implementation of the other instruments (water permits and charges) to signal it to users.

Aside from the management instruments defined in the Water Law, there are other possibilities to introduce conjunctive use strategies in Brazil, for example, with payment for environmental services (PES). The main purpose of PES is to compensate those who promote the conservation of environmental services, such as preservation of springs, soil management to reduce erosion, maintenance of riparian forests to regulate river flows, among others. In 2019, the Brazilian parliament approved a project to create a law that regulates PES; now it must be approved by the Senate (Brazil, 2019). The concept of PES is already in practice by the Brazilian Water Agency, which has a program called "Water Producer," which encourages producers to invest in water management while receiving technical and financial support for the implementation of conservation practice (ANA, 2019).

In 2018, a new resolution about conjunctive use was approved by the National Water Resources Council, which has established guidelines for the integrated management of surface and groundwater resources that included the articulation between the Union, the States, and the Federal District with a view to strengthening water management (MMA, 2018). The approval of specific legislation on integrated management is a milestone in Brazil's water resources management system, and it also emphasizes the need for conjunctive use strategies studies and implementation in the country (Mattiuzi and Marques, 2019). The main highlights of the resolution are:

- Conjunctive use strategies should be applied in aquifers and rivers with direct connectivity between surface and groundwater;
- Recharge processes and aquifer contributions on rivers must be evaluated;
- Water availability and its different uses must be estimated, and hydrometeorological networks should support the generation of data to subsidize the development of conjunctive use strategies; and

- Collaboration between users, managers, and authorities should be strengthened when considering the development of conjunctive use strategies.

From the mechanisms mentioned above, it is possible to address the rich framework of water management instruments in Brazil and the way it could support conjunctive use strategies, contributing to making the water systems more robust, flexible, and adaptable. However, it is important to highlight the importance of establishing coordination mechanisms between administrative spheres and the users, since the fragmentation of responsibilities has been recognized as one of the biggest obstacles in the implementation of management mechanisms (World Bank, 2006). Water resources management should include strong public participation and the technical support of the responsible bodies (Lopes and Freitas, 2007). Strategies for conjunctive use span geographic and administrative scales, requiring coordination between groundwater and surface water policies, as well as soil management and energy planning (Ross and Martinez-Santos, 2010).

16.5 CONCLUSIONS

Conjunctive use operations of surface water and groundwater can improve reliability and sustainability and reduce the vulnerability of water systems, especially in rural regions. In these regions, poverty is strongly related to the lack of access to natural resources, such as water. There are many targets on the UN 2030 Agenda SDGs that focus on reducing poverty and improving rural population's lives, and when assessing rural population access to water resources, there are several challenges to be addressed such as water availability, quantity, quality, storage, among others. Two study cases in rural areas were discussed in this chapter, and each one presented a group of conjunctive use strategies to resolve these challenges in very different scales: local and regional.

In the first study, the authors analyzed alternative operational arrangements to mitigate the excess of fluoride on rural water systems that were supplied by groundwater and reduce endemic fluorosis. It was found feasible alternative operation of a system that included wells, a rainwater harvesting system, and RO to limit the fluoride concentration. By combining the operation of local and neighbor wells with the reversed osmosis treatment, this operational arrangement increased the flexibility of the local rural water supply system by reducing water costs and reliance on external neighbor wells, thus minimizing potential conflicts with users.

In the second study, the authors assessed the economic water allocation and the potential of conjunctive use operations at an irrigated agricultural region in Southern Brazil. Results indicated some conjunctive use strategies such as using groundwater to complement reservoir storage during months with lower rains are expected to reduce scarcity costs. To implement those strategies, the users can adopt soil management to improve recharge, while water managers could establish water credits and design variable water charges, along with shifting water allocation between users.

When designing conjunctive use strategies, we need to assess the technical aspects of the physical environment, as groundwater and surface water withdrawals can impact negatively the whole water system. It is also important to integrate these strategies with the current water resources management institutions and instruments. In Brazil, there are several opportunities to develop conjunctive use programs based on the management water instruments within the National Water Resources Policy (water permits, charges, and water resources plans) and other specific legislation, such as the one dealing with the PES and with managed aquifer recharge.

Through integrated water resources management and conjunctive use, it is possible to avoid conflicts between water users, reduce water availability uncertainties and increase economic benefits. The examples, experiences, and ideas explored here indicate that groundwater is a key element in securing water for food, economic development and the environment, and the implementation of conjunctive use strategies is a powerful tool to build more resilient livelihoods and equitable growth.

REFERENCES

Abdolvandi, A.F.; Parsamehr, A.; Abazadeh, H.; Eslamian, S.; Hosseinipour, Z. (2014) Conjunctive Use of Surface and Groundwater Resources Using System Dynamics Approach (Case Study: Namroud Dam). *World Environmental and Water Resources Congress*, pp. 323–334.

Agência Nacional de Águas – ANA. (2005) Disponibilidade e Demandas dos Recursos Hídricos no Brasil. Caderno de Recursos Hídricos, Brasília/DF, 134p.

Agência Nacional de Águas – ANA. (2010) Atlas Brasil: Abastecimento Urbano de Água – Panorama Nacional, Volume 1. Brasília/DF, 72p.

Agência Nacional de Águas – ANA. (2017) Atlas Irrigação: Uso da Água na Agricultura Irrigada, Agência Nacional de Águas. Brasília/DF, 86p.

Agência Nacional de Águas - ANA. (2018) Conjuntura dos Recursos Hídricos no Brasil 2018: Informe Anual. Brasília/DF, 72p.

Agência Nacional de Águas – ANA. (2019) Programa Produtor de Água. Available at: https://www.ana.gov.br/programas-e-projetos/programa-produtor-de-agua.

Aleixo, B.; Rezende, S.; Pena, J.L.; Zapata, G.; Heller, L. (2016) Human Right in Perspective: Inequalities in Access to Water in a Rural Community of the Brazilian Northeast. *Ambiente & Sociedade* 1, 63–84.

Alley, W.M. (2016) Drought-proofing groundwater. *Groundwater* 54, 309.

Aquino, J.R.; Gazolla, M.; Schneider, S. (2018) Dualismo no Campo e Desigualdades Internas na Agricultura Familiar Brasileira. *Revista de Economia e Sociologia Rural*, 56(1), 123–142.

Arsky, V.; Santana, V. (2012) Acesso à Água na Zona Rural: O Desafio da Gestão. Conselho Nacional de Segurança Alimentar e Nutricional, Presidência da República. Published in 21/03/2012, last modified in 29/06/2017. Acessed in: 1/7/2020. Available at: http://www4.planalto.gov.br/consea/comunicacao/artigos/2012/acesso-a-agua-na-zona-rural-o-desafio-da-gestao.

Bertolo, R.; Hirata, R.; Conicelli, B. (2014) Situação das Reservas e Utilização das Águas Subterrâneas na Região Metropolitana de São Paulo. Centro de Pesquisas de Águas Subterrâneas – USP, 7p.

Bierkens, M.F.P.; Wada, Y. (2019) Non-Renewable Groundwater Use and Groundwater Depletion: A Review. *Environmental Research Letters* 14, 063002.

Boltz, F.N.; Poff, N.L.; Folke, C.; Kete, N.; Brown, C.M.; Freeman, S.S.G.; Matthews, J.H.; Martinez, A.; Rockström, J. (2019) Water Is a Master Variable: Solving for Resilience in the Modern Era. *Water Security* 8, 100048.

Brazil. (1997) Lei N 9.433 de 8 de janeiro de 1997: Institui a Política Nacional de Recursos Hídricos, Cria o Sistema Nacional de Gerenciamento de Recursos Hídricos. Brasília/ DF.

Brazil. (2019) Câmara Aprova Projeto Que Prevê Pagamento Por Serviços Ambientais. Available at: https://www.camara.leg.br/noticias/579925-camara-aprova-projeto-que-preve-pagamento-por-servicos-ambientais/.

Campos, A.P.; Villani, C.; Davis, B.; Takagi M. (2018) Ending Extreme Poverty in Rural Areas: Sustaining Livelihoods to Leave No One Behind. Food and Agriculture Organization of The United Nations (FAO).

Carlevaro, F.; Becerra, G.; Manuel, C. (2011) Costing Improved Water Supply Systems for Developing Countries. *Water Management* 164(3), 123–134.

Carlos, E. (2020) Saneamento Rural: Um Enorme Desafio Para o Brasil – PORTAL DO SANEAMENTO. Instituto Trata Brasil. Available at: http://www.tratabrasil.org.br/saneamento-rural-um-enorme-desafio-para-o-brasil--portal-do-saneamento. Accessed in 1/7/2020.

Castañeda, A.; Doan, D.; Newhouse, D.; Nguyen, M.C.; Uematsu, H.; Azevedo, J.P. (2018) A New Profile of the Global Poor. *World Development* 101, 250–267.

Companhia de Pesquisa de Recursos Minerais – CPRM. (2008) Hidrogeologia – Conceitos e Aplicações. 3ª edição, 812p.

Companhia De Saneamento Ambiental Do Maranhão – CAEMA. (2012) Relatório Anual Da Qualidade da Água Produzida pelo Sistema de Abastecimento de Água de São Luís e São José De Ribamar. São Luís/MA, 2p.

Dai, C.; Cai, Y.P.; Lu, W.T.; Liu, H.; Guo, H.C. (2016) Conjunctive Water Use Optimization for Watershed-Lake Water Distribution System Under Uncertainty: A Case Study. *Water Resources Management* 30, 4429–4449.

Das Gupta, S.K.; Onta, P.R. (1997) Sustainable Groundwater Resources Development. *Journal of Hydrological Sciences – Journal Des Sciences Hydrologiques* 42(4), 565–582.

Datta, P.S. (2005) Groundwater Ethics for Its Sustainability. *Current Science* 89(5), 812–817.

Draper, A.J.; Jenkins, M.W.; Kirby, K.W.; Lund, J.R.; Howitt, R.E. (2003) Economic-Engineering Optimization for California Water Management. *Journal of Water Resources Planning and Management* 129, 155–164.

Evans, W.R.; Evans, R. (2011) Thematic Paper 2: Conjunctive Use and Management of Groundwater and Surface Water. In: *Groundwater Governance: A Global Framework for Country Action.* Available at: https://www.groundwatergovernance.org/resources/thematic-papers/en/.

Food and Agriculture Organization – FAO. (1995) Land and Water Integration and River Basin Management. *Proceedings of FAO Informal Workshop*, Rome/Italy, 31 January -2 February. Available at: http://www.fao.org/3/v5400e/v5400e00.htm

Foster, S.; Van Steenbergen, F.; Zuleta, J.; Garduño, H. (2010) Conjunctive Use of Groundwater and Surface Water: From Spontaneous Coping Strategy to Adaptive Resource Management. GW-Mate Strategic Overview Series 2, World Bank, Washington DC.

Freeze, R.A., Cherry, J.A. (1979) *Groundwater.* Prentice Hall, Inc.

Gaspar, M.T.P. (2006) Sistema Aquífero Urucuia: Caracterização Regional E Propostas De Gestão. Tese De Doutorado/UNB, 204p.

Gheyi, H.R.; Paz, V.P.S.; Medeiros, S.S.; Galvão, C.O. (2012) Recursos Hídricos em Regiões Semiáridas, Campina Grande/PB: Instituto Nacional do Semiárido, Cruz das Almas/BA: Universidade Federal do Recôncavo da Bahia. 258p.

Giordano, M.; Villholth, K. (Eds.) (2007) *The Agricultural Groundwater Revolution: Opportunities and Threats to Development.* Wallingford: CABI. 419p. (Comprehensive Assessment of Water Management in Agriculture Series 3)

Gleeson, T.; Wada, Y.; Bierkens, M.F.; Van Beek, L.P. (2012) Water Balance of Global Aquifers Revealed by Groundwater Footprint. *Nature* 488, 197–200.

Groundwater Governance Global Framework for Action - GGGFA. (2015) Global Framework for Action. Available at: http://www.groundwatergovernance.org/fileadmin/user_upload/groundwatergovernance/docs/general/gwg_framework.pdf.

Hirata, R., Conicelli, B.P. (2012) Groundwater Resources in Brazil: A Review of Possible Impacts Caused by Climate Change. *Anais da Academia Brasileira de Ciências* 84, 297–312.

Hirata, R.; Montenegro, S.; Petelet, E.; Wendland, E.; Marengo, J.; Martins, V.; Bertolo, R.; Cary, L.; Medeiros, E.; Franzen, M.; Pierre, D.; Aquilina, L.; Giglio-Jacquemont, A.; Batista, J. (2012) Coqueiral: Uma Proposta Metodológica Para Solucionar o Problema de Salinização do Sistema Aquífero da Planície do Recife (PE). *XVII Congresso Brasileiro de Águas Subterrâneas*, Bonito/MS.

Hirata, R.; Suhogusoff, A.V.; Villar, P.C.; Marcellini, L. (2020) A Revolução Silenciosa das Águas Subterrâneas no Brasil: Uma Análise da Importância do Recurso e os Riscos pela Falta de Saneamento. Instituto Trata Brasil. Available at: http://www.tratabrasil.org.br/images/estudos/itb/aguas-subterraneas-e-saneamento-basico/Estudo_aguas_subterraneas_FINAL.pdf.

Howitt, R.E.; Medellín-Azuara, J.; Macewan, D.; Lund, J.R. (2012) Calibrating Disaggregate Economic Models of Agricultural Production and Water Management. Environmental Modelling & Software, n. 38, p. 244–258.

Jakeman, A.J. et al. (2016) Integrated Groundwater Management: An Overview of Concepts and Challenges. In: Jakeman A.J., Barreteau O., Hunt R.J., Rinaudo J.D., Ross A. (eds.) *Integrated Groundwater Management*. Springer, Cham. doi:10.1007/978-3-319-23576-9_1.

Jenkins, M.W.; Lund, J.R.; Howitt, R.E.; Draper, A.J.; Msangi, S.M.; Tanaka, S.K.; Riztema, R.S.; Marques, G.F. (2004) Optimization of California's Water Supply System: Results and Insights. *Journal of Water Resources Planning and Management* 130, 271–280.

Kløve, B.; Ala-Aho, P.; Bertrand, G.; Gurdak, J.J.; Kupfersberger, H.; Kværner, J.; Muotka, T.; Mykrä, H.; Preda, E.; Rossi, P.; Uvo, C.B.; Velasco, E.; Pulido-Velázquez, M. (2014) Climate Change Impacts on Groundwater and Dependent Ecosystems. *Journal of Hydrology* 518, 250–266.

Kløve, B.; Allan, A.; Bertrand, G.; Druzynska, E.; Ertürk, A.; Goldscheider, N.; Henry, S.; Karakaya, N.; Karjalainen, T.P.; Koundouri, P.; Kupfersberger, H.; Kværner, J.; Lundberg, A.; Muotka, T.; Preda, E.; Pulido-Velazquez, M.; Schipper, P. (2011) Groundwater Dependent Ecosystems. Part II. Ecosystem Services and Management in Europe Under Risk of Climate Change and Land Use Intensification. *Environmental Science and Policy* 14, 782–793.

Larkin, R.G.; Sharp Jr, J.M. (1992) On the Relationship between River-Basin Geomorphology, Aquifer, Hydraulics, and Ground-Water Flow Direction in Alluvial Aquifers. *Geological Society of America Bulletin*, 104, 1608–1620.

Lopes, A.V.; Freitas, M.A.S. (2007) A Alocação de Água Como Instrumento de Gestão de Recursos Hídricos: Experiências Brasileiras. *Rega* 4, 5–28.

Machado, A.V.M.; Santos, J.A.N.; Quindeler, N; Alves, LMC. (2019) Critical Factors for The Success of Rural Water Supply Services in Brazil. *Water* 11. doi:10.3390/W11102180.

Marinho, F.M.; Marques, G.F.; Velásquez, L.M.N. (2007) Estudo do Planejamento de Fontes de Produção de Água Subterrânea em Área com Alta Incidência de Flúor Natural. *Anais do XVII Simpósio Brasileiro de Recursos Hídricos*, São Paulo/SP.

Marques, G.F.; Lund, J.L.; Leu, M.R.; Jenkins, M.W.; Howitt, R.; Harter, T.; Hatchett, S.; Ruud, N.; Burke, S. (2006) Economically Driven Simulation of Regional Water Systems: Friant-Kern California. *Journal of Water Resources Planning and Management* 132(6), 468–479.

Marques, G.F.; Lund, J.R.; Howitt, R.E. (2010) Modeling Conjunctive Use Operations and Farm Decisions with Two-Stage Stochastic Quadratic Programming. *Journal of Water Resources Planning and Management* 136, 386–394.

Massoud, M.A.; Tareen, J.; Tarhini, A.; Nasr, J.; Jurdi, M. (2009) Effectiveness of Wastewater Management in Rural Areas of Developing Countries: A Case of Al-Chouf Caza in Lebanon. *Environmental Mentoring Assessment* 161, 61–69.

Mattiuzi, C.D.P.; Marques, G.F. (2019) Gestão Integrada dos Recursos Hídricos: Avaliação dos Benefícios do Uso Conjunto de Águas Superficiais e Subterrâneas em uma Região no Sul do Brasil. *Revista Águas Subterrâneas* 33(4), 340–353.

Mattiuzi, C.D.P.; Marques, G.F.; Medellín-Azuara, J. (2019) Reassessing Water Allocation Strategies and Conjunctive Use to Reduce Water Scarcity and Scarcity Costs for Irrigated Agriculture In Southern Brazil. *Water* 11, 1140.

Medeiros, J.F.; Lisboa, R.A.; Oliveira, M.; Silva Junior, M.J.; Alves, L.P. (2003) Caracterização das Águas Subterrâneas Usadas para Irrigação na Área Produtora de Melão da Chapada do Apodi. *Revista Brasileira de Engenharia Agrícola e Ambiental* 7, 469–472.

Ministério do Meio Ambiente - MMA. (2018) Resolução Nº 202 de 28 de junho de 2018 do Conselho Nacional De Recursos Hídricos. Brasília/DF, 3p.

Oliveira, F.M.; Marques, G.F., Velasquez, L.N.M. (2007) Estudo do Planejamento de Fontes de Produção de Água Subterrânea em Área com Alta Incidência de Flúor Natural. *XVII Simpósio Brasileiro de Recursos Hídricos*, 19p.

Organization for Economic Co-operation and Development - OECD. (2015) Water Resources Governance in Brazil, OECD Studies on Water, OECD Publishing. Available at: http://www.oecd.org/brazil/water- resources-governance-in-brazil-9789264238121-en.htm.

Organization for Economic Co-operation and Development - OECD. (2017) Water Charges in Brazil the Ways Forward. OECD Studies on Water. Available at: http://www.oecd.org/brazil/water-charges-in-brazil-9789264285712-en.htm.

Pulido-Velazquez, M.; Marques, G.F.; Harou, J.J.; Lund, J.R. (2016) Hydro-economic Models as Decision Support Tools for Conjunctive Management of Surface and Groundwater. In: Jakeman A.J., Barreteau O., Hunt R.J., Rinaudo J.-D., Ross A. (eds.), *Integrated Groundwater Management. Concepts, Approaches and Challenges*. Ch. 27, pp. 693–710. Springer International Publishing. ISBN 978-3-319-23575-2.

Pulido-Velazquez, M.; Marques, G.F.; Jenkins, M.W.; Lund, J.R. (2003) Conjunctive Use of Ground and Surface Water: Classical Approaches and California Examples. *Proceedings of 11th World Water Congress*, International Water Resources Association, Madrid, Spain.

Raul, S.K.; Panda, S.N. (2013) Simulation-Optimization Modeling for Conjunctive Use Management under Hydrological Uncertainty. *Water Resources Management* 27, 1323–1350.

Rezaei, F.; Safavi, H.R.; Zekri, M. (2017) A Hybrid Fuzzy-Based Multi-Objective Algorithm for Conjunctive Water Use and Optimal Multi-Crop Pattern Planning. *Water Resources Management* 31, 1139–1155.

Riegels, N.; Pulido-Velazquez, M.; Doulgeris, C.; Valerie, S.; Jensen, R.; Moller, F.; Bauer-Gottwein, P. (2013) A Systems Analysis Approach to the Design of Efficient Water Pricing Policies Under the EU Water Framework Directive. *Journal of Water Resources Planning and Management* 139, 574–582.

Ross, A. (2017) Speeding the Transition Towards Integrated Groundwater and Surface Water Management in Australia. *Journal of Hydrology* 567, p. 1–10.

Ross, A.; Martinez-Santos, P. (2010) The Challenge of Groundwater Governance: Case Studies from Australia and Spain. *Regional Environmental Change* 10, 299–310.

Sahuquillo, A. (1985) Groundwater in Water Resources Planning: Conjunctive Use. *Water International* 10, 57–63.

Sahuquillo, A. (2009) La Importancia de Las Aguas Subterráneas. *Revista Real Academia De Ciencias Exactas, Físicas Y Naturales* 103, 97–114.

Sieber, J.; Yates, D.N.; Purkey, D.; A. Huber-Lee. (2005) Weap, A Demand, Priority, And Preference Driven Water Planning Model: Part 1, Model Characteristics. *Water International* 30(4), 487–500.

Silva, F. C. (2007) Análise Integrada de Usos de Água Superficial e Subterrânea em Macro-Escala numa Bacia Hidrográfica: O Caso do Alto Rio Parnaíba. Dissertação de Mestrado, Instituto de Pesquisas Hidráulicas, Universidade Federal do Rio Grande do Sul, 188p.

Silva, F.O.E.; Heikkilab, T.; Filho, F.A.; Silva, D.C. (2012) Developing Sustainable and Replicable Water Supply Systems in Rural Communities in Brazil. *Water Resources Development*, 1–14. doi:10.1080/07900627.2012.722027.

Singh, A. (2016) Optimal Allocation of Resources for Increasing Farm Revenue under Hydrological Uncertainty. *Water Resources Management* 30, 2569–2580.

Taylor, R. G.; Scanlon, B.; Döll, P.; Rodell, M.; Van Beek, R.; Wada, Y.; Konikow, L. (2013) Groundwater and Climate Change. *Nature Climate Change* 3, 322–329.

Velásquez, L.M.N.; Fantinel, L.M.; Costa, W.D.; Uhlein, A.; Ferreira, E.F.E.; Castilho, L.S.; Paixão, H.H. (2003) Origem do Flúor na Água Subterrânea e sua Relação como os Casos de Fluorose Dental no Município de São Francisco, Minas Gerais. Belo Horizonte: Fapemig. (Relatório Cra 294/99), 138p.

Villar, P.C. (2016) Groundwater and the Right to Water in a Context of Crisis. *Ambiente & Sociedade* 19, 85–102.

Wada, Y.; Bierkens, M.F.P. (2014) Sustainability of Global Water Use: Past Reconstruction and Future Projections. *Environmental Research Letters* 9(10), 104003.

World Bank. (2006) Conjunctive Use of Groundwater and Surface Water. World Bank, Washington. Available at: http://documents.worldbank.org/curated/en/387941468041438440/pdf/370310ard0notes1issue601public1.pdf.

World Water Assessment Programme - United Nations. (2018) The United Nations World Water Development Report. 139p. Available at: www.unwater.org/publications/world-water-development-report-2018/.

Chapter 17

The role of groundwater in rural water supply

The case of six villages of Taunggyi District, Southern Shan State, Myanmar

S.Y. May
University of Yangon

K.K. Khaing
Pathein University

J.S.T. Ward
British Geological Survey

CONTENTS

17.1 Introduction 316
    17.1.1 Purpose 316
17.2 Study area: Taunggyi District, Southern Shan State 317
    17.2.1 Location and administrative boundaries 317
    17.2.2 Geography, climate, and geology 317
17.3 Methodology 320
17.4 Results and discussion: the role of groundwater in rural water supply
    of Taunggyi District 320
    17.4.1 Quantity and distribution of springs across Taunggyi District 320
    17.4.2 Case studies 324
        17.4.2.1 Case study 1: Loikun Village, Kakku Village Tract,
            Taunggyi Township 324
        17.4.2.2 Case study 2: Boung Kyaungnar Village, Naung Pi
            Village Tract, Pinlaung Township 324
        17.4.2.3 Case study 3: Nwng Mun Village, Nwng Mun Village
            Tract, Hsihseng Township 327
        17.4.2.4 Case study 4: Taung Kham Village, Kyaukgu Village
            Tract, Indaw Sub-township, Lawksauk Township 328
        17.4.2.5 Case study 5: Naungya Saing Village, Kone Kying Village
            Tract and Nan Kun Village, Paung Lin Village Tract,
            Hopong Township 328
        17.4.2.6 Case study 6: Inn Nge Village, Inn Nge Village Tract,
            Pindaya Township 331
17.5 Overall discussion 334
17.6 Conclusion 334
References 334

DOI: 10.1201/9781003024101-17

17.1 INTRODUCTION

Recently, global efforts have been made to improve water, sanitation, and hygiene (WASH) services within broader integrated water resource management (IWRM), through Sustainable Development Goal 6 (SDG 6) (Carrard, 2019). Water security is a great challenge in developing countries and Myanmar is no exception (Sakai et al., 2013; UNICEF and WHO, 2019). Access to safe water is a significant issue in many developing countries (Sakai et al., 2013). Seven hundred and eighty-five million people globally do not have access to basic water supply services (UNICEF and WHO, 2019). Climate change impacts are expected to cause significant changes in temperature and rainfall patterns, with consequences for the quality and quantity of water available. Myanmar, a Southeast Asian country with a population of 55 million people (Department of Population, Myanmar, 2021), has various types of water sources that vary in quality. Groundwater is of vital importance in Myanmar, particularly because it provides up to 80% of drinking and irrigation supplies in some areas (Visossanges et al., 2017) and water demand is increasing (Kattelus et al., 2017; Pavelic et al., 2018). However, widespread understanding of groundwater hydrochemistry is much more limited as compared with other countries in the South/Southeast Asia region (Tun, 2003; Smedley, 2005).

In 2018, 82% of the Myanmar population was using at least basic drinking water services (FAO, 2018) and the majority of people lived in rural areas (70%). The rural areas of Myanmar have lagged behind urban centers in accessing safe and affordable water supply services. Even in most urban areas, residents have to make effort to get water because they do not have water supply services.

This study focuses on Taunggyi District, in Southern Shan State, which is one of the most populated areas in Shan State and comprises ten townships. Nearly 10% (5.83 million) of the total population of Myanmar live in Shan State, which is the largest state in Myanmar and is located in the eastern highlands (Department of Population, Myanmar, 2021). This territory, Shan State is divided into three areas (Shan East, Shan North, and Shan South) and it comprises thirteen districts. Taunggyi District has a population of approximately 1,701,000, of which 73% live in rural areas (DWIR, 2021). Within Shan State, approximately 890,000 households are located in rural areas. Approximately 55% of these rural households have access to safe drinking water services and about 45% of households rely on unsafe drinking water access (Department of Population, Myanmar, 2017). In rural areas of Taunggyi Township, 42% of households use water from unimproved sources for drinking water (Department of Population, Myanmar, 2017). The Department for Rural Development (DRD) stated that around 70% of people in Southern Shan rely on springs, around 20% on tube wells and the remainders rely on dug wells and rainwater harvesting (Snoad, 2014). Rainwater harvesting ponds are not common and are only constructed by the DRD if no suitable spring or groundwater can be identified (Snoad, 2014). Sustainable development and management of groundwater resources are essential for ensuring long-term water supply in rural Taunggyi District, especially in places that depend mostly on groundwater.

17.1.1 Purpose

The main aim of this study was to present an overview of the current conditions of domestic water use in the rural areas of Taunggyi District, to provide a foundation for better understanding rural water supply, and to support the future development of water resource management plans. The main objectives of this study are as follows:

- To understand the importance of groundwater in rural water resources and its supply in Taunggyi District, Shan State;
- To investigate the water use practices and perceptions of local people regarding water resource availability and water supply in Taunggyi District; and
- To understand any limitations of groundwater availability for water supply in Taunggyi District, Shan State.

17.2 STUDY AREA: TAUNGGYI DISTRICT, SOUTHERN SHAN STATE

17.2.1 Location and administrative boundaries

Myanmar is divided into fifteen states. Shan State is located in eastern Myanmar and comprises thirteen districts. Taunggyi District, in Southern Shan State, is the study area for this research and includes ten townships (comprising cities, towns, and village tracts in an area) and 244 village tracts (Figure 17.1). Village tracts are groups that include several villages in an area. Taunggyi District is administratively bounded by Loilem District in the east, Langkho District in the southeast, Kayah State in the south, Mandalay State/Region in the west, Naypyitaw State/Region in the southwest, and Kyaukme District in the northern part (Figure 17.1).

17.2.2 Geography, climate, and geology

Myanmar is geographically divided into four main divisions: the Western Mountain Ranges, Eastern Highland and their continuation southward, the Central Basin, and Rakhine Coastal Being in the Eastern Highland area, the entire Shan State is composed of mountains.

Taunggyi District, located in Southern Shan State, comprises rolling upland plains, hills, and basins. The altitude ranges from 300 to 1,400 m across the district, with the mountain range generally running in a north–south direction. Between these mountain ranges, there are wide highland basins, valleys, and plains namely; Tharmakhan Plain, Lonpo Plain, NyaungShwe Valley, Moebye Valley, Heho Basin, and Hopone Valley, where intensive agriculture is practiced. Agriculture is the main occupation of rural people in Taunggyi District. The main cash crops are maize, rice, tea, and fruits. Vegetables are grown in lowland fertile areas. Ethnic groups include Shan, Bamr, Pa-O, Intha, and Danu.

Myanmar has a humid subtropical climate, where the temperature and rainfall vary with topography and altitude. In Taunggyi District, Pekon Township receives the lowest rainfall, with average annual total rainfall of 900 mm and Pinlaung Township receives the most rainfall, with an average annual total rainfall of 2,300 mm. Many natural springs can be found in this area.

The regional geology of Taunggyi District comprises Paleozoic and Mesozoic clastic and carbonates form as basement rocks followed by the Tertiary to Quaternary lacustrine deposits (Figure 17.2). The lower Paleozoic litho-stratigraphic unit is well

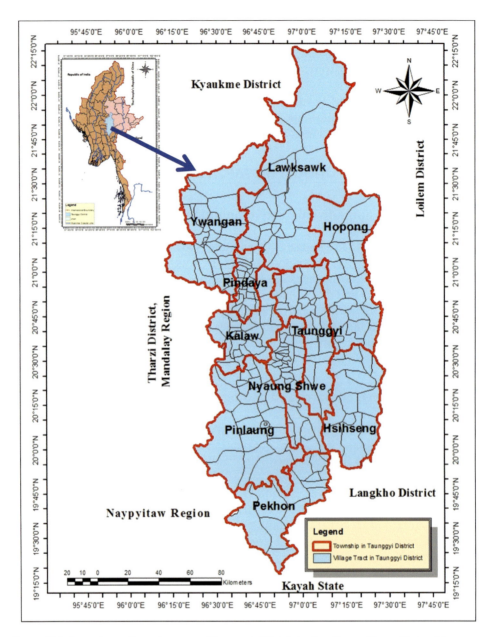

Figure 17.1 Location of study area: Taunggyi District, Southern Shan State, Myanmar. The inset shows the location of Shan State and Taunggyi District within Myanmar. (Outline Sources from Myanmar Information Management Unit, 2019.)

exposed in Pindaya anticline and surrounding area, Kalaw, Pinlaung, Pindaya, and Yaksawk areas. Late Tertiary to Quaternary lacustrine deposits of loosely cemented, and unconsolidated sand, silt, and clays are widely covered in Shwenyaung Plain. South of the study area, the Plateau Limestone Group and Kalaw Red Bed Formation

Groundwater in rural water supply, Myanmar 319

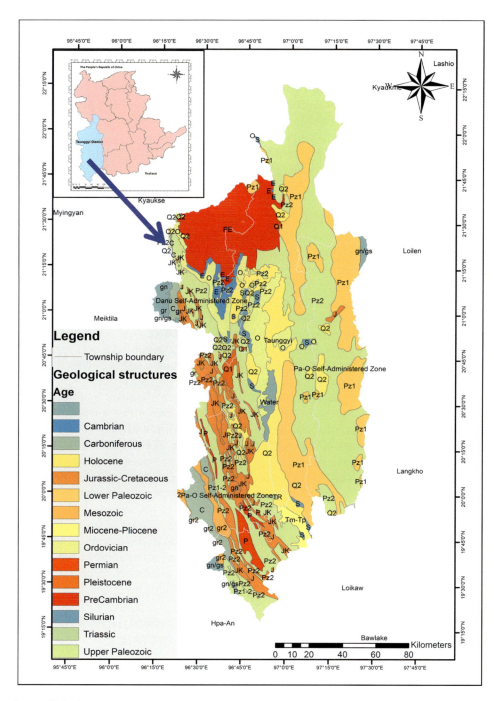

Figure 17.2 Geological map for Taunggyi District, Southern Shan State, Myanmar. The inset shows the location of Taunggyi District within Shan State. (Geology Department, University of Yangon, Myanmar.)

320 Groundwater for Sustainable Livelihoods

are well exposed. North of the study area, the Devonian unit is found in Yebyu area, east of Yakasawk and Taungyi area.

The eastern part of limestone and dolomite limestone and some Precambrian Chaungmagyi groups of phylite, schist, and gneiss are present in Pindaya area. The limestone aquifer is karstic and therefore sinkholes and springs are found throughout the Shan State (Bender, 1983; Zaw and Than, 2017).

Due to the dominance of fracture flow permeability in limestone, access to geophysical equipment to detect the presence of groundwater in the fracture network is needed for the best chance of drilling a successful borehole. A lack of access to this equipment has led to a lot of boreholes failing to produce a water supply.

17.3 METHODOLOGY

Secondary data on the quantity, distribution, and yield of springs in Taunggyi District were collected from the Directorate of Water Resources and Improvement of River System (DWIR). There are three main sources of water supply in Southern Shan State: rainwater harvesting, surface water (water from lakes, ponds, and rivers), and groundwater (from springs and wells). The sources used by individual village tracts depend on the proximity to different water sources (e.g. a nearby lake or borehole) and also seasonal availability (e.g. rainwater harvesting can only be used in the wet season). In rural Taunggyi District, the main type of groundwater source used is natural springs, due to their proximity to villages and reliability of water supply. Natural springs do not have the risks and costs that are associated with drilling boreholes, which require locating a suitable site for drilling and the cost of abstracting the water.

Village tracts that are dependent on groundwater were identified by interviewing local and regional community leaders. Following this, seven villages from six townships were selected as case studies to explore the role of groundwater resources and water supply in rural areas of Taunggyi District (Figure 17.3). In each village, three community leaders were interviewed using a semi-structured interview method, therefore eighteen interviews were completed in total. The age and gender of interviewees are listed in Table 17.1. Interview questions explored the types of groundwater sources used in each village, the number of sources, the number of households depending on each source, perceptions of the community leaders on the amount of water discharge from each source (including seasonal change), water use, and perceptions on water quality, quantity and the limitations of water resources and water supply in each area. Each interview took about 30 minutes. Translators were used where necessary to understand local ethnic minorities' languages.

17.4 RESULTS AND DISCUSSION: THE ROLE OF GROUNDWATER IN RURAL WATER SUPPLY OF TAUNGGYI DISTRICT

17.4.1 Quantity and distribution of springs across Taunggyi District

In Southern Shan State, which includes Taunggyi District, groundwater is a vital source of rural water supply and the majority of the population relies on springs and tube wells (Source: field observation and key informant interview survey). Surface

Groundwater in rural water supply, Myanmar 321

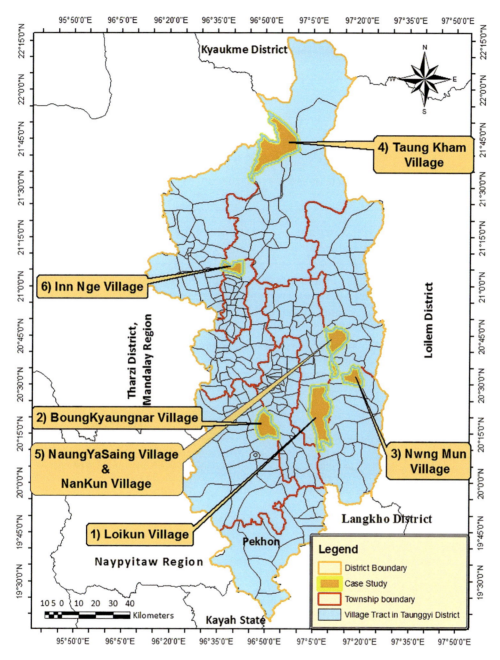

Figure 17.3 Locations of selected villages as case study in Taunggyi District, Myanmar where the local people depend on groundwater resources. The inset shows the distribution of springs within Taunggyi District. (Based on the Database from Myanmar Information Management Unit (MIMU), 2019.)

322 Groundwater for Sustainable Livelihoods

Table 17.1 Interviewee profiles for each village

Village	Loikun	Boung Kyaungnar	Nwng Mun	Taung Kham	Naungya Saing and Nan Kun	Inn Nge
Village tract	Kakku	Naung Pi	Nwng Mun	Kyaukgu	Kone Kying and Paung Lin	Inn Nge
Township	Taunggyi	Pinlaung	Hsihseng	Lawksauk	Hopong	Pindaya
Numbers of interviewee	3	3	3	3	3	3
Gender	2 male and 1 female	3 male	2 male and 1 female	3 male	3 male	1 male and 2 female
Age	64, 46, and 50	60, 49, 60	64, 48, and 62	35, 56, 47	49, 42, 58	58 and 57, 38

water and harvested rainwater also provide additional water sources. Analysis of the data from DWIR (2019) indicated that there are 1,521 natural springs in Southern Shan State, with approximately 80% of these located in Taunggyi District. The 1,206 natural springs found in Taunggyi District form an important part of the local water supply (Figure 17.4). The number of natural springs and the volume of discharge from springs in each township are presented in Table 17.2.

Table 17.2 Number of springs and volume of discharge from natural springs in Taunggyi District in 2019

District name	Township	No. of springs	Spring discharge (m^3/s)	
			For rainy season	For dry season
Taunggyi	Taunggyi	76	0.011	0.0003
	Pindaya	12	0.283	0.213
	Nyaung Shwe	29	0.386	0.217
	Pekhon	103	0.151	0.107
	Lawksawk	113	35.212	41.283
	Pinlaung	467	511.119	249.188
	Hsihseng	27	0.765	0.708
	Hopong	175	0.004	0.002
	Kalaw	104	9,471.136	6,844.748
	Ywangan	100	0.142	0.085
Total		1,206	10,019.209	7,136.551

Source: Data from Directorate of Water Resources and Improvement of River System (DWIR), Taunggyi, Shan State, Myanmar.

In Taunggyi District, the largest number of natural springs can be found in Pinlaung Township ($n = 467$) and the lowest number is found in Pindaya Township ($n = 12$). For all springs, the discharge in the rainy season is higher than in the dry season, with some large ranges in seasonal discharge. The volume of discharge is not always directly related to the number of springs in an area, for example, Kalaw Township ($n = 104$ springs) receives the highest amount of water discharge ($9471.136 \, m^3/s$) from natural springs, but Ywangan ($n = 100$ springs) receives only $0.142 \, m^3/s$.

Groundwater in rural water supply, Myanmar 323

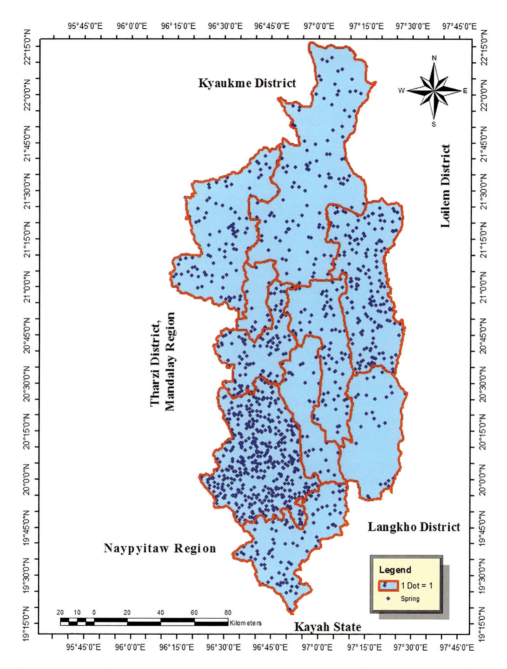

Figure 17.4 Distribution of natural springs source in Taunggyi District, Southern Shan State, Myanmar in 2019. The red line boundaries indicate the Township boundaries. (Based on Data from DWIR and GIS Method was used to portray the distribution of springs within Taunggyi District.)

324 Groundwater for Sustainable Livelihoods

17.4.2 Case studies

Six case studies are presented, which summarize the results from the key informant interviews and provide details on the role of groundwater in rural water supply in Taunggyi District and the challenges the local communities are facing. Table 17.3 summarizes the rural communities, groundwater sources, and limitations to water supply, with further details for each case study provided in the relevant section. The case studies have a range of population size and distance to water sources. This results in different requirements in terms of number of water sources and limitations of the water supply systems. Community perceptions of water quality, and how it influences decisions on water use, is also an important consideration for water supply management.

17.4.2.1   Case study 1: Loikun Village, Kakku Village Tract, Taunggyi Township

The main water supply for Loikun Village is from two springs. In addition, two dug wells supplement the spring water supply. Pipelines transport the spring water from both springs to two separate surface concrete tanks located in the middle of the village. From there, the spring water is distributed via a small pipeline system to each household. In addition, Yakan Spring is located 1,609.3 m from the village, is visually clear and therefore used for drinking water. Kyaungpan Spring is located 4,828.032 m from the village and is visually turbid. Turbidity increases further in the rainy season, therefore the local community has decided this is not suitable for drinking and instead it is used for all domestic purposes and watering house gardens. The two dug wells, approximately 1.83 and 3.66 m deep, respectively, are only used for drinking water. The villagers understand that the water in this area is alkaline due to being located in limestone region, although they are only able to visually assess the water quality routinely.

The main challenge that the community is currently facing is the decrease in groundwater discharge year after year, especially in the dry season due to forest depletion around the spring. Loikun Village totally depends heavily on these two springs, with the two wells only providing supplementary supply. Surrounding villages also use these springs. Increasing population increases the pressure on the water supply system (Figures 17.5 and 17.6).

17.4.2.2   Case study 2: Boung Kyaungnar Village, Naung Pi Village Tract, Pinlaung Township

The highest number of springs ($n = 467$) in Taunggyi District is found in Pinlaung Township. Therefore, almost all the village tracts depend on spring water for water supply, especially in the dry season.

Boung Kyaungnar village depends mainly on spring water, used for drinking and domestic purposes. The spring is located 1,609.3 m from the village and is transported via a pipeline system to surface storage tanks. Small pipelines from the tanks supply each household. At household level, spring water is collected with a small tank for domestic purposes. The spring discharge decreases in the dry season and a tube well

Table 17.3 Summary of results from key informant interviews regarding rural water supply within Taunggyi District, Southern Shan State

Case study	Loikun Village	Boung Kyaungnar Village	Nwng Mun Village	Taung Kham Village	Nan Kun Village and Naungya Saing Village	Inn Nge Village
Population household	550 pop/ 212 HH	200 pop/ 50 HH	9643 pop/ 1607 HH	492 pop/ 120 HH	500 and 400 pop 50 and 104 HH	2,182 pop/ 492 HH
Sources of water	2 springs 2 dug wells	1 spring 1 tube well	1 spring 70 dug wells 100 tube well	1 spring 1 dug well	2 spring 80 dug wells	1 spring
Distance to water resource	1,609.3– 4,828.0 m	1,609.3 m	11,265.4 m (from spring)	11,265.4 m (from spring)	1,207.0 m	16,093.4 m
Water uses	Domestic, drinking, agriculture house gardening	Domestic, drinking	Domestic, drinking, agriculture	Domestic, drinking from well	Domestic, drinking, agriculture	Domestic, drinking, agriculture
Water supply Perception on water quality	Pipeline system Turbid in one spring, Alkaline	Pipeline system Clean	Pipeline system Turbidity	Pipeline system High turbidity from spring water, organic matter and sediment in rainy season	Pipeline system Clean (already tested) for spring Well water is more alkaline	Pipeline system Clear and clean (visually)
Limitation	Decreasing amount of water discharge	Infrastructure (storage capacity)	Water quality (turbid, organic matter)/ distance	Water quality (turbidity, organic matters matters)/ distance	Still no problem	Water supply system due the distance between source of water and village

Source: Key Informant Interview Survey to individual Community Leader in 2020.

Figure 17.5 Yaykan spring is one mile from the village (a), the water is transported via a pipeline and is used for a domestic purpose (b) as well as for agriculture via a small channel (c). (Photos taken by the authors in Field Observation, 2019.)

Figure 17.6 Water from Yakan and Kyaungpan springs is collected and stored in surface concrete tanks (a), from the tank, water is distributed via a small pipeline system to households (b), the local people use the water for drinking and domestic purposes (b). (Photos taken by the authors in Field Observation, 2019.)

Figure 17.7 Natural spring sources where water comes from the crack (a), and rural water supply in Boung Kyaungnar Village, Pinlaung Township where the spring water is stored at concrete tank and is delivered to the village with pipelines. In the figure, twenty-four pipelines are delivered to the village and are extended to one or more household levels from one pipeline (b), and household-level water collection concrete tank (c). (Photo taken by the authors in Field Observation, 2019.)

provides an additional drinking water source if required. In the wet season, rainwater harvesting is used for domestic purposes and agriculture.

Storage capacity is an issue for water supply in Boung Kyaungnar Village. The storage tanks do not provide sufficient capacity to supply the whole village throughout the dry season. It is important to increase the storage capacity to increase the water security of the village. As part of upgrading the water supply system, it was suggested in the interviews that the storage concrete tanks should be placed an equal distance from each household, where possible (Figure 17.7).

17.4.2.3 *Case study 3: Nwng Mun Village, Nwng Mun Village Tract, Hsihseng Township*

Nwng Mun Village is one of the biggest villages in Hsihseng Township. It is situated in the Northernmost part of Hsihseng Township and is famous for its special rice species. Previously, seventy dug wells (2–3 m deep) provided the local water supply in Nwng Mun Village. However, since 2016 the water supply from dug wells has failed to meet demand, due to decreasing groundwater levels and an increase in water demand from a growing population. Therefore, one hundred tube wells are now also used to provide water supply, with each tube well supplying ten or more households, but these run dry in the dry season. Tube wells and purified drinking water or bottled water are used for drinking.

From 2018, a pipeline has provided an additional source from Naung Loi Spring, located 11,265.4 m from the village tract. Spring water is collected in a concrete storage tank and then distributed to tap stands, which supply six households each. The charge for water supply is 100 Myanmar Kyats per unit (which is equal to $1\,m^3$). Currently, the spring sustains the village throughout the year although spring discharge decreases in the dry season.

The spring water is only used for domestic purposes, especially for bathing and washing because it is turbid. Spring water is also used to supplement agricultural water supply, which is also sourced from nearby streams and lakes. The main water supply issue in this village is regular blocking of the pipeline system used to transport the spring water to the village, due to sediment and organic disposal. The pipeline is difficult to maintain due to the distance between spring and village (11.3 km) (Figures 17.8 and 17.9).

17.4.2.4 Case study 4: Taung Kham Village, Kyaukgu Village Tract, Indaw Sub-township, Lawksauk Township

In Lawksauk Township, Taung Kham Village falls on the mountain area which is located at an altitude of approximately 1,700 m above sea level. Therefore, water supply is difficult to compare with other villages within this township because it is located in a different environment, which brings different challenges and opportunities. Groundwater is difficult to obtain; only one tube well (about 6.1 m deep) exists in the foothills of the mountain, which is used for drinking water as well as domestic purposes.

In addition to the tube well, Mongton Spring (11,265.4 m from the village) also provides water supply via a pipeline and is stored in a surface concrete tank. Water supply is piped to households from the storage tank.

The spring water is slightly alkaline and more turbid than the well water, especially in the rainy season due to sedimentation and organic debris. Turbidity increases following heavy rainfall. Therefore, the community uses the spring water for domestic purposes only. The wells are used for the drinking water supply (Figure 17.10).

Water supply issues in Taung Kham Village, include the limited capacity of the storage tank and blocking of the spring water pipeline with sediment and organic matter in the rainy season. Increased storage capacity and regular maintenance of the pipeline system are needed to ensure a sufficient water supply for every household. The pipeline is difficult to maintain due to its length and topography (Figure 17.11).

17.4.2.5 Case study 5: Naungya Saing Village, Kone Kying Village Tract and Nan Kun Village, Paung Lin Village Tract, Hopong Township

The main water source of Naungya Seing Village is Loikar Mong Spring, which is located approximately 1,207 m away. This spring is also shared with other villages for water supply, the spring water is distributed by a pipeline and is used for drinking, domestic and agricultural purposes.

Figure 17.8 Different types of well in Nwng Mun Village domestic dug well (a), dug well, under the tree, with concrete wall (b), and tube well in a house compound (c). (Photo taken by the authors in Field Observation, 2019.)

Villagers stated that they test the spring water quality and the results indicate that it is acceptable for drinking. The spring is protected from animals by fencing. Although the output of water has decreased in recent years, the yield currently still supports the village sufficiently throughout the year. There is also one tube well in Naungya Saing Village, used as a reserve water source, but the quality is more alkaline compared with the spring water.

Figure 17.9 Concrete surface storage tank where water is collected from the Naung Loi Spring for water supply (a), connected with pipeline to water tap, distributing to every six households (b), and an individual household tap (c). (Photos taken by the authors in Field Observations, 2019 and 2020.)

Figure 17.10 Mongton spring, source of spring water which is originated at the middle of the forest (a), trapping spring water at small reservoir, water is turbid (b), and surface water storage concrete tank for water supply which is connected with small reservoir (c). (Photos taken by the authors in Field Observations, 2019 and 2020.)

Figure 17.11 Only one dug well for drinking water purposes (a), household-level water supply (b), water quality is degraded with organic decomposition (c). (Photos taken by the authors in Field Observation, 2019 and 2020.)

In Nan Kun Village, there are about eighty dug wells approximately 4.6 m deep. The local people mostly use well water for drinking and domestic purposes. Some of the wells are shared by two or three households. Currently, the yield from the wells is still enough to supply the village, although the groundwater level has decreased over the past 20 years and there are concerns for future water security.

Ganing reservoir is a small reservoir which is only used for agricultural water supply, about 1.6 km from the village. The reservoir is fed by small mountain streams and is connected to the cultivation area by a channel. The amount of water supply from the reservoir is also decreasing over time (Figures 17.12 and 17.13).

17.4.2.6 Case study 6: Inn Nge Village, Inn Nge Village Tract, Pindaya Township

The main water resource of this village is a spring located approximately 16 km from the village. Over the last 30 years, the village has depended on only this spring, which originates from within a forest on a high mountain. The spring water is collected in

Figure 17.12 Source of the Loikar Mong spring water (a), surface concrete storage tank (b), at the individual household tap (c). (Photos taken by the authors in Field Observations, 2019 and 2020.)

Figure 17.13 Different types of dug wells for household level, a good condition dug well (a) dug well is shared with three households (b), fair condition of dug well. (Photos taken by the authors in Field Observations, 2019 and 2020.)

1.2 m² shaped concrete tanks located 5 km from the village and is then transported to a circular concrete storage tank, 30 ft in radius, located at the entrance of the village. From this place, the water is distributed throughout the village by a pipeline system, with between 1 and 10 households sharing access to a tap. For households with large gardens, there is a charge of 100 Myanmar Kyats per unit (which is equal to 1 m³) to fund the maintenance of the water supply system (Figure 17.14).

Figure 17.14 Spring water near the source (a), the water is trapped with 4 ft² shaped concrete tanks between water source and storage tank (b), at concrete storage tank which is 30 ft radius. (Photos taken by the authors in Field Observations, 2019 and 2020.)

Figure 17.15 Water supply system at domestic faucet where water distribute from storage tank (a), household-level water storage system (b), household-level water storage system. (Photos taken by the authors in Field Observations, 2019 and 2020.)

In terms of water quality, the spring is regarded as a good quality because the visual appearance is clear. It is used for drinking and domestic purposes. Currently, despite a decrease in discharge in the dry season, the spring still provides sufficient water supply throughout the year (Figure 17.15).

17.5 OVERALL DISCUSSION

The case studies, incorporating the key informant interview results confirm that natural springs, dug wells, and tube wells are important for the local water supply in rural Taunggyi District. Spring water is typically piped from the source to the village and is stored in tanks and distributed to households. Groundwater is used for drinking, domestic, and agricultural purposes and therefore supports the community in many aspects. This not only supports progress towards SDG 6, but also other SDGs focused on ending poverty, zero hunger, and improving health.

Spring discharge and groundwater levels are higher in the wet season than in the dry season, but springs and some wells continue to provide water supply throughout the dry season. This is important, especially as seasonal rivers run dry and rainwater harvesting ceases until the following wet season. The stored water supply from rainwater harvesting can also continue to provide water into the dry season. During the wet season, rainwater harvesting, and surface water sources including ponds, lakes, and reservoirs all supplement groundwater supply for domestic and agricultural use. In village tracts with a small spring discharge, rainwater harvesting and surface water sources are more heavily relied on, for example, in Nyaung Shwe Township and Hsihseng Township (Table 17.2).

17.6 CONCLUSION

Groundwater is a vital resource for rural water supply in Taunggyi District, with many communities depending on springs and wells all year round. These sources will continue to be important for water security while facing pressures from population growth and climate change. To ensure the long-term sustainability of water supply in rural Myanmar, effective water resource management will be important. This should consider both the quantity and quality of available sources and the capacity to meet future demand and cope with climate change. Visual assessment of water quality is important for the communities, who allocate more turbid water sources for domestic and agricultural use and reserve other sources for drinking, where possible. There are several limitations to current rural water supply systems, including decreasing spring discharge, insufficient storage capacity, and water quality issues with turbidity and organic debris that reduce quality and can block long pipelines. It is important that these issues are addressed, in order to maintain water supply and meet demand. Increasing storage capacity, developing new sources, and maintaining existing pipelines are all options that should be considered as part of integrated water resources management to improve the security of rural water supply.

REFERENCES

Bender, F. (1983) *Geology of Burma*. Beitrage Zur Regionalen Geologie Dererde, p. 293. ISBN 978-3-443-11016-1.
Carrard, N. (2019) Groundwater as a source of drinking water in Southeast Asia and the pacific: a multi-country review of current reliance and resource concerns.
Department of Population, Myanmar (2017) The 2014 Myanmar population and housing census, Ministry of Labour, Immigration and Population. Taunggyi Township Report, SHAN STATE, TAUNGGYI DISTRICT.

Department of Population, Myanmar (2021) The Republic of the Union of Myanmar, Ministry of Labour, Immigration and Population. https://www.dop.gov.mm/en (Accessed 14.3.2021).

DWIR (Directorate of Water Resources and Improvement of River System) (2021) Taunggyi, Shan State, Myanmar.

FAO (2018). Myanmar country indicators - FAOSTAT [WWW Document]. http://www.fao.org/faostat/en/#country/28. (Accessed 15.3.2021).

GAD (General Administrative Department) (2016) Reports from facts about each and individual township, Taunggyi District, Shan State, Myanmar.

Kattelus, M., Rahaman, M.M., Varis, O. (2014) Myanmar under reform: emerging pressures on water, energy and food security. *Natural Resources Forum* 38, 85–98. doi:10.1111/1477-8947.12032.

Pavelic, P., Johnston, R., McCartney, M., Lacombe, G., Sellamuttu, S.S. (2018) *Groundwater Resources in the Dry Zone of Myanmar: A Review of Current Knowledge.* Springer, Singapore, pp. 695–705. doi:10.1007/978-981-10-3889-1_41.

Sakai, H., et al. (2013) Quality of source water and drinking water in urban areas of Myanmar. *The Scientific World Journal* 2013. Article ID 854261.

Smedley, P.L. (2005) Arsenic Occurrence in East and South Asia.

Snoad, C. (2017) Water Access Analysis, South Shan, Myanmar, Project Report, part of the Livelihood and Food Security Trust (LIFT) funded Making Vegetable Markets Work (MVMW) project.

Tun, T.N. (2003) Arsenic contamination of water sources in rural Myanmar. In: *29th WEDC International Conference.* Towards the Millennium Development Goals. Abuja, Nigeria, pp. 219–221.

UNICEF and WHO (2019) Progress on household drinking water, sanitation and hygiene 2000–2017, Special focus on inequalities. New York.

Viossanges, M., Johnston, R., Drury, L. (2017) Ayeyarwady state of the basin assessment (SOBA) 2A: groundwater resources. International Water Management Institute (IWMI).

Zaw, T., Than, M.M. (2017) Climate change and groundwater resources in Myanmar. *Journal of Groundwater Science and Engineering* 5, 59–66.

Chapter 18

Groundwater-driven paddy farming in West Bengal

How a smallholder-unfriendly farm power policy affects livelihoods of farmers

M. Shah, T. Shah, and S. Daschowdhury
International Water Management Institute

CONTENTS

18.1 Background 337
18.2 Introduction... 338
18.3 Literature review... 339
18.4 Analytical background for the pilot 340
18.5 Noises and limitations in the experiment 341
18.6 Results 342
    18.6.1 Power consumption, profits, and prices.. 342
    18.6.2 Understanding the dynamics being water market players................... 343
        18.6.2.1 Collusion for price setting.. 343
        18.6.2.2 Belief in stickiness of prices 343
        18.6.2.3 Informal agreement on command area................................. 344
        18.6.2.4 Dependence for other services 344
    18.6.3 Highly oligopolistic market structure.. 344
    18.6.4 Billing cycle and late payment surcharges 344
18.7 Conclusion... 345
Note.. 346
References... 346

18.1 BACKGROUND

Sitting on one of the world's best alluvial aquifers, the Indian State of West Bengal is predominantly water abundant. It receives an annual rainfall of 1,500–2,000 mm and has an annual replenishable groundwater resource of 30 billion cubic meters (BCM), of which, 11 BCM is drafted for various purposes (CGWB, 2017). With average cropping intensity of 176%, West Bengal is the largest rice-producing state in the country (ICAR, 2017). There are about 7 million farm families in the State, of which 96% are small and marginal farmers, the average landholding in the State being 0.77 ha (GoWB, 2020). Almost all the farmers grow paddy at least in one season, while many of them grow it in two or three seasons – *Aus* (autumn), *Aman* (winter), and *Boro* (summer), mainly because it ensures their food security throughout the year. *Boro* paddy which requires intensive irrigation but ensures high and certain yield continues to be an important harvest for most farmers in the State.

DOI: 10.1201/9781003024101-18

338　Groundwater for Sustainable Livelihoods

As per the last Minor Irrigation Census 2013–2014, the State had about 4,29,000 groundwater extraction structures, the majority of which were shallow tube wells (STWs) run on electricity or diesel or both (MoWR, 2017). These tube wells were responsible for transforming the State from a perennial rice-deficit to a rice-surplus state after the 1980s (Rogaly et al., 1999). By providing reliable irrigation at the time of need, these groundwater extraction mechanisms have boosted the farm productivity and income of farmers in the State, especially through the additional summer crop of *Boro* paddy. Most of these are owned privately, and through informal water markets have been able to provide benefits of irrigation even to the poorest farmers and tenants and have come to the rescue of the government to alleviate social and economic inequities.

18.2 INTRODUCTION

Until rural electrification picked up pace in the late 1990s, West Bengal's rice economy was held to ransom by high and soaring diesel prices which made *Boro* rice irrigation extremely costly especially for water buyers condemned to cope with oligopolistic water markets. Electricity-run STWs gradually increased in number in the State as rural electrification picked up pace in the 1990s. West Bengal State Electricity Distribution Company (WBSEDCL), formerly West Bengal Electricity Board, charged tube well owners high flat tariff until 2008. To cover their high fixed costs, electric submersible owners competed fiercely to sell more, and in the process, offered their farmer–clients high-quality irrigation service at a lower price compared to diesel STW owners. The pump density, however, stagnated or even declined in 2000s, as the electricity utility started charging full cost of additional infrastructure needed to set up the connection, which was not affordable for most farmers.

In 2008, WBSEDCL made two major changes: (a) it liberalized granting of electricity connection for tube wells to spare farmers from the rising diesel cost and (b) replaced flat tariff by a Time-of-Day (ToD) metered tariff for all farm connections at rates approaching commercial tariff. Under flat-tariff regime, buyers used their strong bargaining power to secure lower price, deferred payment facilities, etc. which had profound redistributive effects since the water buyers are often small and marginal farmers belonging to lower castes. But with the consumption-based billing of electric tube wells at near-commercial rates, water markets hardened. Now that water sellers were no longer under pressure to sell, they began to experience and exercise stronger bargaining power with water buyers in this new sellers' water market. Prices for irrigation services rose overnight. Earlier, STW owners happily sold water for a fixed rate for the whole *Boro* season. Now, pump owners started demanding land for lease at fixed rates during *Boro* season in exchange for irrigation in other seasons as their gains were much higher in cultivating leased land than in selling irrigation for it. This practice is driving small and marginal farmers, especially tenants, out of agriculture and has emerged as a threat to their livelihoods and food security.

In 2011, when the State made it easier to apply for a farm connection by abolishing certain permit requirements, pump density increased in villages. The policy had the potential to transform oligopolistic into a competitive water market as barriers to entry lowered for irrigation sellers. But the parallel policy of metering connections at high tariffs continued to make the market oligopolistic. Hence, just increasing pump

Groundwater and paddy farming in W Bengal   339

density without changing the 2008 tariff structure could not unleash the buyers' market in the State. Moreover, owning an electric STW remains elusive for farmers with small, fragmented parcels and tenants without own land. These are the most vulnerable since they remain permanently dependent on purchased irrigation.

In this chapter, we argue that there is an electricity pricing option, which can cover most or all of the cost of farm power supply and yet recreates the pro-poor character of West Bengal's water markets during the flat-tariff era. It will explore the effects of an alternate regime of a flat-cum-metered tariff structure on water market dynamics; and on irrigation service providers and small farmers, who purchase irrigation services from them. Can such a tariff structure for farm power enhance bargaining power of small and marginal farmers, provide them access to groundwater for irrigation at affordable prices, and ensure profitable and sustainable agriculture for them? We bring some insights from a pilot experiment conducted in Manoharpur village of Birbhum district, West Bengal.

18.3   LITERATURE REVIEW

Like majority farmers across the country, especially since the mid-1980s, the farmers in West Bengal have also come to depend heavily on groundwater for irrigation throughout the year. When Green Revolution took off in the Northern states post-independence, West Bengal was still struggling with its food security issues. Later in the 1980s, the State saw a widespread increase in agricultural production owing to the diesel STWs, which spread swiftly across the region. It witnessed a 6% annual growth rate in agriculture during the decade on the back of rapid expansion in STW-driven irrigation of pre-summer *Boro* rice crop. However as diesel prices began their ascent in the 1990s squeezing the profitability of *Boro* cultivation, the slowdown of West Bengal's agrarian ascent began with growth rate decelerating to 1.2%–2.0% per year (Sarkar, 2006). Mukherji et al. (2012) estimated that during 2000–2008, the index of cost of labor and fertilizer went up from 100 to 136 and 115, respectively, while that for irrigation costs increased from 100 to 223 at 1999–2000 constant prices, a direct result of farmers' dependence on expensive diesel for pumping groundwater and low rates of rural electrification.

Despite the rising cost of irrigation, West Bengal witnessed growth of vibrant and pervasive water markets just like the rest of South Asia, especially in areas not serviced by government canals. The pump owners, who had enough spare capacity to pump water after irrigating their own fields, sold irrigation services to their neighbors who were willing to cover the variable costs of energy (diesel or electricity) and make some contribution to the overhead. Mukherji (2007) found that smallholders benefitted in the informal water markets not only as water buyers but also, in several cases, as entrepreneurial pump owners. She found that, on an average, 77% of all the water pumped and 69% of the area irrigated by any groundwater extraction system was for the benefit of the buyers. The shallow aquifers in the region ensured a round-the-year water availability and helped boom the irrigation service markets and the agrarian economy through an additional summer rice crop. However, the energy policies in the State created several bottlenecks to transform the irrigation service market into a more competitive market in favor of smallholders, as explained in the previous section.

The ToD metered tariff structure provided water sellers with higher bargaining power to lease land from smaller farmers for summer paddy. Daschowdhury (2015)

340 Groundwater for Sustainable Livelihoods

observed that pump owners increasingly preferred leasing in land only for *Boro* rice cultivation at a fixed rental of 3 quintals of paddy or a cash rental of ₹ 5,000 per acre. A pump owner can earn ₹ 3,750 by selling irrigation to an acre of *Boro* paddy but can make a net surplus of ₹ 10,560–26,000 by leasing in an acre on a fixed rental and accounting for all other input costs. Daschowdhury (2015) also reported that many pump owners committed *Amon* irrigation only on the condition that the buyer leases out all or a portion of his land to him for *Boro* cultivation.

Shah and Daschowdhury (2017) also noted that water market dynamics were slightly different in villages where farmers were issued temporary connections for *Boro* season for 120 days, for which they are required to pay an advance amount as security deposit, which is adjusted against the bill generated at the end of the season. They reported that pump owners were apprehensive of their deposit being reimbursed, hence, attempted to sell water to more buyers and cover the advance paid to the utility. It is also interesting to note that not all blocks in the State have temporary farm connection options; villages with high pump density are not issued temporary connections, and pump owners have to bear the cost of temporary poles and cables – something that small farmers might not be able to afford.

18.4 ANALYTICAL BACKGROUND FOR THE PILOT

A year-long pilot was launched in Manoharpur village to test our hypothesis that West Bengal can maximize the benefits to the farmers equitably by adopting a flat-cum-metered tariff structure, where a tube well owner is required to pay a fixed tariff per month per horsepower of the attached motor in addition to a lower consumption linked rate (metered tariff).

We expected the following outcomes in response to the pilot:

i.   Pump owners will be encouraged to sell enough more water to cover the fixed flat component of the tariff, moving the market towards being more competitive, while also avoiding unnecessary extraction of groundwater, as they would no longer be burdened by excessive flat tariff (as before 2008).

ii. Low-metered tariff will help water buyers negotiate better rates for the irrigation services, which clubbed with (i) will cause irrigation cost to lower.

iii. With better bargaining power, water buyers in the market will be able to influence terms of irrigation, including resisting leasing-in of land in *Boro* season by STW owners or negotiating more favorable terms of leasing.

We posited that such a competitive irrigation water market is likely to ensure an equitable benefit of the farm power policy to all the farmers, help utilities reduce power theft and other illegal practices and boost the round-the-year rice cultivation in the State, ensuring resilient livelihoods for small and marginal farmers and optimal utilization of groundwater.

Manoharpur village, located around 18 km away from the town of Shantiniketan in Birbhum district of West Bengal, is home to 273 families, most of whom are dependent on farming as a source of income. The village has nineteen electric STWs and the average price paid per unit of farm power is ₹ 5.18. The average landholding of a water buyer is 1 acre while that of a pump owner in the village is 4.3 acres. None of the STW owners of the village apply for temporary farm connection in *Boro* season, hence, the pump density is constant in all seasons.

All pump owners of Manoharpur participated in the experiment, where the monthly flat rate was set using analysis of historical data. The average units consumed by the pumps in the village were used to set the cutoff point. Additional units consumed were subsidized by 70% during the experiment period, mimicking a high flat rate-low-metered tariff structure. The experiment began at the onset of autumn paddy in July 2017 and ended at the end of summer paddy season in May 2018. Based on the bill generated by the utility every month, refund for each pump owner was calculated, and excessive consumption beyond the set flat tariff was reimbursed.

18.5 NOISES AND LIMITATIONS IN THE EXPERIMENT

Baseline data from Manoharpur showed that almost half the area cultivated by pump owners during *Boro* season was leased-in from water buyers. During the experiment, it would have made economic sense for pump owners to maximize their pump usage, which could be done by either leasing in more land from their water buyers to accrue personal gains from subsidized power or selling irrigation services to larger area of existing water buyers or by getting new buyers to purchase irrigation from them. The first and the economically-more sound option for the pump owners, however, could unfold as a rumor of Operation Barga[1] re-implementation spread amidst local election campaigns in the area. With the fear of losing their land to lessors, only a few landowners leased out their land in the subsequent *Boro* season. Additionally, all pump owners have designated sellers and defined area of operation with an informal agreement within the group to not encroach on area of another pump owner, making the other options infeasible for them to maximize pump usage.

Power pilferage has been a challenge to WBSEDCL, especially as more and more pump owners started drawing power illegally from transmission cables to run their pumps to keep their power bills affordable. In Manoharpur as well, calculations show that only about half the power consumed is billed to the pump owners on an average (refer Table 18.2). But halfway through the pilot, the utility decided to curb power theft and converted all overhead transmission cables to tamper-proof cables. Both irrigation sellers and buyers who benefitted from power pilferage lost out on accruing them during the study year.

The study relied on data from bills generated by the utility to calculate pump usage apart from primary data collected through survey. But we could not depend on the same for triangulation of pump usage data as power consumption and pump-use patterns would have been influenced by the inability to pilfer power during the study year.

While the study tries to capture changes in economic behavior of water market players, it is important to note that all participants were aware of the duration of the pilot being 1 year, which also impeded permanent changes in the groundwater market. Cognizant of the noise and limitations of the pilot, we made several observations and collected data as planned during the study to get deeper insights to test our hypothesis and gather the necessary evidence to argue for sustainable, profitable, and resilient livelihoods for the small and marginal farmers in the State.

342 Groundwater for Sustainable Livelihoods

18.6 RESULTS

18.6.1 Power consumption, profits, and prices

Power consumption data for *Boro* season at the end of the pilot shows that on an average, the pump owners' power consumption increased in the study year (see Table 18.1) while the area irrigated by their pumps did not increase proportionately (Table 18.2). This could be the direct result of installation of tamper-proof insulation cables in the village replacing the older low-tension transmission cables. Table 18.1 also shows the net profit made by pump owners, who are basically levying cost of their own irrigation from their water buyers by charging them 60% more than their total cost of service delivery. There-fore, an average farmer in Manoharpur pays ₹ 2,726 per acre more than the most com-petitive price for irrigating their field – a fairly high amount for a farmer growing paddy for subsistence – while their counterparts in other states have access to free irrigation.

The cartel of water sellers in the area meet at the end of a season and decide a uniform price to be charged from all their water buyers. Even though they receive their bills monthly, it was reported by pump owners that the actual consumption is not reflected in the monthly bills, which are generated using historical data and adjusted at the end of the season when the utility sends out personnel to read meters. It is also when the bill payment camps are set up in the villages, and irrigation service provid-ers collect dues from their water buyers. To pay their dues, these water buyers must sell their harvest at the prevailing market price and forgo the profit they can get by postponing the sale of harvest. Thus, changing the farm power policy to make it more pro-poor can not only ensure that small farmers are able to get competitive prices for irrigation but also maximize their profits from sale of harvest.

Table 18.1 Comparison of pricing and profit in the base year and study year

Parameter	2016–2017	2017–2018
Average units of power consumed (kWh)	3,503	4,588
Cost of power at ₹ 5.18/unit	₹ 18,146	₹ 23,765
Average pump maintenance cost in *Boro* season	₹ 5,156	₹ 3,495
Operating cost (₹ 3,000/month)	₹ 15,000	₹ 15,000
Total cost of service delivery over 5 months	₹ 38,300	₹ 42,260
Buyers' area irrigated by one pump (acre)	10	9.8
Break-even price per acre (excluding own/ leased area)	₹ 3,830	₹ 4,350
Actual price charged (seller-reported)	₹ 3,815	₹ 4,350
Actual price charged (water buyer-reported)	₹ 4,035	₹ 4,495
Cost of irrigating own/ leased-in land	₹ 7,737	₹ 10,059
Net profit	25.5%	28.05%

Table 18.2 Comparison of estimated power consumed and billed

	2016–2017	2017–2018
Average area irrigated by a pump (including own land)	14.4	15.5
Average pump running hours (at 70 hours/acre)	1,008	1,086
Power consumed by 6.5 HP pump for given hours	4,888	5,269
Units consumed at 75% efficiency	6,501	7,008
Percentage of actual consumption billed	53.9%	%

18.6.2 Understanding the dynamics being water market players

Noises in the experiment and billing cycle issue resulted in the pilot not being able to achieve the desired results. However, it revealed some key characteristics of the irrigation market in the study village.

18.6.2.1 Collusion for price setting

It is not only the farmers of Manoharpur who stick together to fix the price of irrigation for a particular season but the farmers of all adjacent villages, who meet and decide a common price for irrigation every season. Hence, despite offering subsidies on power in one village, the pump owners did not lower prices and kept it comparable to other villages in the area.

18.6.2.2 Belief in stickiness of prices

Knowing that the subsidy under the pilot would only last 1 year and without any anticipation of such a tariff structure being implemented by the utility, the pump owners did not lower prices even during the study period. Had they done so; they would open themselves up for negotiation from water buyers in the following seasons as prices in this market are sticky. To make the most of the subsidy, instead, they filled in farm ponds, expanded their cropped area in one or more seasons, but rarely chose to lower prices to attract more buyers (Figure 18.1).

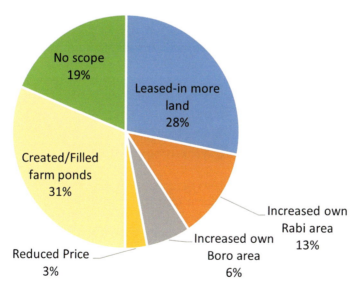

Figure 18.1 Strategies of pump owners to maximize gains from the subsidy under the pilot.

18.6.2.3 Informal agreement on command area

Purchasing a pump and getting farm connection requires a substantial investment, which was even higher a decade ago. No pump owner would invest in one without guaranteed command area for selling water. Thus, they have an informal agreement not to encroach into each other's designated area set during pump installation itself and they continue to follow that even today.

18.6.2.4 Dependence for other services

Many pump owners also own other agricultural equipment like tractors, threshers, etc. which gives them more power over their water buyers, who require these machines in addition to irrigation services – mostly all at credit till they sell off the products or in exchange for crop. Based on the study of interlinked factor markets in agricultural production in Eastern India, Bardhan (1990) pointed out that interlinking of transactions in different markets is a very effective way for the dominant party (owner of land and capital) to charge high prices in some markets or dictate terms for a scarce resource when prices are inflexible. Swain (2000) found higher incidence of interlinked factor markets in irrigated villages compared to non-irrigated villages. Such interlocking could increase the exploitative powers of stronger elements and through the interpenetration of markets, increase the quantum of surplus extraction (Chakrabarti, 1998). Manoharpur is a case in point.

18.6.3 Highly oligopolistic market structure

The Herfindahl–Hirschman Index shows that the Manoharpur irrigation services market had a score of 806 at baseline and 740 at end-line, indicating monopolistic competition in the market.

Pump owners, who are significantly more well-off than the water buyers, have no incentive to offer competitive prices when gain from the oligopolistic market structure benefits them much more than it would by expanding their business. The smallholder water buyers will continue to be at a disadvantage until systemic changes are brought in through a pro-poor farm power policy.

18.6.4 Billing cycle and late payment surcharges

Fourteen out of the nineteen pump owners reported that they clear their electricity bills seasonally after collecting payment from their water sellers. As their monthly bills accumulate over the months, they end up paying anywhere between 15% and 30% more as of late payment penalties. This amount is added to their bills and in the end, is charged from the poor water buyers.

The utility charged high flat tariff before 2008, which reduced bargaining power of pump owners; hence, it is the least preferred option for farm power tariff for them even now. In fact, majority of them shared that lower electricity bills were the only way for them to reduce prices in this market (Figure 18.2).

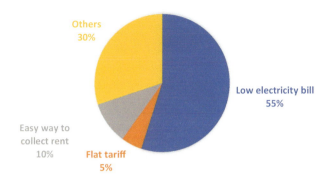

Figure 18.2 Motivating factors for pump owners to offer competitive irrigation service price to water buyers.

18.7 CONCLUSION

Shah and Chowdhury (2017), based on their exploration of *Boro* paddy situation in the State, reported multiple instances of farmers giving up *Boro* paddy cultivation altogether due to disconnected electricity connection as they were unable to pay bills and had accumulated large penalties. Small and marginal farmers, for whom *Boro* paddy is prime to food security, have been forced to migrate for work.

Data from Manoharpur show that only 2.7% of water buyers have Kisan Credit Card or any formal source of credit, compared to 60% of pump owners. Could one imagine the difficulty faced by the water buyers to pay for irrigation in case of crop failure in any season? The probability of these farmers getting caught in a vicious cycle of borrowing from moneylenders at exorbitant prices to pay their irrigation service providers so that they can get irrigation for the next season is quite high. Hence, the sustainable solution is to create policy shifts to enable affordable access to groundwater irrigation for these farmers.

The utility incurs ₹4.84 for an average unit of power sold but earns only ₹4.07 per unit from farm consumers (PFC, 2016). When temporary connections for *Boro* were more prevalent in the State, farmers paid an advance deposit upfront to the utility, which was adjusted, based on their final metered consumption. The pump owners as well as water buyers enjoyed cheaper irrigation services then.

A flat-cum-metered structure with low-metered tariff billed seasonally can be a win-win for all stakeholders involved and can create equitable opportunities for profit marking. For the utility, it assured upfront payment from pump owners for the flat component and reduces the cost of meter reading every month. It also reduces the incentive to pilfer power and hence, reduces unaccounted losses of the utility. The pump owners would not have to pay the late payment surcharges and would be motivated to sell at competitive prices to cover the cost of flat components in the tariff. It might also create space in the market for new pump owners, which will subsequently increase competition in the market.

Most importantly, a competitive water market will ensure affordable and economically feasible irrigation service for the small and marginal farmers of the State, which can go a long way in helping them lease-in more land and increase their income from agriculture. Groundwater, which pulled the State out of food insecurity, can once again be a blessing for the State's farmers and provide them with an opportunity for resilient livelihoods doing what they do best – cultivating paddy throughout the year.

NOTE

1 Launched in 1978, Operation Barga was a land-reform movement to create official record of sharecroppers with an aim to protect their rights against forceful eviction by land owners, ensure fair share of produce, and ultimately convert some to landowners.

REFERENCES

Bardhan, P.K. (1980). Interlocking factor markets and agrarian development: A review of issues. *Oxford Economic Papers*, 32(1), 82–98.

CGWB (Central Ground Water Board). (2017) *Ground Water Yearbook - India 2016–17.* New Delhi: CGWB.

Chakrabarti, D. (1998). *Agrarian backwardness and interlocking of product and factor markets in agriculture: A study of Cooch Behar district in West Bengal.* University of North Bengal. Report number 126689.

Daschowdhury, S. (2015). *Report on impact of liberalized electrification.* Unpublished report. IWMI-Tata Water Policy Program. Anand.

GoWB (Government of West Bengal). (2020). *West Bengal state portal: Agriculture.* [Online] Available from: https://wb.gov.in/departments-details.aspx?id=D170907140022669&page=Agriculture [Accessed 30 December 2020].

ICAR (Indian Council of Agricultural Research). (2018). *State specific chapter: West Bengal.* [Online] Available from: https://icar.org.in/files/state-specific/chapter/125.htm [Accessed 2 April 2019].

MoWR (Ministry of Water Resources). (2017). *5th Census of Minor Irrigation Schemes Report 2013–14.* New Delhi: MoWR.

Mukherji, A. (2007). *Political economy of groundwater markets in West Bengal, India: Evolution, extent and impacts.* PhD thesis, University of Cambridge, United Kingdom.

Mukherji, A., Shah, T. and Banerjee, P.S. (2012). Kick-starting a second green revolution in Bengal. *Economic & Political Weekly*, 47(18), 27–30.

PFC (Power Finance Corporation). (2016). *The Performance of State Power Utilities for the years 2013–14 to 2015–16.* New Delhi: PFC.

Rogaly, B., Harris-White, B. and Bose, S. (1999). *Sonar Bangla? Agricultural Growth and Agrarian Stagnation in West Bengal and Bangladesh.* New Delhi: Sage Publications.

Sarkar, A. (2006). Political economy of West Bengal: A puzzle and a hypothesis. *Economic & Political Weekly*, 41(4), 341–48.

Shah, M. and Daschowdhury, S. (2017). *Causes and consequences of decline in boro paddy area in West Bengal.* Unpublished report. IWMI-Tata Water Policy Program. Anand.

Swain, M. (2000). Agricultural development and interlocked factor markets. *Indian Journal of Agricultural Economics*, 55(3), 308–316.

Chapter 19

Assessment of options for small-scale groundwater irrigation in Lao PDR

P. Pavelic, D. Suhardiman, O. Keovilignavong, C. Clément,
J. Vinckevleugel, S.M. Bohsung, K. Xiong, L. Valee,
M. Viossanges, S. Douangsavanh, T. Sotoukee,
and K.G. Villholth
International Water Management Institute

B.R. Shivakoti
Institute for Global Environmental Strategies

K. Vongsathiane
Department of Irrigation, Vientiane, Lao PDR

CONTENTS

19.1 Introduction 347
19.2 Description of the study area 349
19.3 Research methodology 351
      19.3.1 Approach 351
      19.3.2 Methods 351
19.4 Results 353
      19.4.1 Open dugwells 353
      19.4.2 Deep boreholes 354
19.5 Discussion 356
      19.5.1 Factors contributing to successful groundwater irrigation 356
      19.5.2 Comparative exploration of the two options 358
      19.5.3 Governance considerations for scaling up 359
19.6 Conclusions 360
Acknowledgments 361
Notes 361
References 361

19.1 INTRODUCTION

Lao PDR is a landlocked country of around 7 million people situated in the heart of the Mekong region. Socioeconomically, the country's level of development is significantly lower than that of neighboring China, Thailand, and Vietnam but comparable to Myanmar and Cambodia.[1] Farming is the primary source of income of almost 80% of all households in Lao PDR. Agriculture remains largely subsistence-based and is thus of high importance for food security and livelihoods at the national level (MAF, 2012).

DOI: 10.1201/9781003024101-19

Lao PDR is also one of the fastest-growing economies in the region with an average GDP growth consistently exceeding 6% for at least the past decade.[2] Poorly developed, water-rich countries such as Lao PDR have historically paid the most attention to surface water resources, with limited consideration to groundwater. Irrigated agriculture, is critical for addressing poverty alleviation, food security, and climate change.

Although Lao PDR is well endowed with water resources ($49,000\,m^3$/capita/year in 2017[3]), rainfall is almost entirely concentrated within the wet season. Drought-like conditions characterize the extended dry season and the wet season when rainfall is poor or breaks in rainfall are long. As a result, most fields lie fallow for much of the year due to a lack of sufficient infrastructure for surface water irrigation. Irrigation would allow farmers to advance beyond a single paddy crop each year to include shorter duration, cash crops that produce multiple harvests per year and offer higher and steadier income streams (Suhardiman et al., 2016). Despite the major efforts put in irrigation development over the past three decades, only around 20% of farming households nationally have access to irrigation water, with lack of access being a major constraint to agricultural development (MAF, 2012).

Groundwater is used for domestic consumption in most villages and some towns across upland and lowland parts of the country. However, there has only been limited exploitation of groundwater for agricultural purposes to date (Pavelic et al., 2014; Vote et al., 2015). Farmers in villages across the most productive lowland areas, in particular, would benefit from agricultural water supply options to maintain and enhance their livelihoods as settlements are prone to drought and often situated remote from perennial rivers and streams where the costs of lifting and transporting water are high.

Studies have long recognized that groundwater could be more widely used for agriculture (e.g. Johnson, 1986), and help make important inroads toward meeting Lao PDR's commitments to Sustainable Development Goals such as "poverty reduction" (SDG 1) and "zero hunger" (SDG 2). However, extremely poor knowledge of the aquifer systems and limited institutional and human capacity for planning, development, and management of these resources has prevented wider use for irrigation (Pavelic et al., 2014). Groundwater potentially offers Lao smallholder farmers scope to flexibly enhance and regulate water supply on their farms to enable diversification beyond wet season paddy to improve their livelihoods and resilience. It can potentially allow dry season cash cropping, support livestock and fisheries production, and provide supplementary irrigation during the wet season or in times of drought.

This chapter focuses on evaluating and comparing the performance and scalability of two complementary groundwater irrigation options, taking into account their associated opportunities and constraints. In doing so, it also seeks to establish what role these options may have to offer for livelihood improvement in the lowlands of Lao PDR, while placing it within the context of livelihood resilience and more equitable growth.

Considerable research has emphasized the important role that groundwater may play for poverty reduction and livelihood enhancement for smallholder farmers in developing countries and particularly Sub-Saharan Africa, East Asia, and South Asia (e.g. Villholth, 2013; Wang et al., 2007; Shah et al., 2006). With irrigation, farmers generate greater agricultural production and economic surpluses that enable the expansion of assets and capital that in turn, help them to rise above poverty traps (Moench, 2003). Ensuring that livelihood benefits of groundwater irrigation are achieved relies

Small-scale groundwater irrigation in Lao 349

on getting the mix of external factors such as policies, institutions, and markets right (Mukherji and Shah, 2005). The Southeast Asia region has been little studied in this regard, with some exceptions (Kwanyuen et al., 2003) and hence this study seeks to contribute to this effort.

19.2 DESCRIPTION OF THE STUDY AREA

This study was carried out at Ekxang village, situated on the Vientiane Plain (VP) around 55 km north of Vientiane, the national capital (Figure 19.1). The poverty level for Vientiane Province, which takes in much of the VP,[4] was 5.3% in 2018/2019, considerably lower than the national average of 18.3%. Aided by steady economic growth nationally, poverty levels in rural areas have been in gradual decline over the past three decades (Lao Statistics Bureau and World Bank, 2020). The VP is one of the major "food bowls" of the country, which together with its relatively good infrastructure and accessibility, make it an appropriate focal area for this research. Within the upper (northern) zone of the VP, where Ekxang village is located, large surface water irrigation systems are generally absent and past investigations suggest good prospects for developing the groundwater resources (JICA, 1993).

Lao PDR has a tropical monsoon climate with a distinct wet season from May to October, a relatively cool dry season (winter) from November to February and a hot dry season (summer) in March and April. The district-averaged rainfall of the study area is approximately 2,400 mm/year and the potential evaporation is 800 mm/year (Department of Meteorology and Hydrology data for 2009–2015).

The Ekxang population of 1,280 people comprises 236 households made up of three ethnic groups: the majority Lao Lum (lowland Lao) (95%) together with minority Hmong (5%) and Khmu (<1%) (Suhardiman et al., 2018). About 90% of households in the village rely on agriculture for their livelihoods (Suhardiman et al., 2016). Farmers focus on lowland rice, seasonal or year-round vegetable farming (e.g. long bean, cucumber, lettuce, morning glory, and herbs) and other high-value cash crops (e.g. dragon fruit and watermelon), with some also raising livestock (cattle and poultry) and aquaculture. Rice is sold mostly to local millers. Vegetables, livestock, and fish are sold at nearby markets or to local traders. Farm sizes range from <1 ha for small farmers to >2.5 ha for large farmers.

In geological terms, the VP sits on the northern margin of the Khorat Plateau which extends over much of northeast Thailand. On the VP, the predominantly Mesozoic marine sandstones are overlain by Pliocene to recent sand and clay infill originating from fluvial transport from the surrounding uplands (Viossanges et al., 2018; Perttu et al., 2011). Highly variable thickness of the more productive alluvial layers, and patchy and uncertain presence of connate salt deposits impact on the success of drilling efforts. The Ekxang area is situated in the central part of the plain where alluvial deposits are thick and the overlying freshwater layer above brackish or saline groundwater is extensive. The village is underlain by around 30 m of alluvial sediments, with aquifer transmissivity of around 300 m²/day. Well yields can reach up to 10 L/s. Groundwater levels fluctuate annually from a depth of 3 m below ground level late in the wet season to a depth of 6 m late in the dry season (Clément et al., 2018). Groundwater is of an acceptable quality for irrigation and drinking purposes (Brindha et al., 2017).

350 Groundwater for Sustainable Livelihoods

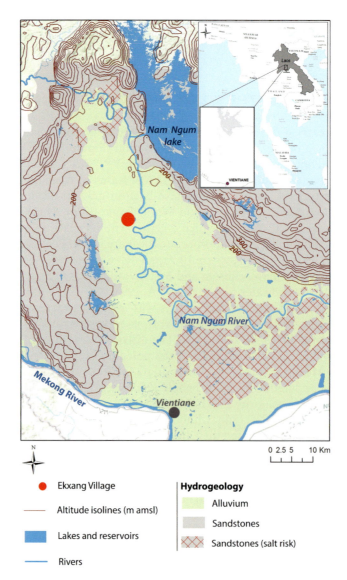

Figure 19.1 The location of Ekxang village situated on the Vientiane Plain.

The village has access to three distinct sources of water: local ephemeral rivers/streams, canal water from a small reservoir in a neighboring village (Phonthan), and shallow groundwater. Surface water is unavailable throughout the entire dry season and usually ceases by March. Groundwater is accessed by wells distributed throughout the built-up part of the village and within the farming land.

Groundwater is used throughout Ekxang village for domestic purposes. As many as 96% of households in Ekxang village own shallow, lined wells, whereas the remaining 4% depend on community wells established through donor intervention (Suhardiman et al., 2018). Groundwater is also commonly used for irrigation. The source of water for irrigation is usually from shallow lined ringwells for home gardening, and from large shallow open dugwells for cash-crop production. Water is withdrawn from ringwells using a bucket or a surface pump that may attain a pumping rate of 0.5–1 L/s. For dugwells, water is pumped using a two-wheel tractor as a generator and a mobile surface centrifugal pump, ensuring a pumping rate as high as 8–10 L/s until the capacity of the dugwell is exhausted, after which time the well is allowed to recover. Around 50% of households in the village own an open dugwell.

19.3 RESEARCH METHODOLOGY

19.3.1 Approach

This study considers two distinct types of groundwater irrigation options:

1. Open and unlined dugwells (DW), typically four-sided with side dimensions of 3–5 m and excavated in fields to depths of up to 10 m (Table 19.1). Groundwater is lifted by mobile surface pumps powered by a diesel motor fitted on a two-wheel tractor, delivering water through a 2-inch diameter pipe system to the field, where it is typically distributed through unlined furrows. They serve individual farmers by providing supplemental irrigation during the wet season and/or dry season.
2. Relatively deep boreholes (BH) around 30 m drilled and fitted with electric submersible pumps to irrigate adjacent paddy lands to support dry season cropping. Boreholes offer a more reliable irrigation water supply than shallow wells that are prone to drying out in the dry season thereby affecting crop production. This is a first-of-its-kind pilot trial established with the participation of the community and made possible through the financial assistance associated with a research project.[5] The research team worked closely with local communities and village authorities in developing the community-based groundwater irrigation trial. The trial was also supported and endorsed by the district and provincial governments.

19.3.2 Methods

The assessment of the two options was carried out independently.

Suhardiman et al. (2016, 2018) evaluated dugwells in Ekxang village through farming systems analysis comprising: (a) a socioeconomic survey of eighty randomly selected households; (b) five focus group discussions with different farmer typologies and (c) in-depth semi-structured interviews with twenty farm households. Fieldwork was carried out in 2013/2014, while planning for the borehole pilot trial was underway. The broad-based data collected by Suhardiman et al. (2016, 2018) helped to compliment the more intense, field monitoring approaches described below.

Bohsung et al. (2015) carried out a detailed evaluation of two dugwell sites over one dry season (November to February) in 2014/2015. Monitoring was done with the participation of two farmers (*a* and *b*) who grew similar areas of watermelon (*a* irrigated

Table 19.1 Overview of the two types of groundwater irrigation options practiced on the Vientiane Plain

Option / Characteristics	Shallow dugwells (DW)	Deeper boreholes (BH)
Depth (m)	4–10	32
Pump type	Diesel pumps fitted on two-wheel tractors	Electric submersible (power transmission line brought to the site)
Purpose	Cash crops in dry season and rice in wet season	Cash crops in dry season (and potentially supplemental irrigation in wet season)
Irrigated area	~1 ha	6 ha (design command area for the two borehole system)
Construction	By the landowner	Shared between the research project team and the local community
Management	Individual farmers, sometimes involving water-sharing agreements amongst farmers	Community via a water user group
Example		

Small-scale groundwater irrigation in Lao 353

1.2 ha and farmer *b* irrigated 1.3 ha). They each recorded water levels within their dug-wells and their irrigation schedule as well as the quantity of rainfall on a daily basis. Semi-structured interviews were also carried out with each farmer to gain a better understanding of their agriculture practices. The interviews also provided quantitative information on expenses and returns associated with watermelon cultivation.

Clément et al. (2018) designed, set up, and monitored the operation of the groundwater irrigation pilot trial using boreholes. The monitoring covered a detailed set of biophysical and socioeconomic parameters over two successive dry seasons in 2015/2016 and 2016/2017. It included climatic variables, groundwater levels, water quality, farming practices, system and plot-level water use, investment costs, seasonal input and output costs, groundwater user group (GWUG) establishment, and functioning. Some of the farmers participating in the trial assisted with data collection.

19.4 RESULTS

19.4.1 Open dugwells

Irrigation management differed between the two field sites which are situated less than 1 km apart. Although the fields were watered daily in both cases, farmer *a* was able to irrigate the entire crop, whereas farmer *b* was only able to irrigate about one-third of the crop due to labor shortages, hence covering the area through rotation. Pumping was only interrupted during limited heavy rainfall events. Farmer *a* irrigated an estimated 382 mm of groundwater (4,582 m^3 on 1.2 ha), whereas farmer *b* irrigated 160 mm (2,082 m^3 on 1.3 ha) (Table 19.2). Thus, water used by farmer *a* was more than twice that of *b*. The difference is likely due to different irrigation application methods rather than intrinsic differences in site conditions.

Farmer *a* irrigated within the Food and Agriculture Organization (FAO) crop water requirement for watermelon, whereas farmer *b* under-watered the crop by around half of the recommended amount. Yields from farmer *b* were not negatively affected compared to the yields from farmer *a*. Watermelon prices secured by the two farmers were different by almost 50% (LAK[6] 7,300 vs. 4,000 per kg) as farmer *a* was able to sell his produce to local traders in January, a month before farmer *b* when produce supply was higher.

Net incomes of farmer *b* for the 2014/2015 winter season studied was LAK 19 million ($US 2,300).[7] Input and operational costs associated with seeds, fertilizers, pesticides, fuel and external labor amounted to around LAK 5–6 million ($US 600–700).

Based on additional information provided from interviews with the two farmers, gross incomes for the entire dry season cultivation over the period from 2011 to 2015 were estimated to be in the range of LAK 15–45 million ($US 1,800–5,500). High short-term market price volatility, weather conditions, pests and diseases, and differences in input costs between winter and summer crops all affected farmer revenues.

The investment cost for constructing an open dugwell is around LAK 4 million ($US 490). The internal rate of return (IRR) on this investment for a conservative net income of LAK 8 million ($US 980) per annum is high (>100% within the first year, taking into account only the direct benefits). This clearly demonstrates the profitability of investments in open dugwells.

354 Groundwater for Sustainable Livelihoods

Table 19.2 Summary of the key results for both groundwater irrigation options

	Shallow dugwells (DW)		Deeper boreholes (BH)	
	Farmer a	Farmer b	2015/2016 Dry season	2016/2017 Dry season
Participating farmer(s)	1	1	4	1
Area irrigated (ha)	1.2	1.3	1.52	0.67
Irrigation method(s)	Furrow irrigation	Furrow irrigation	Furrow irrigation, with some experimentation using raised sprinkler and drip irrigation	Raised sprinkler
Crop(s) grown	Watermelon	Watermelon	Rice, watermelon, sweet corn, gourd, pumpkin, mixed vegetables and herbs	Rice, sweet corn, peanut, mixed vegetables and herbs
Water use (mm)	382	160	75–930	255–302
Water fees	Farmer pay for fuel	Farmer pays for fuel	Research project pays for water use	Water user pays electricity bill
Net incomes per farmer (Mill. LAK/$US)	-	19/$US 2,300	9.2/$US 1,100 (w/out subsidy)	2.1/260

Financial analysis for a model 0.67 ha farm with the three major crop types shows that watermelon is by far the most profitable crop per unit area, generating a 2.7-fold higher profit than paddy rice and 3.8-fold higher profit than long bean (Suhardiman et al., 2018).

19.4.2 Deep boreholes

During the first evaluation period in the dry season of 2015/2016, four farmers participated in the trial, irrigating ten crop types over a total area of 1.5 ha (Table 19.2). Pumping and some maintenance costs were subsidized by the research project for the first year, to encourage adoption. In the second evaluation period of 2016/2017, when the subsidy on pumping was removed, only one farmer who had also participated in the previous year used the system, irrigating a total area of 0.7 ha. Before the start of the second year, concerted efforts were made to involve the poorest farmers including landless farmers from Ekxang and neighboring villages. Although seasonal leasing of land was theoretically possible according to local rules, it appears likely that socio-cultural factors limited the opportunity for farmers from neighboring villages to participate (Clément et al., 2018).

Dry season cash-crop cultivation was found to generate profits (even modest profits without the subsidy in the first year). Rice cultivation, which required the heaviest water demands, was only profitable through the subsidy. Farmers cultivating a range of cash crops in 2015/2016 would have generated net profits over the dry season ranging from LAK 1 to 7 million ($US 120–860) without the subsidy. One farmer cultivating dry season rice on a 0.8 ha plot would have generated an unsubsidized loss of LAK 2 million ($US 250), but instead, made a subsidized profit of LAK 0.7 million ($US 80).

During the second evaluation period of 2016/2017, a total of seven crop types including a mix of cash crops (57%) and rice (43%) were cultivated. The net income of the farmer over the dry season was LAK 2.1 million ($US 260). This is a significant boost over the net income of LAK 4.8 million ($US 590) gained by the same farmer from the wet season production of glutinous rice cultivated over the same area. Despite its high water requirement, rice was cultivated in the dry season primarily for household consumption and not for income generation. In proportional terms, the most profitable crop was peanut,[8] a crop which the village had no previous experience in growing but was started through advice and seeds provided by agricultural extension services.

The groundwater irrigation scheme has a design area of 6 ha and the capacity to irrigate around 6–9 ha, depending on the cropping patterns and irrigation methods chosen by farmers. Financial analysis shows the system is economically viable, particularly if there is a high uptake by farmers. Assuming a 6 ha irrigated area and full adoption, the investment cost would be around LAK 18 million/ha ($US 2,200/ha), and the IRR would be as high as 45% (Clément et al., 2018). Even with 1.5 ha adoption, an IRR of 15% would be achieved. Calculated values from both scenarios exceed the 12% discount rate that multilateral organizations typically charge for irrigation project investments (Savva and Frenken, 2002).

Groundwater level monitoring showed no evidence of enhanced drawdown due to the trial. Total groundwater withdrawals from the trial were just 0.2%–1.1% of the estimated annual irrigation demand in the village.

To encourage good water management practices, the GWUG was provided with limited quantities of sprinkler and drip irrigation equipment; technologies that were new to the farmers, together with adequate training on how to use them. Farmers thus irrigated with a mix of sprinkler and drip, as well as the traditional furrow irrigation. Irrigation efficiencies, representing the gross water held in the root zone relative to evapotranspiration requirements and accounting for percolation losses, was highest for rice (95%) and furrow-irrigated pumpkin and sweetcorn (95%–96%); values well above sprinkler irrigated watermelon and small vegetables (41%–55%), and drip-irrigated watermelon (47%). Lack of experience with sprinkler and drip irrigation, compounded by the waiver on pumping costs in the first year, probably led the farmers to over-irrigate. Since furrow irrigation generally relied on the farmers being present to undertake the activity manually, the water application rate was found to be less.

The irrigation system was developed jointly with the community. The Ekxang GWUG was composed of seven farmers, including one female (Deputy Head). The participants in the trial are members of the group who have land within the command area and earn most of their income through on-farm activities. This alternative institutional setup (i.e. collective instead of individual use) provided some employment opportunities for the poorer farmers (e.g. landless, tenant farmers and laborers). There is scope for farmers who might not be able to invest in wells and pumps to join the GWUG to pursue cash-crop farming and thereby contribute toward more growth that is equitable.

356 Groundwater for Sustainable Livelihoods

Participation of farmers in this new irrigation technology was not easily forthcoming due to concerns of the GWUG on the reliability and cost of the new technology on top of other constraints related to farming such as labor shortages, soil fertility, pest attack, and market uncertainty. Despite the lack of participation, the trial served to overcome the initial doubts of the farmers, who were generally convinced that the system could function effectively. As farmers gained experience with the high-performance electric pumps, pipe distribution systems and other equipment, their overall trust in the system grew, and they came to have a sense of ownership over the system. Incrementally, the system is being more valued, particularly in drought conditions when there are no alternative water sources.

Despite the modest numbers of farmers involved in the trial, which is not totally unexpected in this new setting, follow-up observations of the site after the trial in 2018 revealed ongoing utilization of the system by farmers on an unsubsidized basis. Discussions revealed that the GWUG is planning to invest their own funds to purchase alternative pumping technologies. This shows that farmers are interested in using borehole systems provided that the technologies are well-tailored to their needs. It also highlights some of the inherent difficulties in conducting interdisciplinary participatory research. Even with the best intention of the research team to engage fully with the community and to pursue bottom-up approaches wherever possible, this was not as successful as it could have been.

19.5 DISCUSSION

19.5.1 Factors contributing to successful groundwater irrigation

Our results suggest that the scope for farmers to engage effectively in groundwater irrigation will be strongly dependent on a number of factors, both biophysical and socioeconomic in nature. They are summarized as follows:

1. *Hydrogeological conditions*: Adequate groundwater availability and quality is an essential prerequisite for groundwater irrigation. Sites should be selected with adequate aquifer yields and low salinity water. Resource availability alone does not guarantee beneficial outcomes to farmers from irrigated agriculture.
2. *Household labor*: Cultivation of cash crops is a labor-intensive activity and requires considerably more effort than rice cultivation. According to the farmer interviews conducted by Suhardiman et al. (2018), typically about three laborers per household are needed to profitably farm, although this depends highly on farm sizes. By this measure, the village demographics are such that around 60% of households would have a labor shortage if pursuing cash-crop cultivation. The labor shortage is further compounded due to proximity of the area to Vientiane Capital where off-farm employment opportunities are higher.
3. *Livelihood alternatives*: A well-established practice throughout the rural parts the country is the migration to urban centers within Lao PDR or neighboring countries with stronger economies to pursue wage labor in construction, or other sectors. Seasonal or longer-term migration directly competes with agriculture, bringing about labor shortfalls and increased costs for farm labor.

4. *Cropping choices*: With water pumped from a borehole or well being limited, and abstraction adding to the cost of agricultural production, using water efficiently and profitably is of great importance to farmers and water resource managers more broadly. Groundwater used for dry season rice is likely to be marginally economic and increases the risk of groundwater overexploitation (Clément et al., 2018). Groundwater is best used for growing high-value, low-water use crops, rather than dry season rice. There may be merit, however, in utilizing groundwater for supplementary irrigation of wet season rice in periods of drought (ACIAR, 2017).
5. *Market access*: Whereas farmers of the past produced rice and vegetables mainly for home consumption, nowadays there is a strong commercialization trend in which most households in the village produce crops and livestock for market (Suhardiman et al., 2018). Ekxang village's proximity to both local markets and the national capital supported by relatively good roads make it ideally located.
6. *Knowledge and tools*: The success of individually or community-managed groundwater irrigation relies heavily not only on farmer acceptance of new technologies, but also on other factors impacting farming practices such as cropping choices, access to micro-credit, markets, and knowledge and support provided by government extension and advisory agencies. Experience from the irrigation subsector in Lao PDR shows that irrigation systems with strong user groups operate and manage the infrastructure most profitably and incur the lowest levels of debt (Bouahom et al., 2016).
7. *Institutional backing*: Under the local administrative structure of Lao PDR, the role of the village leader is crucial for the uptake of any new technology. The formation of the GWUG, which enabled wider participation of farmers, was facilitated through the village head. Besides involvement of formal and informal local institutions, such as the District Agriculture and Forestry Office, Women's Group, or Vegetable Farmer Group, are also important to ensure proper alignment of interests and establishing the most profitable uses for groundwater.

Overall, the profitability of groundwater irrigation for farmers is determined not just by favorable hydrogeological conditions, soil types present, crops cultivated and agricultural practices applied, but also by the market conditions and a range of other factors. Farmer awareness on all of these topics is thus of importance and negative impacts due to any single factor (e.g. pest infestation, poor market conditions, or labor shortages) could easily be detrimental to the adoption of groundwater for irrigation. Targeted advocacy of groundwater irrigation development for specific settings and agricultural commodities stands the best prospect of success, in particular, when they are coupled with other relevant practices such as Good Agricultural Practice (GAP).

Clearly, further development will depend not just on local hydrogeology, but also on a the social and economic factors that influence farmers' decisions. Adoption of groundwater irrigation is dependent on the willingness to engage individuals through larger groups of farmers and supportive institutional mechanisms to link and integrate individual farmers to the wider economy. Potential changes in household priorities and alternative livelihood options that may influence the level of adoption must be clearly understood in advance. Similarly, measures to increase attractiveness toward commercial farming and enhancement of income generation at the local level are equally important to address out-migration of labor from farming.

358   Groundwater for Sustainable Livelihoods

Introducing technological innovations to make groundwater more affordable, less labour-intensive and sustainable is also essential to overcome local barriers. The emergence of new technologies such as solar pumps with their prospects to reduce farmers' production costs could also make a real difference to the attractiveness of groundwater irrigation. The Lao solar industry has just started and solar pumping systems are beginning to be used for domestic water supply.[9] Even more recently, a solar powered demonstration trial for irrigated agriculture has been established in Vientiane.[10]

19.5.2   Comparative exploration of the two options

While both of the options presented herein fulfill the basic goal to provide an additional source of water for irrigation, there are substantial differences in how this goal is delivered. An examination of the standalone and comparative opportunities and constraints, as summarized in Table 19.3 and detailed below, is therefore of value.

Dugwells offer a simple, small-scale technology that individual farming households have spontaneously adopted with apparently little or no external support. This option may prove lucrative for participating farmers where the minimum set of conditions are met – sufficiently shallow groundwater levels (i.e. maximum depths of 6 m) available during the time of cultivation, sufficient land area and tenure security, and access to markets. It follows then that dugwells are not well suited to areas where water

Table 19.3 Opportunities and constraints of the two options for groundwater irrigation

	Opportunities	Constraints
Shallow dugwells (DW)	• Scope for crop diversification • Shallow water table areas with suitable water quality • Simple operation and management • Easily replicated • Limited capacity for overexploitation as only shallow aquifers are tapped • Relatively lower investment cost	• Limited to small-scale development • Farmers with sizable land holdings and access to family or paid labor • Shallow depth of groundwater accessed limits buffering capacity to rainfall variability and thus increases vulnerability to climate risks • Risk of polluted surface runoff impacting on groundwater quality
Deeper boreholes (BH)	• Areas where good aquifers overlay with good infrastructure and supply chains to enable commercial agriculture and crop diversification • Areas where surface water development is inappropriate • Scope for incremental development, bringing about large-scale development • Opportunities to engage small farmers and landless • Least affected by intra- and inter-annual climate variability	• Requires external financial and technical support to operate and manage sustainably (at least in the initial stage) • Potential competition with other water users if aquifer is overexploited • May require strong regulatory mechanisms to ensure sustainable resource use

levels are deep or widely fluctuating across seasons (such as in the case of Ekxang village), where farmers lease or own small or dispersed landholdings, or where markets are remote. Dugwell irrigation may also leave behind households with lowest incomes or limited to no access to land. On the VP up to one-third of the area may have biophysical conditions potentially suited to establishing dugwells (inferred from field data reported by Brindha et al., 2019). Although easily replicated, dugwells are small in scale and are an inappropriate option for larger-scale irrigation development. Open dugwells pose a threat to groundwater quality from inputs of agrochemicals associated with farming activities.

Borehole irrigation, when managed by groups of farmers, is founded on the principle of being inclusive of larger numbers of farmers, which may in theory include marginal or landless farmers. Borehole systems require higher investment costs, collective management, and some specialist knowledge to maintain infrastructure but allow relatively lower unit cost for irrigation on each farmer due to the economy of scale. Boreholes provide more water and more stable supply, thus enabling more area under irrigation and more income which may provide dividends over the longer term as the modeling results given earlier suggest.

Whereas the reliability of groundwater supplies from dugwells with an average depth of around 5 m is limited since they skim only the upper saturated layer, boreholes that are 30 m in depth offer more reliable, year-round supplies. Borehole systems offer greater capacity to buffer against supply constraints within years (e.g. late in the dry season) and between years (e.g. when monsoon rainfall for one or more years has been poor). Boreholes may also be constructed in a wider range of hydrogeological environments than dugwells.

19.5.3   Governance considerations for scaling up

The potential to develop groundwater irrigation is best revealed through groundwater resource assessments. The resources of the VP are better understood than other parts of Lao PDR due to a series of crosscutting studies carried out on topics that include:

1. Broadscale hydrogeological investigation (Batelaan et al., 2018; Viossanges et al., 2018; Perttu et al., 2011; JICA, 1993);
2. Recharge estimation (Lacombe et al., 2017);
3. Domestic groundwater use estimation (Heang and Keovongdy, 2016);
4. Water quality surveys (Brindha et al., 2017, 2019);
5. Groundwater network establishment, monitoring, and management plan development (ACIAR, 2016) and
6. Numerical groundwater modeling (NREI, 2017).

In countries with weak institutions and legal frameworks such as Lao PDR, there is little or no outside control over resource utilization, regardless of whether that use lies in the hands of individual farmers or groups of farmers (Pavelic et al., 2014). Nor is it advisable to limit the prospects of groundwater use for improving crop production and enhancing resilience out of the fear of potential overexploitation, as observed in other parts of the world such as India. Establishing a system to govern and manage

360 Groundwater for Sustainable Livelihoods

groundwater in a sustainable manner cannot be delayed, as leaving groundwater resources unmanaged runs the risk of future over-exploitation in response to increased water demand. In the Lao context, a balanced approach of groundwater development based on triangular coordination among the Ministry of Natural Resources and Environment (MONRE), Ministry of Agriculture and Forestry (MOAF), and Ministry of Energy and Mines (MEM) is needed to manage the potential trade-offs between environment, energy and crop production (ACIAR, 2016). Appropriate systems of management by the establishment of a Groundwater Department under MONRE, are being put in place in the upper VP but it remains too early to establish their effectiveness (ACIAR, 2016). In other areas where groundwater has a role in accelerating poverty alleviation it is recommended that until the resource is better understood and systems of management strengthened, development should commence with pilot demonstrations. The pilot demonstration helps identify the pros and cons of groundwater irrigation in a timely manner. Further, it avoids prospects of unplanned upscaling of groundwater and associated risks due to increased awareness and experiences on groundwater irrigation. Further, the use of groundwater for irrigation (or industry) should not compromise the delivery of water for domestic and environmental purposes, which must be maintained as a priority (ACIAR, 2017).

19.6 CONCLUSIONS

This study has evaluated the use of shallow dugwells managed by individual farmers as well as deeper boreholes managed by the community. Both irrigation options provide a viable means for farmers in Lao PDR to irrigate cash crops during the dry season and improve food security, while also contributing to more resilient livelihoods and equitable growth, especially for farmers lacking access to irrigation water from canals or ponds. Individual farmer-managed groundwater irrigation with dugwells is a successful, demand-driven approach that is simple to apply and requires low investment and maintenance costs. It is therefore easily scalable but does require very specific biophysical and socioeconomic conditions. Community-managed irrigation through boreholes is new to the country and was established as part of a larger research project. The community-managed irrigation through boreholes ensures higher resilience compared to dugwells, which are prone to seasonal water level fluctuations and drying. The trial demonstrates the challenges in convincing farmers to apply a new technology that would at first seem potentially risky due to higher upfront investment and operational challenges. Participation by farmers covered 25% of the design command area in the first year with project subsidies in place, and only 12% in the second year with subsidies removed.

Despite limitations in adoption, the results provide the basis for identifying the drivers and conditions that incentivize and enable groundwater irrigation to generate positive development outcomes for smallholder farmers.

As Lao PDR seeks to strengthen food security and rural livelihoods while adapting agriculture to climate change impacts, the potential for groundwater irrigation can only grow. The potential to develop groundwater for irrigation in the lowlands of Lao PDR is high. Whether or not farmers can make the best use of the available groundwater resources depends upon overcoming economic incentives and barriers faced by the farmers. It also requires ensuring there is a clearly defined need (when and how much

Small-scale groundwater irrigation in Lao   361

to use), and gaining a better understanding of the hydrogeological systems at the local level, defining more precisely the suitability for groundwater irrigation development, empowering farmers through knowledge and technical support, institutional backing, and accounting for the local culture and other contextual factors. Further adoption of groundwater irrigation needs to be flexible to adapt to new opportunities (such as solar irrigation) and governance challenges. These prerequisites would provide wider opportunities for agricultural groundwater development in Lao PDR. Lessons learned from the existing and piloted options evaluated should prove useful for helping to realize this potential.

ACKNOWLEDGMENTS

This work was supported by the Australian Centre for International Agricultural Research through research project LWR/2010/81 and the CGIAR Research Program on Water, Land and Ecosystems (WLE). The authors thank the wider group of project team members from the Department of Irrigation, Department of Water Resources, Natural Resources and Environment Institute, National University of Laos (Lao PDR), Khon Kaen University (Thailand), Institute for Global Environmental Strategies (Japan), and International Water Management Institute for their contributions to this research. Particular thanks go to the enterprising farmers of Ekxang village for generously sharing their time and experience with the project team. Peer review by Dr Robyn Johnston (ACIAR) helped improve the quality of this chapter.

NOTES

1 A UN designated *'least developed country'* (https://www.un.org/development/desa/dpad/least-developed-country-category/ldcs-at-a-glance.html).
2 https://data.worldbank.org/country/lao-pdr.
3 http://www.fao.org/aquastat/en/.
4 Vientiane Province does not include the national capital, which has lower poverty levels.
5 ACIAR Project LWR/2010/081: *Enhancing the resilience and productivity of rainfed dominated systems in Lao PDR through sustainable groundwater use* (https://www.aciar.gov.au/node/13071).
6 LAK is an abbreviation for the national currently, the Lao Kip.
7 The exchange rate used is $US 1 = LAK 8,150 (based on the Bank of Lao PDR official exchange rate in November 2016; http://www.bol.gov.la/english/exchrate.html).
8 This farmer did not chose to grow high-value watermelon which is typically grown on a 3-year rotational basis.
9 http://www.oecd.org/dev/asia-pacific/saeo-2019-Lao-PDR.pdf.
10 https://www.researchgate.net/publication/332570912_A_Solar_Power-Based_Groundwater_Irrigation_System_for_the_NAFRI_Agriculture_and_Forestry_Learning_Garden.

REFERENCES

ACIAR (Australian Centre for International Agricultural Research) (2016) *Enhancing the resilience and productivity of rainfed dominated systems in Lao PDR through sustainable groundwater use.* Final Report for ACIAR Project LWR/2010/081. Available at: http://aciar.gov.au/publication/fr2016-35.

ACIAR (Australian Centre for International Agricultural Research) (2017) *Groundwater for irrigation in Lao PDR: Promoting sustainable farmer-managed groundwater irrigation technologies for food security, livelihood enhancement and climate resilient agriculture*. ACIAR Policy Brief. Available at: http://aciar.gov.au/publication/fs2017.

Batelaan O., Banks E., Hatch M., Douangsavanh S., Sithiengtham P., Enemark T., Pavelic P., Xayavong V. and Xayviliya O. (2018) Geophysics to enhance agricultural productivity and livelihoods of smallholder farmers through improved groundwater management of the Vientiane Plain, Lao PDR. *SEG International Exposition and 88th Annual Meeting*, pp. 2501–2505. doi:10.1190/segam2018-2998321.1.

Bohsung S.M. et al. (2015) *Small-scale groundwater irrigation in the Vientiane Plain: A case study on dug wells*. International Water Management Institute draft report, unpublished.

Bouahom B., Pavelic P., Makin I., McCartney M., Suhardiman D., Sellamuttu S.S., Venot J-P. and Vongsathiane K. (2016) Performance review and policy options for sustainable irrigation development in Lao PDR. *Proceedings of the 2nd World Irrigation Forum (WIF2)*, 6–8 November 2016, Chiang Mai, Thailand.

Brindha K., Pavelic P. and Sotoukee T. (2019) Environmental assessment of water and soil quality in the Vientiane Plain, Lao PDR. *Groundwater for Sustainable Development*, 8, 24–30.

Brindha K., Pavelic P., Sotoukee T., Douangsavanh S. and Elango L. (2017) Geochemical characteristics and groundwater quality in the Vientiane Plain, Laos. *Exposure and Health*, 9(2), 89–104.

Clément C., Vinckevleugel J., Pavelic P., Xiong K., Valee L., Sotoukee T., Shivakoti B.R. and Vongsathien K. (2018) *Community-managed groundwater irrigation on the Vientiane Plain of Lao PDR: Planning, implementation and findings from a pilot trial*. Colombo, Sri Lanka: International Water Management Institute (IWMI). IWMI Working Paper 183, 52p. doi:10.5337/2018.230.

Heang V. and Keovongdy P. (2016) *Estimating Groundwater Use from 4 Villages: Ekxang, Phousan, Luk52 and Viengkham villages in Phonhong District, Vientiane Province*. BSc Student Thesis, National University of Laos Faculty of Environmental Science, Lao PDR (written in Lao language).

JICA (Japan International Cooperation Agency) (1993) *Basic Design Study Report on the Project for Groundwater Development in the Vientiane Province in Lao PDR*. Chiyoda: JICA, 199p.

Johnson J.H. (1986) *Preliminary appraisal of the hydrogeology of the Lower Mekong Basin*. Report of the Interim Committee for coordination of investigations of the Lower Mekong Basin, 147p.

Kwanyuen B., Mainuddin M. and Cherdchanpipat N. (2003) Socio-ecology of groundwater irrigation in Thailand. *IWMI Southeast Asia, Kasetsart University, Seminar on Scientific Cooperation*, Bangkok, Thailand, 26 March 2003, pp. 23–41.

Lacombe G., Douangsavanh S., Vongphachanh S. and Pavelic P. (2017) Regional assessment of groundwater recharge in the lower Mekong basin. *Hydrology*, 4(4), 60. doi:10.3390/hydrology4040060.

Lao Statistics Bureau and World Bank (2020) Poverty profile in Lao PDR: Poverty report for the Lao Expenditure and Consumption Survey 2018–2019. Available at: http://pubdocs.worldbank.org/en/923031603135932002/Lao-PDR-Poverty-Profile-Report-ENG.pdf.

MAF (Ministry of Agriculture and Forestry) (2012) *National Agricultural Census 2010/11 Highlights*. Report prepared by the Steering Committee for the Agricultural Census, Agricultural Census Office, Government of Lao PDR.

Moench M. (2003) Groundwater and poverty: Exploring the connections. In: R. Llamas & E. Custodio (Eds.), *Intensive Use of Groundwater: Challenges and Opportunities*, pp. 441–455. Lisse: A.A. Balkema.

Mukherji A. and Shah T. (2005) Groundwater socio-ecology and governance: A review of institutions and policies in selected countries. *Hydrogeology Journal*, 3(1), 328–345. doi:10.1007/s10040-005-0434-9.

NREI (Natural Resources and Environment Institute) (2017) *Groundwater model development in Vientiane Plain. Ministry of National Resources and Environment, Lao PDR.* Draft Report (written in Lao language).

Pavelic P., Xayviliya O. and Ongkeo O. (2014) Pathways for effective groundwater governance in the least-developed-country context of the Lao PDR. *Water International*, 39(4), 469–485.

Perttu N., Wattanasen K., Phommasone K. and Elming S.Å. (2011) Characterization of aquifers in the Vientiane Basin, Laos, using magnetic resonance sounding and vertical electrical sounding. *Journal of Applied Geophysics*, 73(3), 207–220.

Savva A.P. and Frenken K. (2002) *Financial and economic appraisal of irrigation projects.* Module 11 of Irrigation Manual: Planning, development, monitoring and evaluation of irrigated agriculture with farmer participation. Rome: Food and Agriculture Organization of the United Nations (FAO).

Shah T., Singh O.P. and Mukherji A. (2006) Some aspects of South Asia's groundwater irrigation economy: Analyses from a survey in India, Pakistan, Nepal Terai and Bangladesh. *Hydrogeology Journal*, 14, 286–309. doi:10.1007/s10040-005-0004-1.

Suhardiman D., Giordano M., Bouapao L. and Keovilignavong O. (2016) Farmers' strategies as building block for rethinking sustainable intensification. *Agriculture and Human Values*, 33(3), 563–574.

Suhardiman D., Pavelic P., Keovilignavong O. and Giordano M. (2018) Putting farmers' strategies in the centre of agricultural groundwater use in the Vientiane Plain, Laos. *International Journal of Water Resources Development*. doi:10.1080/07900627.2018.1543116.

Villholth K.G. (2013) Groundwater irrigation for smallholders in Sub-Saharan Africa – A synthesis of current knowledge to guide sustainable outcomes. *Water International*, 38, 369–391. doi:10.1080/02508060.2013.821644.

Viossanges M. Pavelic P., Rebelo L.-M., Lacombe G. and Sotoukee T. (2018) Regional mapping of groundwater resources in data-scarce regions: The case of Laos. *Hydrology*, 5(1), 2. doi:10.3390/hydrology5010002.

Vote C., Newby J., Phouyyavong K., Inthavong T. and Eberbach P. (2015) Trends and perceptions of rural household groundwater use and the implications for smallholder agriculture in rain-fed Southern Laos. *International Journal of Water Resources Development*, 31(4), 558–574.

Wang J., Huang J., Rozelle S., Huang Q. and Blanke A. (2007) Agriculture and groundwater development in northern China: Trends institutional responses and policy options. *Water Policy*, 9(1), 61–74.

Index

agriculture
  productivity 4, 35
  rainfed 3, 20, 41, 53
aquifer 297
  basement crystalline 96, 102, 202, 216
  coastal 82
  fractured 258, 320
  governance 114, 197, 199, 215
    managed aquifer recharge 115
  sandstone 36, 141, 242, 258
  shallow 102, 106
  specific capacity 259
  systems 255
  transboundary 38
  transmissivity 96, 169, 259
  unconsolidated 39, 103, 162, 201, 337
    underground dam 108

boreholes 97, 129, 284, 351, 358
  abandoned 291
  artesian 43, 257
  cost 68, 167
  drilling 206
    standards 71
    training 209, 203
  gamma ray log 162
  maintenance 166
  success rate 163, 173
  test pumping 140
  yield 67, 96, 142

climate change 25, 28, 90, 136, 153, 169,
  267, 276
  rising water level 90
community participation 55, 114, 161,
  268, 356
conjunctive use 153, 296
  strategies 303, 308
crop diversification 128, 132

data gathering 279, 298, 353
  interviews 66, 69, 277, 320, 325
  monitoring 136, 225, 269
  questionnaire 219
disaster
  mitigation 31
  recovery 236
disease water borne 203, 299
drought 26, 236, 276
  coping strategy 27, 277
  early warning systems 27
  prolonged 28, 44
    desertification 102
    salinisation 116, 235
  resilience 107

education 267
  awareness rising 90
El Niño 153

financial analysis 354
  cost benefit 280, 290
  net present value 280, 289
flow
  diffuse 5, 6
  flownet 111
  groundwater 297
  modelling 115, 168
  turbulent 6
food security 30, 33, 37, 55, 136, 215, 278, 348

gender 227, 279
GRACE 124
groundwater
  access 1, 4, 28, 69, 98, 199, 215, 316
    collection time 69, 86, 289
  capacity 10
  dependency 218
  development 10, 239
  exploration 9

366 Index

groundwater (*cont.*)
  governance 71, 199, 215, 222, 254
  hydrograph 88, 105, 136, 143, 150
  levels 28, 42, 43, 145, 235
    declining 124, 142, 200, 242, 328
  management 78, 220, 249, 359
    resource inventory 10
  mineralisation 30, 37
    (*see also* hydrochemistry)
    quality 8, 9, 138
    salinity 142, 146
  piezometry 167, 170
  recharge 37, 88, 102, 144, 168, 181, 246,
    258, 298
    indirect 103
    water balance 169, 171
    zone 153
  scarcity 11
  spring discharge 324
    use 98

hydrochemistry
  chemical type 184, 202, 258
  graphic Piper Diagram 185
  inorganic 184
    arsenic 27, 37
    fluoride 27, 37, 303
    guidelines 8, 9
    nitrate 37, 90, 203
    pH 257
    salinity 89, 124, 171
    specific electrical conductance 6, 8, 89,
      183, 202
    total dissolved solids 142, 236, 242
  organic
    pesticides 124

institutional
  capacity 221
  building 224
irrigation 102, 104, 131
  coastal 87
  drip 54, 106
  security 52
  small scale 42, 57
    water markets 339
IWRM 220, 316
  application 114

karst
  caves 3
    tourism 16
  dolines 3
  epikarst 4, 6
  porosity 5
  recharge 5
  risings 180–193

sinks 2
spring discharge 183, 186

land degradation 79, 136
landscape restoration 143, 144
livestock prices 15

migration 15, 26, 125

peri-urban, see Urban
policy guidance 28
  technical 221
pollution
  acid mine drainage 203
  air 125
  anthropogenic 28, 89, 216
  bacterial 9, 90, 203
    coliform faecal 9, 89
  pesticides 124
  protection zones 91, 198
  vulnerability 102
population density 10, 19
poverty
  alleviation 30, 348
  rural 26, 102, 277, 296
pump
  hand 165, 216, 276, 285
  solar 166

rainfall
  harvesting 54, 235
  unpredictable 26, 46, 136
rural livelihoods 31
  strategy 35
  sustainable framework 31, 33
  zoning 35, 45

sanitation 79, 90, 198, 202, 267, 316
seawater intrusion 90, 243
stable isotopes 169
stakeholder engagement 219, 268
suspended sediments 9, 17
sustainable Development Goals 95, 254, 296,
  334, 348

Tracer test 5

urban and peri-urban development 30, 63
  growth 81
  water supply 87, 197, 204

Vision 2030 95

water
  consumption 18
  credits 309
  demand 65, 83, 136

evaporation 247
harvesting 136, 148
integrated use 114, 138
law 40, 71, 128, 197, 220, 308
management 27, 268, 302
metering 68, 247
price 7, 17, 18, 71, 81, 86, 98, 167
    ability to pay 175
safe 28, 79
scarcity 10, 28, 49, 98, 297

supply
    coverage 64, 70
    peri-urban 63–65
    vending 74, 98
wealth creation 78, 297
    human development index 10, 48
wells
    duck billed 111
    hand dug 41, 97, 129, 141, 153, 163, 284, 351, 358